T0141672

Caring for Eeyou Istchee

Caring for Eeyou Istchee

PROTECTED AREA CREATION
ON WEMINDJI CREE TERRITORY

**EDITED BY
MONICA E. MULRENNAN, COLIN H. SCOTT,
AND KATHERINE SCOTT**

UBCPress · Vancouver · Toronto

28 27 26 25 24 23 22 21 20 19   5 4 3 2 1

Printed in Canada on FSC-certified ancient-forest-free paper
(100% post-consumer recycled) that is processed chlorine- and acid-free.

Cataloguing-in-publication data for this book is available from
Library and Archives Canada.

ISBN 978-0-7748-3858-0 (hardcover)
ISBN 978-0-7748-3860-3 (PDF)
ISBN 978-0-7748-3861-0 (EPUB)
ISBN 978-0-7748-3862-7 (Kindle)

Canada

UBC Press gratefully acknowledges the financial support for our
publishing program of the Government of Canada (through the Canada Book Fund),
the Canada Council for the Arts, and the British Columbia Arts Council.

This book has been published with the help of a grant from the
Canadian Federation for the Humanities and Social Sciences, through the
Awards to Scholarly Publications Program, using funds provided by the
Social Sciences and Humanities Research Council of Canada.

Printed and bound in Canada by Friesens
Set in Univers Condensed and Minion by Artegraphica Design Co. Ltd.
Copy editor: Deborah Kerr
Proofreader: Kristy Lynn Hankewitz
Indexer: Noeline Bridge
Cartographer: Eric Leinberger
Cover designer: Lara Minja
Cree orthography and syllabics: Frances Visitor and Theresa Georgekish

UBC Press
The University of British Columbia
2029 West Mall
Vancouver, BC V6T 1Z2
www.ubcpress.ca

ᐊᕆᕠᑕᑭᓄᐅᐃᐤᒡ ᐊᕝᕆ ᑭᕐ"ᐤ ᓄᐱ ·ᐊᕝᓄᕆ"ᔟ ᑭᕐ"ᐤ ᓄ ᕆ·
ᐃ"ᕆ"ᐴᐱᒡᐢᓄᕝ ᕆᐞᓄᕝᐅᕝ ᐴ ᐸᕆ ᕐᓄ·ᐊᐞᔟ"ᔟ ᑭᕝ"ᐤ ᐊᕝᕆ ᕝᕆ"ᐴᕝ ᐅᕝ ᐊᕝᕆᕝᐤx
ᐊᕆᕆ ᒑᕝᕆᓄ·ᐊᐞᔟ"ᔟ ᑭᕝ"ᐤ ᐅᕆᕝᑯᓕᐢ·ᐃᓄ·ᐊᔟ"ᔟ ᐊ"ᕆᕝ

·ᐊᕆ"ᐢ·ᐊᐞᕝ"ᑕᕝ ᐊᓄ·ᕝᐣx

Dedicated to the lands and waters of Wemindji territory
and to the elders who have loved and cared for it, protected it,
and shared their knowledge through the generations.

Contents

Figures and Tables

TABLES

Foreword

STAN STEVENS

In the late nineteenth century, the settler countries of Canada, the United States, Australia, and New Zealand were in the vanguard of what became a global movement to conserve nature through the creation of uninhabited national parks and other protected areas. Very often these were on Indigenous lands. In these countries, conservationists and government officials typically saw Indigenous peoples as threats to conservation, a view that rationalized their displacement and legitimized the appropriation of their territories as protected areas, through nationalization and state governance. Initially, these exclusionary protected areas focused on preserving scenic wonders and charismatic species for the public enjoyment and benefit of settler societies, but later they were pre-eminently dedicated to conserving biological diversity. Through much of the twentieth century, most conservation organizations and influential bodies such as the International Union for Conservation of Nature (IUCN) promoted this "fortress conservation" approach to establishing exclusionary, uninhabited, "Yellowstone model" national parks and other protected areas. In recent decades, however, human rights advocates, international treaty monitoring mechanisms, courts, and many conservationists and conservation organizations – including the IUCN – have joined Indigenous peoples in rejecting many of the core assumptions, policies, and practices associated with fortress conservation. They are calling for far-reaching rethinking of conservation, including the reform of protected area establishment, design, governance, and management.

New thinking about Indigenous peoples, conservation, and protected areas has reshaped international policies and standards. In 2003 the IUCN Vth World Parks Congress in Durban, South Africa, announced a "new protected area paradigm." This drew on recent international experience with protected areas that are governed by and with Indigenous peoples and on evolving IUCN rights-based conservation policy. The new paradigm promotes respect for Indigenous peoples' knowledge, values, collective tenure, stewardship, customary sustainable use and management of bio-diversity, rights and responsibilities, full and effective participation in protected area governance, and their informed consent to decisions that affect their traditional lands and waters, their well-being, and their self-determination. Since 2003, despite some controversy and skepticism, the IUCN and the Parties to the Convention on Biological Diversity (the primary international treaty that focuses on protected areas) have re-affirmed and elaborated these principles in international environmental law and policy. Most recently, in 2016, both the IUCN and the Parties to the Convention called for recognition of what the IUCN refers to as "the territories and areas conserved by Indigenous peoples and local commun-ities (ICCAs) that are overlapped by protected areas." This affirms that Indigenous peoples' and local communities' continuing self-governance, stewardship, and sustainable use of their territories should be a founda-tion for protected area governance and management that respects their worldviews, knowledge, values, institutions, and practices.

Although the new paradigm has yet to be implemented in many coun-tries, protected areas have been established or are in development in several that embody these principles to varying degrees – including Canada. These include the two protected areas that are the focus of this book, the prov-incial Paakumshumwaau-Maatuskaau Biodiversity Reserve in Quebec (established in 2008) and the proposed federal Tawich (Marine) Con-servation Area. Both are initiatives of the Cree Nation of Wemindji, which envisions Cree-led protected areas as a means of safeguarding the ecological integrity of key parts of its territory of Eeyou Istchee in eastern James Bay from unwanted mining and other impacts. In alignment with the latest international law and policy, the Wemindji Cree will share governance of these protected areas with the provincial and federal governments, while ensuring that their management will be grounded in Cree knowledge, values, institutions, and practices. A key aspect of this is basing protected area management on recognition of the continuing authority of traditional hunting territory leaders for management of extended family hunting territories.

This landmark book presents findings and analysis from an extraordinary collaborative research program, the Wemindji Protected Areas Project. The project brought together Cree leaders and community members, faculty and graduate students from four Canadian universities and a spectrum of academic fields, an environmental organization, and staff from federal and provincial government agencies who shared the Wemindji Cree protected area vision. This research collaboration was remarkable in many ways, including its diversity of partners, the co-leadership of Chief Rodney Mark and anthropologist Colin Scott, the number and breadth of the projects and their ten-year span, and the degree to which knowledge exchanges between Cree and non-Indigenous academics and the co-creation of new knowledge informed the strategizing, planning, and negotiation of the two protected areas. The resulting book provides an exceptionally in-depth and detailed documentation of the importance of protecting these specific terrestrial and marine areas of Wemindji Cree territory, the historical and political contexts and conditions within which the two protected areas are being developed, and a nuanced analysis of how their creation draws on the perspectives of the Cree and their diverse partners.

The Paakumshumwaau-Maatuskaau Biodiversity Reserve and the plan for the Tawich Conservation Area exemplify much of what the IUCN's new protected area paradigm seeks to promote. This paradigm shift could fundamentally reform conservation and the relationship between protected areas and Indigenous peoples. Differently envisioned, governed, and managed protected areas can become an important means of recognizing Indigenous peoples' self-governance, use, and conservation of their territories according to their own values and practices. Such areas may become key sites of social justice and reconciliation as well as biodiversity. The experiences and insights shared in this book will be an important source of inspiration and guidance for Indigenous peoples, academics, conservationists, human rights defenders, and government officials in Canada and beyond who seek to envision and create protected areas such as those established and planned by the Wemindji Cree and their partners.

Preface

This volume is an outcome of the Wemindji Protected Areas Project, which began in 2001 as a partnership between the Cree Nation of Wemindji (Eeyou Istchee) and a multi-disciplinary research team from McGill University, Concordia University, the University of Manitoba, and the University of British Columbia. It was co-directed by Chief Rodney Mark of Wemindji and anthropologist (and principal investigator) Colin Scott of McGill University. The research team involved more than a dozen co-investigators and nearly twenty graduate students from a range of disciplines, including anthropology, archaeology, biology, forest ecology, geography, philosophy, and political science. The partnership also included the Grand Council of the Crees (Eeyou Istchee), the Cree Regional Authority (now the Cree Nation Government), the Cree Trappers Association, the Tawich Development Corporation, the Ministère du Développement durable, de l'Environnement et des Parcs du Québec, Parks Canada, and la Société pour la nature et les parcs du Canada, which is the Quebec chapter of the Canadian Parks and Wilderness Society.

The primary source of funding was the Social Sciences and Humanities Research Council of Canada, specifically its Community-University Research Alliance (CURA) and Northern Aboriginal Research programs. The purpose of the CURA was to support the creation of alliances between community organizations and universities whose ongoing collaboration and mutual learning would foster innovative research, training, and the generation of new knowledge in areas of importance for the social, cultural, or economic development of Canadian communities. The Northern Aboriginal Research program encouraged Indigenous participation and leadership in research through the promotion of partnerships between academics and Indigenous communities.

The catalyst for this particular partnership was community-level concern about environmental protection in Wemindji Cree territory, where the pressures for resource development were mounting. The objective was to identify strategies for Cree-led environmental protection that would respond to priorities for both cultural continuity and development on locally defined terms. These priorities included the protection of lands, waters, and resources upon which traditional activities depended, thereby contributing to the well-being of community members. They also entailed the establishment of a regime of protection that built on Cree institutions of knowledge and environmental stewardship in support of enhanced local-level authority. Strengthening local knowledge resources and building local capacity to engage state-level conservation agencies through innovative approaches and emergent technologies were goals throughout the partnership, including an interaction between Indigenous and scientific knowledge. In addition, the needs of a growing population had to be addressed through negotiated terms on which large-scale resource extraction would be carried out on the territory.

Strategies for Cree-led environmental protection would be developed in ways that were socially and culturally appropriate to the circumstances and interests of Cree people. "Culturally appropriate" strategies built upon Cree institutions for managing land- and sea-based activities and resources, and they contributed insights from scientific knowledge while supporting the authority of Indigenous knowledge and its practices in environmental management.

The Wemindji Protected Areas Project employed a range of methodologies to support this objective. These included lightly structured interviews with local Cree experts, often hunters, on topics related to environmental protection. Workshop sessions on the same topics were organized with elders, community leaders, senior hunters, women, and youth. Participant-observation took place in Cree hunting/fishing camps and at community gatherings. Collaborative field surveys examined geophysical, floral, and faunal environmental components and archaeological sites. Interviews were also conducted with federal, provincial, and regional Cree policy-makers who were engaged in environmental conservation in northern regions. Mapping and modelling of environmental features and changes in the study area relied on a combination of local knowledge, remote sensing imagery, ground-truthing, core sampling, and pollen analysis. Regular community-team meetings provided a forum for reflection and discussion of the project's progress. Students and co-investigators, usually in teams that included community members, participated in workshops and conference symposia, and co-authored chapters in this volume, as well as articles and newsletters to communicate the story of their work.

A primary outcome of the partnership was the Quebec government's acceptance on May 14, 2008, of a proposal from the Wemindji Cree Nation to establish the Paakumshumwaau-Maatuskaau Biodiversity Reserve. The reserve comprises nearly five thousand square kilometres and represents approximately 20 percent of the terrestrial component of Wemindji territory. Official registration of the projected reserve triggered an immediate moratorium on new mining claims in the area and led to the subsequent withdrawal of existing claims in a core zone of the proposed reserve. A second outcome was the development of a proposal to establish the Tawich (Marine) Conservation Area in the adjacent offshore as part of a joint undertaking by east coast James Bay Cree communities, the Grand Council of the Crees of Eeyou Istchee, and Parks Canada under the Parks Canada National Marine Conservation Area program.

This volume brings together the efforts of team members and community partners to support the creation of protected areas on Wemindji territory. It features contributions from twenty-nine authors, including two Indigenous partners, eleven co-investigators, fourteen graduate students, and two provincial government employees. Our ultimate aim is to showcase the experience and efforts of one community to protect a substantial portion of its territory in the midst of growing pressures from large-scale development. We trust that the experience of Wemindji Crees will be a source of inspiration for other Indigenous communities in the Canadian North and beyond.

Acknowledgments

This volume is the work of many hands. Countless people in Wemindji and from far-flung locations have contributed to these chapters in diverse ways. We would like to thank you all from the bottom of our hearts. Your input has made this project possible.

First and foremost, we thank the people of Wemindji for their generosity, hospitality, warm welcome, and willingness to teach us about their beautiful living territory. You patiently shared your knowledge and guided us along the way – from those who mentored and worked with us on the land to those who sat through interviews, who welcomed us into their homes, who gave us wonderful places to stay, and who shared the gifts of the land with us.

Special thanks go to Rodney Mark, co-director of the project; Dorothy Stewart, who acted as our community-university liaison officer; Fred Stewart, who was *uchimaau* (tallyman or hunting boss) of the hunting territory where much of our work took place; and Peter G. Brown, who kept us mindful of the bigger picture.

We thank our university and research colleagues whose work intersected with ours over the years, who attended our research meetings, advised us, and shared their precious time. We also thank community leaders and people of Eastmain, Chisasibi, and Waskaganish, who participated in some of our meetings, for their shared commitment to the goal of protecting the coast and offshore. We also acknowledge the valuable input and support of many individuals who work in regional Cree government entities, as well as in provincial and federal government departments.

The list of individuals that follows is in alphabetical order, and any omissions are deeply regretted, but please know that we appreciated your help.

Elaine Albert, Sylvain Archambault, Fred Asquabaneskum, Leonard Asquabaneskum, Ronnie and Hilda Asquabaneskum, Walter and Juliet Asquabaneskum, Winnie Asquabaneskum, Andrew Atsynia, the late Daisy Atsynia, Sr., Frank Atsynia, Freddie and Queenie Atsynia, Lillian Atsynia, Raymond Atsynia, Sarah Atsynia, Chris Beck, Fred Blackned, the late Jimmy Blackned, Ron Blackned, Sammy Blackned, Valter Blazevic, Nelson Boisvert, Jennifer Bracewell, Jim Chism, Shu-Yi Chu, Matthew Coon Come, Earl and Nancy Danyluk, David Denton, Jessica Dolan, Holly Dressel, Gwilym Eades, Harvey Feit, Margaret Forrest, Bradley A.J. Georgekish, the late Clifford and Emily Georgekish, Dennis Georgekish, Edward Georgekish, Elmer Georgekish, the late Jimmy Georgekish, Johnny T. Georgekish, Johnny "Zack" Georgekish, Melvin Georgekish, the late Roderick Georgekish, Ronnie Georgekish, Theresa Georgekish, the late Samuel Georgekish, Sinclair Georgekish, Willard Georgekish, Vincent Gérardin, Albert Gilpin, the late Billy Gilpin, Christine Gilpin, Frank Gilpin, Sam Gilpin, Tom Heller, Eby Heller, Ernie Hughboy, the late Harry and Emily Hughboy, the late Sam Hughboy, Leslie Kakabat and Rita Atsynia, the late Lot Kakabat, Anne-Elise Keen, Hugo Kitching, George Kudlu and his late wife, Louisa, Jessica Labrecque, Manuelle Landry-Cuerrier, Youcef Larbi, Marty Lechowicz, Johnny Mark, Reggie Mark, Thomas Mark, Nick Marotchnik, Abraham Matches, Alan Matches, Clayton Matches, the late Doris Matches, Duane Matches, Jeremy Matches, Jimmy and Minnie Matches, the late John Matches, Kimberly Matches, Stacy Matches, Beverly Mayappo, the late Michael Mayappo and his wife, Louise, the late Sinclair Mayappo and his wife, Alice, George McCourt, Marina Milligan, Clarence Miniquaken, John and Florrie Mistacheesick, Sinclair and Irene Mistacheesick, William Mistacheesick, Tania Monaghan, Chantal Otter Tétreault, Ieva Paberzyte, Cartter Patten, Alan Penn, Troy Pretzlaw, Genevieve Reid, Jinya Scott, Sean Scott, Marilyn Scott, Raja Sengupta, James and Annie Shashaweskum, the late Lloyd Shashaweskum, Renée Sieber, Bernice Stewart, Bill Stewart, the late Donald Stewart, Fred

Stewart, the late George Stewart and his wife, Ellen, the late Harry Stewart and his wife, Alice, Henry Stewart, Linda Stewart, Roy and Jennifer Stewart and their daughter Cherish, Sarah Stewart, Christine Stocek, Jesslyn Stoncius, Adrian Tanner, Charlie Tomatuk, Maurice Tomatuk, Abel Visitor, Arden Visitor, Elmer Visitor and Clara Stewart, Frances Visitor, Christopher Wellen, Kathleen Woottoen, Christina Zhu, and the McGill School of Environment students in Research in Environment 401 who came to Wemindji in 2003, 2004, and 2005.

Finally, the editors would like to thank James MacNevin, Michelle van der Merwe, Katrina Petrik, and the rest of the team at UBC Press, for their guidance and support. We also greatly appreciate the helpful comments and suggestions that we received from the reviewers of the book.

Caring for Eeyou Istchee

Introduction

RODNEY MARK, MONICA E. MULRENNAN,
KATHERINE SCOTT, and COLIN H. SCOTT

The history of protected area creation in Canada, as in other settler states,[1] has deep colonial roots. Early protected area policies routinely dispossessed Indigenous peoples of their lands, which resulted in widespread suffering and lasting distrust of government land management efforts (Colchester 2004). In recent decades, increased attention to social justice and Indigenous rights and a rethinking of environmental protection approaches have supported a new paradigm of conservation based on greater collaboration with Indigenous peoples and international recognition of their rights (Stevens 2014). However, translating this paradigm into meaningful action has been difficult and uneven. Too often, policy-makers have characterized Indigenous people as stakeholders rather than primary rights-holders. As stakeholders, they must compete with other more powerful interests to have their voices heard and to participate in defining the terms of environmental protection and development on their own lands.

This volume arises from the extraordinary experience of one particular Indigenous community, the Cree Nation of Wemindji, located in Eeyou Istchee (which translates as "the land of the people" and refers to the homeland of the Crees of northern Quebec) on the eastern shore of James Bay. But since "Eeyou" applies not only to Crees or other human beings, but to living beings in general, Eeyou Istchee also connotes "the living land." Wemindji Eeyouch have established a protected area on their traditional territory that is defined by their priorities,[2] and they have done so in the context of state-supported, industrial-scale resource extraction on other

parts of their territory. Wemindji Eeyouch are no strangers to development, having seen extensive portions of their inland territory altered since
the 1970s by the James Bay hydro-electric project and the construction of
ancillary infrastructure such as roads and power lines. In the early 2000s,
a new wave of development in the form of mining exploration raised the
prospect of further threats to the ecological integrity of their territory. In
response, community leaders articulated a vision of the future, featuring
enhanced authority for decisions that affected their lives, to support their
long-held responsibilities as stewards of their lands and waters.

The challenges that Wemindji Eeyouch confronted almost fifty years
ago, and still grapple with today, mirror the complex and intractable challenges faced by many Indigenous communities. They must find a balance
between protecting traditional territory and ensuring that hunting, fishing,
and trapping remain a vibrant and viable way of life, while negotiating
the terms of industrial-scale development on their territory and guiding
community-based development to provide jobs and entrepreneurial possibilities for their youth. In search of solutions and seeking strategies for
environmental and cultural heritage protection, Wemindji Eeyouch formed
a partnership with a team of university researchers, which became known
as the Wemindji Protected Areas Project. Partners agreed to prioritize
local institutions of land and sea tenure, traditional authority in environmental stewardship, ecological and cultural knowledge, and environmental ethics.

Wemindji's achievements in creating a protected area, and thereby
gaining some control over its lands, helped to motivate the regional
Cree government to establish the Eeyou Protected Areas Committee, out
of which emerged the Cree Regional Conservation Strategy (Cree Nation
Government 2014). Recently, the International Union for Conservation
of Nature (IUCN) recognized the Cree Nation Government's innovative
approach to protected area creation and welcomed it as a new member.

There have been obstacles, confrontations, and compromises along
the way – but Wemindji Eeyouch have negotiated benefits from development even as they steadily increased the percentage of their territory that
is off-limits to high-impact forms of industrial resource extraction. Yet
they have continued to cultivate positive relations with development companies on certain parts of their territory. In these efforts, Wemindji provides

a source of inspiration for other Indigenous communities, most especially in achieving some balance between development and environmental protection. Wemindji Eeyouch have also demonstrated the potential of protected areas as a political strategy that redefines relations with government in terms of a shared responsibility to care for land and sea.

This book is one of two that emerged from the research conducted as part of the Wemindji Protected Areas Project. Its focus is on environmental protection in Wemindji territory. The second volume, which is still in development, is dedicated to the exchange of knowledge across cultures: hunters and scientists, local people and academics, northerners and urban southerners, First Nations and settlers.

Rodney Mark, co-author of this chapter, played an essential role as a leader of the Wemindji Protected Areas Project. Chief of the Cree Nation of Wemindji during the project's first six years, he then became deputy grand chief of the Crees (Eeyou Istchee) until 2017. His reflections, which he shares below, show a profound understanding of local needs and aspirations, as well as a deep commitment to Eeyou values that support caring for the entirety of lands, waters, humans, and non-human beings that make up and inhabit Wemindji territory. His account also speaks to the resilience of Wemindji Eeyouch, although Rodney prefers to frame this in relation to "tradition." For him, a "traditional way of life" refers to a way of being on the land that relies mainly on hunting, trapping, and fishing. Sustaining it depends on the inter-generational transmission and adaptation of knowledge, sometimes referred to as traditional knowledge. "Traditional" in this sense does not mean old or no longer relevant. Rather, it is about adjusting to new conditions, changing with the times, keeping what is meaningful, and innovating as required. Thus, the traditional way of life on the land has changed profoundly over the centuries, but its vibrancy and relevance remain undiminished.

A WEMINDJI EEYOU PERSPECTIVE ON ENVIRONMENTAL PROTECTION: REFLECTIONS FROM RODNEY MARK

The real challenge in protecting and preserving a Cree traditional way of life is to ensure that lands and waters are available for that purpose. This requires a balanced and flexible approach to development in our territory while seeking to ensure that land users are comfortable and unhindered in their hunting, fishing, and trapping practices. Today, those who spend most of their time living on the land see the economic potential of natural resource development in the territory. However, they feel strongly – with community support – that this development should not compromise a traditional way of life that is centred on hunting, fishing, and trapping. At the same time, they are well aware of the importance of optimizing benefits for the community and its young people in terms of quality of life. This means finding satisfying work without moving away from family and community, as well as not having to compromise on opportunities for training and advancement. The approach is flexible, but concern for the long-term impacts of development on the environment and its effects on the hunting way of life runs very deep. The bush, the coast, and the offshore are vital resources for traditional land users and animals alike; this co-existence is essential for the survival of both.

Also, for all of us, our understanding of who we are in the world, and our place in the world, informs and motivates us. Our goal and fundamental interest in protecting a certain amount of Wemindji territory is about defending our concept of the world as a network of relations between all living things. Protecting the environment is about protecting the animals, a way of life, and all that nature gives. In our understanding and experience, environment (the land and water) is a living being that we are part of, and it is part of us. Animals and the environment, participating in each other's worlds, in the whole network of relationships of mutual respect and action, are the essence of a traditional way of life. Living well as part of that total community is our ultimate aspiration.

The parts of our territory that we prioritize for protection have historical importance to our communities and our families. There are stories in the land that are stories about us, about where we came from, and where we go from here. A culture that is evolving has a living connection from its past to its present. For us, continuity lies in recognizing the value of life and all living things, so what is called "culture" includes *all* the living things of the land that we protect. Our responsibility remains for all of our lands and waters – not just those that might some day be designated protected areas by provincial or federal governments.

The establishment of state-recognized protected areas is one strategy among many in a proactive approach to nurturing our way of life and the relationships with the land that it depends upon. Family hunting territories are everywhere on our traditional lands and waters. There is no place that does not come under the authority and stewardship of our *nituuhuu uchimaauch* (traditional hunting territory leaders). We do not intend to ignore Cree responsibility for our entire territory, and we will continue to seek the best means to care for all of it. We already have some useful tools. Our hunting territories and the authority of our nituuhuu uchimaauch are recognized under the James Bay and Northern Québec Agreement (JBNQA). Together with other Cree communities in the Grand Council of the Crees (Eeyou Istchee), we have co-management authority for wildlife everywhere on our territory. We also have co-management authority in environmental impact assessment procedures everywhere in our territory. We recognize that we need to avail ourselves of these to help mitigate the negative impacts of development. At the same time, we are experimenting with various technologies, such as hand-held tracking devices and satellite imagery, to enhance our own capacity as "the eyes and ears" of Eeyou Istchee (for details, see Brammer et al. 2016). The Grand Council of the Crees is also negotiating with provincial and federal governments to assert our self-government authority everywhere within our land and sea territory.

How we see ourselves as a Cree nation is reflected in the words "Eeyou Istchee," which refer to the territory represented by the Grand Council of the Crees in our dealings with federal and provincial governments, and with other entities in Canada and abroad. Simply translated, "Eeyou Istchee" means "Cree land," but at a deeper level, it declares awareness of our cultural self. We use the word "Eeyou" to distinguish Crees from other Indigenous peoples. But it also distinguishes Indigenous people in general from non-Indigenous people and humans in general from animals and other living things. At the broadest level, it embraces all living beings – all being is living in some way. So who we are, as Crees, as Eeyouch, is nested within the more and more inclusive connections that make up the living world. "Istchee," which focuses on the idea of land, has a similar extension of meanings. It can mean the living soil, the moss that is the ground in our forests, but it also refers to a Cree family hunting territory. At the broadest level, it encompasses the living world. So, ultimately, Eeyou and Istchee are one and the same. This is what our elders have shown us, and that is how we call ourselves and our land.

We (Wemindji Eeyouch) will develop protected areas, in co-operation with provincial and federal governments, in ways that respect and defend this outlook, which makes our lands and waters available for our children and grandchildren forever so they can experience as real our Cree values that sustain stewardship of land and sea and all the living things that comprise them. At the same time, we acknowledge that we have a responsibility to develop pathways between the experience of the bush and the outlook that requires, and the realities of village life and life in the wider society.

The elders tell us always that we must have balance to have a good life. People want jobs, but they want to live in the bush too. How to achieve balance, then, becomes key. Some of our children will have more opportunities than others to live in hunting and fishing camps, and to experience certain aspects of life on the land. We must not imagine that we can replace time lived on the land, but we can communicate and complement the knowledge of our people on the land by experimenting with new media and innovative modes of documentation and presentation. We are in the process of developing a Wemindji Eeyou knowledge center that provides an interactive virtual landscape of our territory so that the place names and the stories that go with those places can be enjoyed and assimilated by all Crees. The centre will also be a site for communicating our expanding knowledge of the archaeological record. As a research centre, it can support ongoing studies of recent and natural histories of our territory in partnership with researchers from McGill and Concordia, as well as elsewhere. Language teaching and research, cultural programming, mapping, and a library will also have a home there.

We intend that Wemindji's protected area network, our knowledge centre, a public website, and various cultural and ecological tourism ventures under development will together enable us to communicate our knowledge and experience to the wider world. We, the Cree people, have never sought to isolate ourselves, and we do not believe that our situation is unique. Relationship and respect are the basis for addressing social and ecological challenges everywhere. The promotion and protection of culture, and the promotion and protection of our ecological heritage, are one. Whereas Western notions of protection are centred on protecting the land for human good and preserving it by excluding or limiting human disturbance, the Cree view moves toward resolving the tension between these two. That is, we see humans as part of the larger community of life and believe that each form of life has its own, equally essential, role in that community. In short, we're all in it together. We are talking, then, about caring for the whole community of life in Eeyou Istchee. This includes Cree land-based livelihoods and traditional activities, Cree knowledge, and the renewal of Cree knowledge by maintaining these connections.

AA WIICHAAUTUIHKW: PARTNERSHIPS IN ACTION AND PROTECTED AREA CREATION

Rodney's perspective provided a guiding spirit for the protected area project and the research that supported it. Our Cree colleague Dorothy Stewart used the Cree word "*aa wiichaautuihkw*," which can be translated as "they are coming together to walk together," to describe our partnership at the beginning of this journey. The word evokes the priority given to the relational aspects of our partnership since 2001, when our project began. As we look back over almost two decades of working together, we recognize the significance of our partnership in advancing Wemindji's approach to balancing development and environmental protection. Partnerships are what make things happen, and the relationships developed are sustained over decades.

The initial impetus for creating a protected area was very much a coming together of ideas. In July 2000, as Colin Scott and Peter Brown of McGill University took a canoe trip down the magnificent Paakumshumwaashtikw (Old Factory River), they discussed the need to protect it. When they reached Wemindji village, they met up with Deputy Chief Rodney Mark, who had also been thinking about ways to protect the river's cultural, historical, and ecological values. Rodney's concerns were tied to his involvement in an annual canoe expedition on the river for Cree youth and to a growing sense of apprehension in the community regarding the proliferation of mining claims in the region. Next, Colin, Peter, and Rodney raised the possibility of protecting the Paakumshumwaau-Maatuskaau watersheds, speaking informally with local leaders and families whose hunting territories were directly threatened. More formal discussions with Wemindji Band Council members and other locals soon followed. With strong support for the idea, the next move was to build a research partnership. Rodney recalls contacting Colin. They talked and exchanged ideas. According to Rodney, "I think he was thinking about building a protected area, and we were kind of thinking the same thing, and it kind of just fell into place. That made this partnership really successful because we were kind of independent, but we were interdependent at the same time. We complemented each other's objectives, and that added value to both parties."[3]

The first steps in conducting research for the project took place during the summers of 2003, 2004, and 2005, as part of a McGill School of Environment undergraduate field course led by Colin Scott. Student efforts were directed toward gaining an understanding of community needs and issues, and conducting preliminary field survey work. A significant contribution of the latter was basic mapping of biodiversity and other features of the traditional territory that community members identified as important. The contributions of the nituuhuu uchimaauch, as well as expert guides and mentors Leonard and Ronnie Asquabaneskum, Dorothy Stewart, Abel Visitor, and Freddie Atsynia, were particularly valuable in this regard. During these years, Clara Stewart, Fred Stewart, Elmer Visitor, and the Asquabaneskum family worked with Véronique Bussières, a master's student at Concordia University at the time, to document stories and place names in the Paakumshumwaau area (Bussières 2005; Chapter 10, this volume). These early projects built on the more sustained research engagements of Wemindji Eeyouch with Colin Scott and Monica Mulrennan, of Concordia University's Department of Geography, Planning and Environment, that enabled community members to weigh the possible benefits and impacts of an expanded research partnership (for a comprehensive account of the partnership, see Mulrennan, Mark, and Scott 2012).

A trans-disciplinary team of ten co-investigators from the natural sciences, social sciences, and humanities joined the project in 2005. Daisy Atsynia, Fred Blackned, Leslie and Lot Kakabat, Beverly Mayappo, and Fred, Henry, and George Stewart, as well as various family members and many other residents, including administrative staff for the Cree Nation of Wemindji, joined as expert knowledge holders who could guide the research agenda. Many of the two dozen graduate students made significant contributions to both the research and this volume. A few of them, such as Véronique Bussières, Claude Péloquin, Jesse Sayles, and Katherine Scott, who also served as project administrator, worked with us after early undergraduate field courses and continued through to master's and doctoral studies. Dorothy Stewart, whose family hunting territories are in the Paakumshumwaau-Maatuskaau Biodiversity Reserve, also helped shape the project and became an invaluable liaison as translator, interpreter, and cross-cultural adviser.

The research partnership, which followed the principles of community-based participatory research through all its stages (see Mulrennan, Mark, and Scott 2012), was intended to achieve a cross-fertilization of Indigenous and academic perspectives that would yield detailed knowledge of local ecology and innovative approaches to environmental protection and cultural survival. According to Rodney,

> It really validated the hunters' and land users' claims when the scientists basically made the same observations and correlated them back with the trappers' knowledge. To me, that was such a good thing for the trappers and land users. And I want to make sure that we use the protected areas to help get access to further research funds and that all these other things put together add value to Cree knowledge. I think that is basically the biggest goal I have for this project.[4]

Collaborative work presents challenges and opportunities for both communities and researchers, since there are many unknowns from the start. Participants must adapt to working with unfamiliar people and experiment with new ways of working. In the process, one can feel that the experience is much like skating on very thin ice. The outcomes of our work were valued in various ways, but the partners generally agree that the experience was mutually beneficial. Of most immediate value to the community, of course, are the tangible outcomes: the biodiversity reserve itself and the potential for creating the Tawich (Marine) Conservation Area. For the academic partners (as well as some community members), this book is a meaningful result of the partnership, as are the twenty related theses and dozens of papers, reports, and presentations that preceded it.

It is clear to all of us that the most satisfying results are the rich and mutually beneficial engagements and long-term friendships that persist to this day. New projects and new research partnerships continue to grow from the relationships that created this volume, such as the team working on the community's new cultural learning centre and several recent master's and doctoral research projects. Many members of our team feel strongly that a research partnership does not terminate when the funding for a particular project runs out.

WEMINDJI IN REGIONAL CONTEXT

Wemindji is located on the east coast of James Bay, northern Quebec, and is one of ten Cree First Nations in Eeyou Istchee (Figure 0.1). The village site of Wemindji, formerly known as Paint Hills, was established in 1959 at the mouth of the Maquatua River (Mwaakutwaashtihk), following the relocation of the community from Old Factory Bay, fifty kilometres to the south. Wemindji's traditional territory consists of twenty-one family hunting areas bounded by latitudes 52°30'N and 53°10'N and extending inland about three hundred kilometres. The community has a population of 1,444 (Statistics Canada 2017) and a mixed economy comprising formal wage-labour, income subsidies, and subsistence harvesting (Scott 1988, 1996).

Although this volume focuses on the Cree Nation of Wemindji, the community is closely tied through historical and kinship connections to other Cree groups, particularly the neighbouring Cree nations of Chisasibi to the north and Eastmain to the south. Wemindji, which has one-third the population of Chisasibi and twice that of Eastmain, fared comparatively better in terms of the impact from the James Bay hydro-electric project. What is now the Cree community of Chisasibi, located on the island of Fort George at the mouth of the La Grand River, suffered major disruptions during the late 1970s. It was forced to move from Fort George to the present site due to the dramatically increased flow of the La Grande River and the construction of reservoirs and power stations upstream. Eastmain was heavily affected by the massive flooding that occurred when the Eastmain and Opinaca Rivers were diverted and by the lack of provision for flow maintenance. The Eastmain diversion flowed through Wemindji territory en route to the La Grande Complex via Sakami Lake, flooding portions of inland hunting territories but leaving coastal hunting territories untouched (Alan Penn, pers. comm., October 25, 2017). Nevertheless, the majority of Wemindji's hunting territories were directly affected by reservoirs and river diversions.

FIGURE 0.1

Wemindji Cree family hunting territories and the Paakumshumwaau-Maatuskaau Biodiversity Reserve

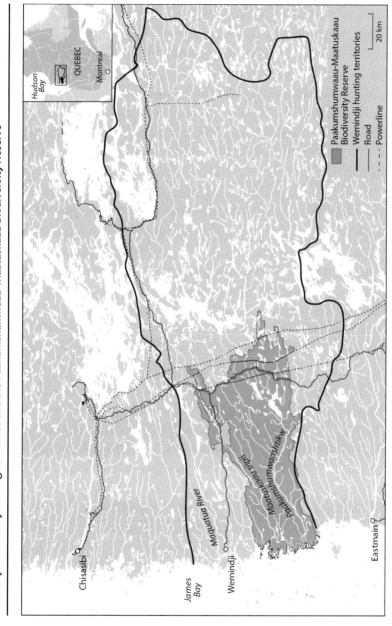

Source: Adapted from a map drawn by Gwilym Eades.

The implementation of the JBNQA in 1975, arising from the James Bay Project, established a host of regional-level provisions that addressed differing categories of land (in relation to landownership, Cree harvesting rights, and provincial rights to develop), economic and financial compensation, environmental protections, and Cree institutions responsible for education, health and social services, and local government. Wemindji's status as a beneficiary of the JBNQA means that it is very much part of a regional Cree political society and the increasingly complex institutional landscape that supports it. This involves working with the Grand Council of the Crees (Eeyou Istchee)/Cree Nation Government, which has a complex social and political agenda at the regional level, shaped by recent governance agreements between the Cree Nation and both Quebec and Canada. This institutional context is relevant to an appreciation of the dynamics between Cree communities and the regional nation, including their shared interests in economic development and environmental protection (Alan Penn, pers. comm., October 25, 2017). Indeed, the fact that Rodney Mark became vice-chair of the Cree Nation Government and deputy grand chief of the Grand Council of the Crees after having served as chief of Wemindji reflects the linkages across various levels of Cree governance.

In October 2015, the Cree Nation Government launched the Cree Regional Conservation Strategy as a framework for the co-ordination of Cree engagement in conservation and protected areas planning. The strategy recognizes community-led initiatives supported at the level of the family hunting territory – such as the Paakumshumwaau-Maatuskaau Biodiversity Reserve – "as the building blocks of a regional conservation areas network" (Cree Nation Government 2014, 11). Co-ordination at the regional-planning scale is intended to ensure that the network takes account of ecosystem processes and ecological connectivity issues as well as the knowledge, circumstances, and priorities of hunting territories and communities (Cree Nation Government 2014, 11). Speaking of his commitment to Lake Bienville (Apishukimimiish) on his home territory of Whapmagoostui, some five hundred kilometres north of Wemindji, Cree elder Andrew Kawapit described the wider context and growing interest in Eeyou Istchee for the creation of protected areas at a public consultation meeting on June 17, 2014 (the following quotation, translated by George Masty at the public meeting, refers to Kawapit as "he"):

> He is talking about certain areas for the Cree that have always been important ... We have a special relation with Lake Bienville. He and his brother grew up there and that's why they still spend time there. They have a relationship with this place and are still most comfortable being there. When you have a special relation with the land, you know that the place can be wounded. He wishes that we do not wound these special places any further. And if we do have developments, then he hopes that the developers are careful of certain areas that must be left undisturbed. This is how they must look at the land. About the land, he is aware, as are many of the elders, that when the land is changed, the animals are affected too ... If something happens to it, we are disturbed too. (Andrew Kawapit, public meeting of community consultations on the Eeyou Marine Region Agreement, Whapmagoostui, June 17, 2014.)

A BRIEF OVERVIEW OF THE VOLUME

The chapters in this book address several major questions, some of which we've mentioned above: What do we mean by environmental protection in the Wemindji Eeyou context? What values, interests, and knowledge enter into advocating or opposing particular visions and projects of environmental protection? What is being protected? How does our understanding of the natural and cultural history of the area inform the design and purpose of environmental protection? What kind of protection is appropriate and worthwhile, and what compromises or trade-offs impinge on environmental protection?

This volume is divided into three parts. The chapters in "Part 1: Context" attempt to situate environmental protection in relation to the politics and power dynamics of Indigenous ecological knowledge, the shifting political landscape of protected areas in relation to northern Indigenous groups, and the legacy of early scientific studies in the James Bay area. "Part 2: What to Protect" examines changes over time in the community of life, both human and non-human, that makes up the Paakumshumwaau-Maatuskaau watersheds and the adjacent offshore. "Part 3: How to Protect" concentrates on the efforts to safeguard the ecological and cultural heritage of the Wemindji area while also exploring options for a balance between large-scale development (specifically mining) and environmental protection.

Part 1 opens with a chapter from Monica Mulrennan and Fikret Berkes, who trace the history of dispossession and marginalization associated with early parks and protected areas, and identify relatively recent efforts to accommodate Indigenous rights and interests in protected areas. Particular attention is given to an emergent trend among many northern Indigenous groups, who (like the Cree Nation of Wemindji) have declared an interest in establishing protected areas as an element of their own governance strategies. The authors explore the thinking behind this phenomenon and identify opportunities and constraints that protected areas offer in a context of relentless expansion of northern resource extraction on Indigenous lands.

In Chapter 2, Wren Nasr and Colin Scott examine the political dimensions of Indigenous ecological knowledge in protected area creation and cultural continuity. They observe asymmetries of power in representation and knowledge production that have underpinned both historical and contemporary resource development and environmental protection projects. Crees, however, have proven remarkably effective negotiators and defenders of their autonomy through reliance on strategies that hinge on their distinctive knowledge and practices of environmental stewardship. As a result, hunting knowledge has acquired new relevance, and environmental protection in the territory has become a tool for reinforcing and enhancing Cree knowledge, values, and practices of caring for the land.

In Chapter 3, Ugo Lapointe and Colin Scott address the politics of mining in Wemindji territory, its relationship to protected area creation, and the search for a balance between development and environmental protection. Their story unfolds around Wemindji's decision to protect the culturally, historically, and environmentally significant Paakumshumwaau (Old Factory Lake), at the heart of the Paakumshumwaau-Maatuskaau Biodiversity Reserve. This decision was at odds with the aspirations of a mineral development company that invested heavily in exploring the potential of gold deposits at the site. Eventually, the conflict was resolved positively, but

protecting valued lands and fostering economic development and livelihoods at the local level involves a potentially risky balancing act of resistance
and accommodation that confounds simple stereotypes about Indigenous
positions regarding development and environmental protection.

The partnership between Wemindji Eeyouch and a team of academic
researchers that led to the creation of the Paakumshumwaau-Maatuskaau
Biodiversity Reserve was initiated and carried out in the context of northern
(particularly subarctic) research, which has a long and not always collaborative history. In Chapter 4, Katherine Scott outlines this history in Eeyou
Istchee, reminding us that researchers in various guises have visited James
Bay since the earliest explorers began collecting new-to-them flora and
fauna to take home for further study. Conversations between Scott and
Cree partners revealed that though local people guided and looked after
visiting researchers, their contributions were not always recognized. Too
often, research findings were not communicated back to them. To support
legal battles against Hydro-Québec in the 1970s, Crees began to employ
academic researchers as consultants and have increasingly taken charge
of their own research agendas and process. These initiatives nourished the
relationships and work presented in this volume.

Part 2 focuses on deciding what to protect, a distinct challenge in one
of the world's most dynamic coastal environments, where human beings
have lived for more than five thousand years. The particularities of this history – in terms of connections between shoreline displacement, ecological
diversity, wildlife resources, and human settlement and adaptation – have
not been documented previously; we begin that task in this volume. In
Chapter 5, Florin Pendea and his colleagues present a synopsis of landscape
evolution in eastern James Bay since the last Ice Age. Their study is based
on the examination of a site at Old Factory Lake. Extensive investigation
by this team of archaeologists, palynologists, and community members
supports a reconstruction of how the coastal landscape changed as the ice
receded, the land rose, and humans entered the area as early as six thousand years ago. Their account not only underscores the length of time that
humans have occupied this land, but argues that they were present in
greater numbers and capable of sustaining more permanent settlements
than previous studies have suggested.

In Chapter 6, Jim Fyles and his colleagues embrace the tradition of storytelling in their intriguing account of the ecological history of Paakumshumwaau's dynamic and diverse land- and seascapes. They join scientific perspectives with Wemindji Eeyouch ways of understanding the diverse and complex patterns in marine and terrestrial plant ecosystems and the varied bio- and geophysical processes that have shaped them.

Whereas Fyles and his team focus on the physical features and vegetation of the area, wildlife biologists Murray Humphries, Jason Samson, and Heather Milligan worked closely with Wemindji hunters and trappers to provide an account of some of the mammals in the Paakumshumwaau-Maatuskaau watersheds. In Chapter 7, they highlight, most notably, the extraordinary co-existence of species at the limits of their northern and southern ranges.

People, of course, have also played an integral part in the changing face of Eeyou Istchee's landscapes and ecosystems. Chapters 8 and 9 discuss human adaptations to changes in the availability of Canada geese, the seasonal hunting of which is the primary subsistence activity for Wemindji Cree hunters. In Chapter 8, Claude Péloquin and Fikret Berkes examine the multiple complexities of change in the goose hunt over recent years. The explain how Eeyou hunters have adapted their hunting practices to changing conditions while at the same time maintaining Cree principles of respect and reciprocity. In Chapter 9, Jesse Sayles and Monica Mulrennan investigate coastal landscape modifications, specifically in relation to the creation of ponds and flyways, as examples of Wemindji hunters' efforts to resist and redirect change so that important goose hunting places may be created and maintained.

In the final chapter of Part 2, Véronique Bussières, Monica Mulrennan, and Dorothy Stewart pick up the narrative thread. They take us from Paakumshumwaau (Old Factory Lake) down the Paakumshumwaashtikw (Old Factory River) and onto the coast and offshore, reminding us of the profound connections and continuities that Wemindji Eeyouch maintain with these lands and waters through memories of lives lived here, place names, and stories.

Part 3 deals with the challenge of how to protect by documenting the journey of the Wemindji Protected Areas Project in relation to the various strategies, obstacles, and opportunities encountered. In Chapter 11, Julie Hébert and co-authors provide a comprehensive account of the Quebec provincial government process that established the Paakumshumwaau-Maatuskaau Biodiversity Reserve. In Chapter 12, Monica Mulrennan and Colin Scott offer a parallel portrayal of efforts by the community and regional Cree leadership, as well as the federal government, to establish the Tawich (Marine) Conservation Area. Both accounts are of value in giving rare behind-the-scenes insights into the process of establishing protected areas. In the Conclusion, Monica Mulrennan, Katherine Scott, and Colin Scott highlight some of the challenges and achievements of the project.

NOTES

1 Although Canada is a settler state, it practises extractive colonialism in the North rather than settler colonialism. This distinction has implications for protected area establishment and conservation (see Chapter 1, this volume).
2 "Eeyouch" is the coastal dialect expression for "our people." "Eeyou Istchee" means "the land of the/our people"). "Eenou" or "Eenouch" are used by inland communities. The words are pronounced "ee-yooch," "ee-nooch," and "ee-you ist-chee."
3 This statement was made during a conversation between Rodney Mark and Katherine Scott, which was recorded in February 2012.
4 Conversation between Rodney Mark and Katherine Scott, recorded in February 2012.

WORKS CITED

Brammer, Jeremy R., et al. 2016. "The Role of Digital Data Entry Participatory Environmental." Special issue on citizen science, *Conservation Biology* 30 (6): 1277–87. https://doi.org/10.1111/cobi.12727.

Bussières, Véronique. 2005. "Towards a Culturally-Appropriate Locally-Managed Protected Area for the James Bay Cree Community of Wemindji, Northern Québec." Master's thesis, Concordia University.

Colchester, Marcus. 2004. "Conservation Policy and Indigenous Peoples: Indigenous Lands or National Park?" *Cultural Survival Quarterly* 28 (1). http://www.culturalsurvival.org/publications/cultural-survival-quarterly/none/conservation-policy-and-indigenous-peoples.

Cree Nation Government. 2014. "Cree Regional Conservation Strategy." Nemaska, QC, Cree Nation Government.

Mulrennan, Monica E., Rodney Mark, and Colin Scott. 2012. "Revamping Community-Based Conservation through Participatory Research." *Canadian Geographer* 56 (2): 243–59.

Scott, Colin H. 1988. "Property, Practice and Aboriginal Rights among Québec Cree Hunters." In *Hunters and Gatherers – Property, Power and Ideology*, eds. James Woodburn, Tim Ingold, and David Riches, Vol. 2, 35–51. London: Berg Publishers Ltd.

–. 1996. "Science for the West, Myth for the Rest? The Case of James Bay Cree Knowledge Construction." In *Naked Science: Anthropological Inquiry into Boundaries, Power, and Knowledge*, ed. Laura Nader, 69–86. London: Routledge.

Statistics Canada. 2017. "Focus on Geography Series, 2016 Census." Statistics Canada Catalogue No. 98-404-X2016001, Ottawa. http://www12.statcan.gc.ca/census-recensement/2016/as-sa/fogs-spg/Facts-csd-eng.cfm?LANG=Eng&GK=CSD&GC=2499050&TOPIC=1.

Stevens, Stan, ed. 2014. *Indigenous Peoples, National Parks, and Protected Areas: A New Paradigm Linking Conservation, Culture, and Rights*. Tucson: University of Arizona Press.

PART 1

Context

1

Protected Area Development in Northern Canadian Indigenous Contexts

MONICA E. MULRENNAN and FIKRET BERKES

Protected areas are a cornerstone of international biodiversity conservation strategies, whose rise to prominence has been shaped by a series of wide-ranging and vigorous debates. The conservation-preservation debate of the early twentieth century (Oelschlaeger 1991) was followed by the parks versus people debate (Western and Wright 1994; Terborgh 1999) and then by the biodiversity protection versus sustainable use debate (Miller, Minteer, and Malan 2011; Shafer 2015). These shifts in perspectives and priorities, and the policies and practices they inform, have had important implications for Indigenous peoples.

In this chapter, we examine recent developments in protected area establishment and management in the Canadian North. Our intention is to provide some context and insight for understanding why northern Indigenous communities, such as the Wemindji Cree Nation of northern Quebec, have come not only to accept protected areas but to embrace them as a central element of their own Indigenous governance strategies. More specifically, we examine the opportunities and constraints that protected areas present for northern Indigenous people.

We begin by overviewing some recent trends that are contributing to a rethinking of conservation and protected area approaches, examining if and how they are being translated into policy developments at international, national (Canada), and provincial (Quebec) levels. Next, we address the question of why some northern Canadian Indigenous groups are espousing protected areas. Finally, we consider the new paradigm of protected areas – articulated in 2003 (Phillips 2003) and refined a decade later (Stevens 2014b) – as an approach that encompasses social, economic, conservation, and recreational objectives. We conclude with a discussion of prospects for wider adoption of protected area development policies in the North.

EMERGENT TRENDS IN PROTECTED AREA DEVELOPMENT

Historically, protected areas have been linked to the dispossession of Indigenous peoples in the pursuit of preservationist goals and imperatives (Spence 1999; Nash 2001; Dowie 2011). An alternative people-oriented conservation took hold in the early 1990s, moving away from strict nature protection to concentrating on the needs, interests, and involvement of local communities (Western and Wright 1994; Wells and McShane 2004; Lockwood and Kothari 2006; Otto et al. 2013). This shift was a response to concerns about the social impacts of protected areas (Neumann 1998; West, Igoe, and Brockington 2006; Adams and Hutton 2007). It also reflected a mounting recognition that top-down, centralized, state-based approaches based on regulation and enforcement – also known as fortress conservation – don't always work (Brockington 2002).

Also influential was a growing appreciation of the fundamental connection between ecosystems and people, particularly the recognition "that healthy landscapes are shaped by human culture as well as the forces of nature, that rich biological diversity often coincides with cultural diversity, and that conservation cannot be undertaken without the involvement of those people closest to the resources" (Brown and Mitchell 2000, 70). Concepts such as cultural landscapes, social-ecological systems, resilience, and biocultural conservation, in which "social and ecological values [and practices] are embedded in the landscape" (Shultis and Heffner 2016, 1229), have helped to build a compelling counter-narrative to historically ingrained hegemonic paradigms. They facilitate an engagement of Indigenous worldviews and livelihoods in revised models for the conservation of environmentally, culturally, and historically important places.

Not everyone is onboard with this rethinking, however. Nature conservationists assert that people-centred approaches undermine conservation goals through their emphasis on poverty alleviation and sustainable use (Brandon, Redford, and Sanderson 1998; Terborgh 1999; Redford,

Robinson, and Adams 2006). They suggest that strictly managed protected areas offer the best hope for the conservation of biodiversity (Locke and Dearden 2005). Others, particularly those who work in the developing world, point to the pervasive reach of neo-liberal conservation, which holds that free markets and the commodification of nature can provide win-win solutions in terms of increased democracy and participation while promoting sustainability. They fear that this form of conservation is further displacing local peoples (Igoe and Brockington 2007). Social conflict (Adams and Hutton 2007) and the sometimes profoundly disruptive impacts of ecotourism and commodification (West, Igoe, and Brockington 2006) are additional consequences.

At the same time, we are seeing possibilities for more viable, socially just approaches to protected areas creation, including legal mechanisms to support decolonization. These are informed by concerns about the social impacts of protected areas on local people and are fuelled by an international-level acknowledgment of the political and human rights of Indigenous peoples (Brosius 2006; Scott and Mulrennan 2010; Stevens 2014a; Tauli-Corpuz 2016). Nevertheless, progress is uneven and unstable for various reasons, one of which is its possible eclipse by what Peter Brosius and Sarah Hitchner (2010, 150) refer to as a "strategic turn in conservation." This they define in terms of the adoption of a discourse of investment and valuation, and a preference for large-scale eco-regional conservation initiatives supported by geo-spatial technologies and other methodological tools that seek to reduce and simplify the complexity of local contexts.

RECENT PROTECTED AREA POLICY DEVELOPMENTS

International

Strategies for worldwide conservation efforts are shaped at the international level by a variety of organizations, including several United Nations agencies and non-governmental organizations. Key among these is the International Union for Conservation of Nature (IUCN), which, through the World Commission on Protected Areas, has supported the establishment of a worldwide representative network of terrestrial and marine protected areas during recent decades. These areas vary considerably in size, purpose, governance, management, and outcomes (Dudley et al. 2010). To facilitate their listing and cataloguing, the IUCN established

TABLE 1.1
IUCN protected area categories

IUCN category		Definition
I(a)	Strict nature reserve	These areas are set aside to protect biodiversity and also geological and geomorphological features. Human visitation, use, and impacts are strictly controlled and limited to safeguard conservation values. Such areas can serve as indispensable references for scientific research and monitoring.
I(b)	Wilderness area	These are usually large unmodified or slightly modified areas, retaining their natural character and influence. They lack permanent or significant human habitation and are managed to preserve their natural condition.
II	National park	These large natural or near natural areas are set aside to protect large-scale ecological processes, as well as their characteristic complement of species and ecosystems. These also provide a foundation for environmentally and culturally compatible spiritual, scientific, educational, recreational, and visitor activities.
III	National monument	These are set aside to protect a specific natural monument, which can be a landform, sea mount, submarine cavern, geological feature such as a cave, or even a living feature such as an ancient grove. They are generally quite small and often have high visitor value.
IV	Habitat/species management area	These aim to protect specific species or habitats, and their management reflects this priority. Many category IV areas need regular active interventions to address the requirements of particular species or to maintain habitats, but this is not a requirement of the category.
V	Protected landscape/ seascape	These are established where the interaction of people and nature has produced an area of distinct character, with significant ecological, biological, cultural, and scenic value, and where safeguarding the integrity of this interaction is vital to sustaining the area and its associated nature conservation and other values.
VI	Managed resource protected area	These conserve ecosystems and habitats, together with associated cultural values and traditional natural resource management systems. They are generally large, with most of the area in a natural condition. A portion is under sustainable natural resource management, and low-level non-industrial use of natural resources compatible with nature conservation is a main aim.

Source: Adapted from Nigel Dudley (2008).

a set of categories that reflect differing management objectives as well as a range of levels of human intervention and management authority (IUCN 1994). The purpose of these categories, shown in Table 1.1, is to promote the importance and scope of protected areas, provide international standards for reporting different types of protected areas, and improve communication and understanding among all stakeholders in conservation (Lockwood 2006).

The World Parks Congress (WPC), organized by the IUCN at decadal intervals, has played a critical role in promoting protected areas and in informing approaches and practices of protected area management (Brosius 2004). Of key importance was the fifth WPC, held in Durban, South Africa, in 2003. According to Adrian Phillips (2003, 19), this meeting endorsed "a new paradigm for protected areas, which contrasts in almost every respect with that which prevailed 40 or even 30 years ago." This new paradigm, which affirmed "a broader way of looking at protected areas" (ibid., 31), included a broader range of actors recognized among those who initiate and manage protected areas; a broader scale, as reflected in the emphasis on ecological networks and bioregional planning; and a broader scope in terms of increased options for governance types and management categories.

The Durban Action Plan, which supported this "sea-change" in approach (Shadie 2006b, 78), identified strategic directions to guide protected area development over the next ten years. These included improved governance of the areas, as well as the increased participation of Indigenous peoples and local communities in their designation and management (Borrini-Feyerabend, Kothari, and Oviedo 2004; IUCN 2005; Berkes 2009). Of particular significance was official recognition of the contribution made to biodiversity by Indigenous and community conservation areas (ICCAs). These were defined as "natural and/or modified ecosystems containing significant biodiversity values, ecological services and cultural values, voluntarily conserved by Indigenous peoples and local communities, both sedentary and mobile, through laws or other effective means" (Borrini-Feyerabend, Kothari, and Oviedo 2004, xv).

The IUCN's protected area category system has since been reviewed and amended to reflect these shifts in priorities and to respond to evolving issues and challenges identified by the international conservation community.

The latter include concerns that the IUCN categories "were being used as an excuse for relocating indigenous peoples from their traditional territories" (Dudley et al. 2010, 486). Revised guidelines on the categories, released in 2008 (Dudley 2008), include some changes that could improve the status of Indigenous peoples while also including others that could be deleterious to their rights and interests (ibid.). For example, the definition of protected areas now lays greater emphasis on nature conservation, whereas ICCAs are given formal recognition as a distinct governance type, but only if their management objectives adhere to the IUCN definition of an ICCA (Dudley et al. 2010). Established in 2008, the ICCA Consortium, which collaborates directly with the IUCN, the Convention on Biological Diversity Secretariat, and other international organizations, has played a major role in promoting appropriate recognition and support for ICCAs, including defending the interests and efforts of individual communities with respect to their ICCAs (ICCA Consortium 2010, 2017).

Indigenous governance was a central theme of the sixth WPC, which was held in Sydney, Australia, in 2014 (Enkerlin-Hoeflich et al. 2015). Participants reported "increased recognition and respect for Indigenous Peoples' and local communities' rights and traditional knowledge by many governments, intergovernmental organizations and international policy processes" (IUCN 2014, 1) during the intervening decade. Significant challenges remain, however. These were acknowledged in "The Promise of Sydney Vision" (IUCN 2014, 2), which articulated a commitment "to redress and remedy past and continuing injustices in accord with international agreements" and to recognize and support through partnerships with Indigenous and local communities their "long traditions and knowledge, collective rights and responsibilities" (ibid.). Recommendations to fulfill this commitment include the creation of a new IUCN category to address "Indigenous territories management, including Indigenous Protected Areas" (IUCN 2014, 2). More recently, at the World Conservation Congress, IUCN members approved a motion for the recognition and respect of ICCAs that overlap protected areas (Stevens, Broome, and Jaeger 2016).

Notwithstanding ongoing debates and the uneven experiences of Indigenous peoples with protected areas across the globe, recent international policy developments affirm a new paradigm of protected areas.

It commits national governments, as parties to the Convention on Biological Diversity (CBD), "to develop participatory, ecologically representative and effectively managed national and regional systems of protected areas, stretching, where necessary, across boundaries, integrated with other land uses and contributing to human well-being" (Shadie 2006a, 704). There is optimism that the United Nations Declaration on the Rights of Indigenous Peoples, adopted in 2007, will further strengthen provisions of the CBD, particularly concerning state obligations to Indigenous peoples (Cittadino 2014).

Canadian North

Although Canada is a settler state, the colonial history of its northern regions is most accurately characterized as "extractive colonialism." Unlike other parts of the country, where colonialism functioned through the replacement of Indigenous populations with an invasive settler society, colonization in the North concentrated on resource exploration and extraction, allowing maintenance of regional majority populations of Indigenous people. The assertion of state bureaucratic control over northern wildlife, forests, and wilderness (rather than the settlement of European migrants) "provided one of the most important administrative vehicles through which the imperial powers attempted to assert authority over subsistence hunters and small-scale agriculturalists in hinterland regions" (Sandlos 2014, 135).

Until recently, the dominant narrative regarding the early twentieth-century Canadian conservation movement focused on the dedicated efforts of certain federal government employees and their earnest pursuit of enlightened wildlife programs (Foster and Hammond 1998). Revisionist accounts suggest a more complex history in which these same programs enabled the expansion of state control over northern Indigenous communities (Loo 2006; Sandlos 2007). According to this view, the implementation of game regulations, the creation of sanctuaries, and the hiring of game officials undermined local authority over wildlife resources and "had the effect of marginalizing local customary uses of wildlife ... [as] part of the colonization of rural Canada" (Loo 2006, 6). John Sandlos (2014) argues that federal government efforts to interfere with the subsistence economy of northern Indigenous hunters reflect a colonial mentality, with respect to northern conservation programs that persisted until recent times. Parks and protected areas are heavily implicated in this colonial history. As David Neufeld (2008, 183) contends, national parks represent "powerful tools in the business of constructing the State" that have until recent years been associated with a misrepresentation of the histories, cultures, and places of Canada's Indigenous peoples.

During the first half of the twentieth century, the guiding principle of national park management was to ensure the "benefit, education and enjoyment" of parks by Canadians (National Parks Act 1930, s. 4; Parks Canada 1964). Aboriginal rights and title were not acknowledged, and little account was taken of the interests of Indigenous populations. Although conservation was not a stated goal, Parks Canada focused on establishing a parks system to protect "representative natural areas" rather than "vital wildlife habitat," which simply further alienated Indigenous groups and undermined their subsistence economies and land-based cultures (Peepre and Dearden 2002; Dearden and Bennett 2016).

A major change in the discourse on Indigenous peoples and national parks occurred in the late 1970s, facilitated by public hearings associated with Justice Thomas Berger's 1977 report on the Mackenzie Valley Pipeline Project, which offered a new perspective on northern development and Indigenous use of wilderness areas (Berger 1977). Parks Canada responded by defining a new relationship between the federal government and Indigenous peoples through the adoption in 1979 of a policy of joint management (Peepre and Dearden 2002). The terms of this relationship were refined by section 35 of the Constitution Act, 1982, which recognized and affirmed the existing Aboriginal and treaty rights of Canada's Indigenous peoples. Several legal decisions subsequently clarified the nature of those rights, resulting in a 1994 overhaul of Parks Canada policies to include provisions for the continuation of subsistence harvesting activities in parks and opportunities for Indigenous participation – though often as "reluctant partners" (Mulrennan 2015, 71) – in park development and management. However, the principle of ecological integrity, defined as "the minimization of human impact on natural processes of ecological change" (Parks Canada 1994, 119), was also endorsed at this time as the first priority of park management.

A federal panel investigation of Canada's national parks, commissioned in the late 1990s, found that they were under serious threat for a variety of reasons. Among these, the panel noted that "Parks Canada has tradition-ally adopted a legalistic approach and position in dealing with Aboriginal issues – which are often referred to as 'problems'" (Parks Canada 2000, 7.3). Panel recommendations included legislative amendments to confirm ecological integrity as the first priority of national parks, alongside the adoption of clear policies to encourage and support the development and maintenance of genuine partnerships with Indigenous peoples in the management of parks (Parks Canada 2000). According to Douglas Clark, Shaun Fluker, and Lee Risby (2008, 157), these recommendations set up a paradox "advocating ecological integrity in a wilderness-normative sense that excludes people, while at the same time promoting stronger relation-ships between Parks Canada and Aboriginal Peoples." Catriona Mortimer-Sandilands (2009) argues that in privileging science to define and value nature, ecological integrity legitimizes the dominance of the federal gov-ernment in managing parks even as it marginalizes the knowledge inputs of Indigenous peoples.

Amendments to the Canada National Parks Act in 2000 brought many changes and led to the creation of seven new national parks, five of which were established through agreements with Indigenous peoples. Others followed; by the end of the decade, about 70 percent of land protected under the national park system was the result of Indigenous partnership arrangements (Langdon, Prosper, and Gagnon 2010). The terms of these partnerships were almost exclusively dictated by park-specific management arrangements laid out in northern comprehensive land claim agreements, including provisions for establishment of co-management boards, trad-itional access rights, and protocols for archaeological work, as well as eco-nomic opportunities (Gladu et al. 2003; Berkes et al. 2005; Environment Canada 2015).

Assessments varied regarding the relationship between Indigenous peoples and the federal government. With respect to the creation of Gwaii Haanas National Park in the 1990s, C. Lloyd Brown-John (2006, 1) com-mended Parks Canada for its willingness and ability to move quickly from its early experience in negotiating collaborative management to embrace "full-fledged collaborative management for the operation of all national parks in the territories." Similarly, Frances Gertsch et al. (2003, 1) applauded the rapid evolution of co-operative management in Quttinirpaaq National

Park in Nunavut, attributing this to "stronger assertion of land owner-ship by Aboriginal Peoples, and the pursuit of participatory democracy by governments."

Others provided less upbeat assessments, particularly concerning Parks Canada's record prior to and during the ascendancy of Stephen Harper's Conservatives, who formed the federal government from 2006 to 2015. For example, Juri Peepre and Philip Dearden (2002, 331, emphasis added) criticize Parks Canada for its failure to establish "*true* partnerships with shared authority and resources" or "*genuine* joint or co-operative decision-making." Vicki Sahanatien (2007) suggests that Parks Canada had not developed policy or direction in sufficient detail to provide guidance on wilderness management for national parks that were tied to northern land claim agreements. Similarly, Eugene Thomlinson and Geoffrey Crouch (2012, 76, 77) observe that, though Parks Canada appeared to be gradually "coming to the realization" that Indigenous relationships to the land could be consistent with Western conservation goals, its engagement with com-munities had "seemingly been on an ad hoc, case-by-case basis with a continuum of degrees of involvement by Aboriginal people." The five-decade process of creating the Torngat Mountains National Park Reserve, which included a legal challenge by Inuit and a lengthy period during land claim negotiations when plans for the park were on hold, illustrates the roller-coaster relationship between Parks Canada and some northern Indigenous peoples (Canadian Parks Council 2008).

Overall, collaboration between the federal government and Indigenous peoples has improved. According to the Canadian Parks Council, three primary factors account for this. These include the leadership role asserted by Indigenous communities in protecting and managing their traditional lands, efforts to develop more meaningful partnerships based on respect and trust, and recognition of the contribution of cultural resources and traditional knowledge to park management (Canadian Parks Council 2008). For their part, Grant Murray and Leslie King (2012) attribute chan-ges in Parks Canada's position to its alignment with international-level commitments, as well as national legal and policy developments. Philip Dearden and Nathan Bennett (2016) note that national parks established since the passage of the Constitution Act, 1982, which affirmed Aboriginal and treaty rights, are associated with improved working relationships with

Indigenous peoples. The greater certainty and assurances provided by legal frameworks, such as comprehensive land claim agreements, also encourage some northern Indigenous groups to take the lead in protected area development. They have become "enthusiastic instigators" of such areas (Mulrennan 2015, 72), which has been the real game-changer in Canada over the past twenty years.

Initiatives such as the Northwest Territories Protected Areas Strategy (NWT-PAS), which is based on a unique community-driven process to establish a network of protected areas across the Northwest Territories, demonstrate the desire of northern communities for new approaches that extend beyond the terms of land claim settlements (NWT-PAS Advisory Committee 1999). In 2009, the community of Deline, in partnership with Parks Canada, established Saoyú-ʔehdacho as the first protected area created under the NWT-PAS. Five new protected areas championed by First Nations were proposed in the Northwest Territories between 2006 and 2011 (Environment Canada 2015). And in 2013, a five-year land withdrawal order was registered, granting interim protection of an important cultural and spiritual gathering place known as Edéhzhíe (*Northwest Territories Lands Act* 2014). Similar developments occurred across the Canadian North (Environment Canada 2015). New designations that seek to accommodate First Nation interests and to support collaborative working relations also emerged, such as traditional use planning areas in Manitoba and a type of conservancy in British Columbia (Stronghill, Rutherford, and Haider 2015). However, except for the tribal park model asserted by the Tla-o-qui-aht and other First Nations in British Columbia, ICCAs have been slow to gain traction in Canada (Herrmann et al. 2012). One explanation for this is that Indigenous participation in Canadian protected areas (especially in the North) tends to occur in the context of modern land claim agreements, treaties, and other agreements, which dictate the terms and range of possibilities for Indigenous engagement.

Between 2006 and 2011, Indigenous peoples contributed to the establishment of "tens of thousands of square kilometres of protected areas designation" in Canada through various forms of collaborative management and benefit sharing (Environment Canada 2015, 7). The emphasis of these new initiatives on cultural values and regional priorities, and the attention given to sustainable resource use and local knowledge over

southern concepts of wilderness, and in some cases over Western know-
ledge, represent a marked departure in Canadian policy (Dearden and
Bennett 2016). Specific initiatives include the collaboration between Parks
Canada and the Dehcho First Nations on a six-fold expansion of Nahanni
National Park Reserve, a commitment to create the Náàts'ihch'oh National
Park Reserve in partnership with the Sahtu Dene and Metis people, and
the establishment in 2010 of Gwaii Haanas National Marine Conservation
Area Reserve and Haida Heritage Site (McNamee 2010). According to
Kevin McNamee (ibid., 149), such progress was possible "precisely because
Parks Canada has taken the time to establish relationships and negotiate
agreements to ensure new parks are cooperatively managed." Obstacles
remain, however. Steven Nitah was the Lutsel K'e Dene First Nation's chief
negotiator for the establishment of Thaidene Nëné National Park Reserve,
Northwest Territories. As he explained, "One of the biggest challenges we
had at the negotiating table was fettering the minister's authority. When
we're talking nation-to-nation discussions, the minister has to be able to
get into a partnership relationship with indigenous governments and be
able to share responsibility and authority over those areas" (Standing
Committee on Environment and Sustainable Development 2016).

Canada currently lags behind other industrialized democracies when
it comes to protecting its lands and fresh waters (CPAWS 2017). A decade
of Conservative government policies was characterized by a determined
weakening of environmental protection in favour of industrial and resource
development, and the deterioration of relationships between Ottawa and
Indigenous communities. At the end of 2017, only 10.5 percent of Canada's
land area and 2.9 percent of its ocean estate was protected (Canadian Council on Ecological Areas 2017a) (Figure 1.1). Prime Minister Justin Trudeau
campaigned on a commitment to change, including a "strong environmental platform and ambitious targets" (Jessen, Morgan, and Bezaury-
Creel 2016, 18). Following its 2015 election, his Liberal government adopted
a suite of objectives, known as the 2020 Biodiversity Goals and Targets
for Canada, that aligned with Canada's international obligations under the
CBD's Strategic Plan for Biodiversity 2011–20 and the Aichi Targets. Key
among these was Aichi Target 11, a commitment to conserve (through
networks of protected areas and other effective area-based measures) at

FIGURE 1.1

Canada's protected area network

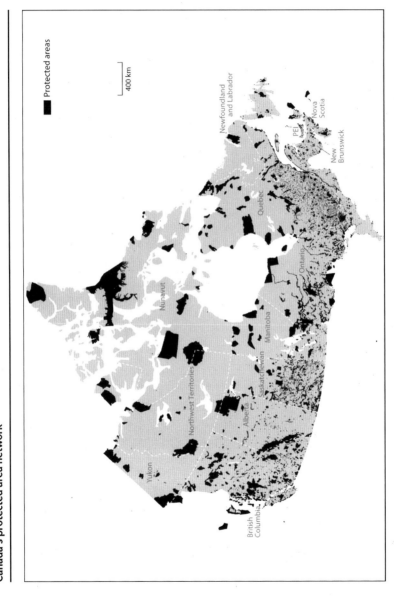

Protected areas

400 km

Newfoundland
and Labrador

Nova
Scotia

PEI

New
Brunswick

Quebec

Ontario

Nunavut

Manitoba

Saskatchewan

Northwest Territories

Alberta

British
Columbia

Yukon

Source: Adapted from the Canadian Council on Ecological Areas (2017b), https://ccea.org/wp-content/uploads/2019/04/CCEA_
CANADA_15M_LETTER_CARTS_IUCN_20171231.pdf.

least 17 percent of terrestrial areas and inland water, and 10 percent of marine and coastal areas of Canada by 2020.

A central tenet of the federal approach to meeting these commitments is a renewed nation-to-nation relationship with Indigenous communities, based on the recognition of Indigenous rights, respect, co-operation, and partnership (Parks Canada 2017b). As a first step in this approach, an Indigenous Circle of Experts was created in 2017 by the federal government to provide Indigenous advice throughout all aspects of the biodiversity initiative. The government is also paying special attention to the development of criteria for Indigenous protected and conserved areas, regarded as essential to its efforts to achieve the 2020 targets (Parks Canada 2017a). These are hopeful signs that Canada has entered a new collaborative process, involving partnerships with Indigenous peoples that will protect more land and sea, and contribute to reconciliation (CPAWS 2017). However, there is no room for complacency. A recent audit highlighted Ottawa's failure to implement its climate change commitments (Gelfand 2017), and when the federal government approved two major pipelines without their consent, First Nations groups in British Columbia felt betrayed. Clearly, sustained political pressure will be needed to ensure that the relationship between protected areas and Indigenous peoples in Canada changes for the better.

Quebec

At the turn of the millennium, less than 3 percent of Quebec – almost exclusively in the southern part of the province – enjoyed protected area status (MDDEP 2000). Since that time, Quebec has demonstrated a strong commitment to protected area development. The adoption of the Stratégie québécoise sur les Aires Protégées (SQAP) in 2000 by the Ministère du Développement durable, de l'Environnement et des Parcs du Québec (MDDEP) was intended to address Quebec's weak record in protected area development (MDDEP 2000). Under the SQAP, 8 percent of Quebec's land mass was to be protected by 2005, with each of thirteen natural provinces represented. This target was reached in March 2009, with the announcement of seventeen newly proposed protected areas, which encompassed a total of 135,326 square kilometres (MDDEP 2009). New strategic guidelines for protected areas were adopted in April 2011, with the objective to expand the network to 12 percent by 2015 (MDDELCC 2011b). As of March 2017, 9.35 percent of Quebec's terrestrial area was under protection (MDDELCC 2017) (Figure 1.1).

The 2002 Natural Heritage Conservation Act (NHCA), in concert with a Strategic Action Plan for Protected Areas, facilitated the establishment of Quebec's protected area network by clarifying the various types of areas and the activities prohibited within each; it also facilitated the devolution of some management responsibilities to local organizations, including regional Indigenous governments (MDDEP 2003a, 2005). These measures, according to MDDEP, were intended to make communities the first beneficiaries of biodiversity conservation (MDDEP 2003b). Biodiversity reserves and aquatic reserves were introduced as two new categories of protected area under the NHCA. These were less restrictive than the other categories, and the Quebec government saw them as more appropriate for the North (see Chapter 11, this volume).

The first park created under the NHCA was Pingualuit National Park, established in 2004 after more than three decades of protracted negotiations between Inuit representatives and the Government of Quebec (MDDELCC 2007). A significant complication was the fact that the park was located on lands that are under the jurisdiction of the James Bay and Northern Québec Agreement (JBNQA). Ultimately, the rights of Inuit as beneficiaries of the JBNQA – which include provisions for Indigenous harvesting rights, land categories, and resource management regimes – prevailed over park regulations. The latter included prohibitions against carrying a firearm and restrictions concerning campfires and the cutting of trees (Environment Canada 2015). Due to provisions in the NHCA, Inuit participation was facilitated through the devolution of park operations and management to the Kativik Regional Government (Société de la faune et des parcs du Québec 2000). Two more provincial national parks, also in Nunavik – Kuururjuaq Park on the Quebec-Labrador Peninsula and Tursujuq on Hudson Bay (the largest national park in Quebec) – were established in 2009 and 2013, respectively, based on a similar set of arrangements for devolution (Canadian Parks Council 2008).

The experience gained in establishing these protected areas "includes the hard-earned lessons of the importance of working with [Indigenous peoples] as equal partners, of the need to achieve mutual respect and trust, of lengthy setbacks, and of tough negotiations and compromise among diverse parties" (Canadian Parks Council 2008, 55). The experience of the Cree Nation of Mistissini (Eeyou Istchee) in negotiating the creation of the Albanel-Témiscamie-Otish Park is illustrative of the extent to which parks policy has evolved. The initial proposal for the park in the early 1990s included a promise from Quebec that policies would be modified "to allow" Crees to practise their traditional subsistence activities there. In response, the Cree Nation of Mistissini pointed out that the "right to harvest" was constitutionally protected under the JBNQA (Morrison 1997). It then set out the core values and guiding principles that should govern the development of the park. In 2008, despite many setbacks along the way, these values and principles were adopted to inform the spirit and intent of the park (MDDEP 2005).

By March 2009, Quebec had met its 8 percent target under SQAP. Later that year the Quebec Liberal government announced Plan Nord, an economic development strategy that committed it to a twenty-five-year, $80 billion investment in large-scale energy, mineral, and forestry resource development on the northern two-thirds of the province (roughly 1.2 million square kilometres). The strategy was accompanied by government assurances that it would also dedicate, by 2035, 50 percent of the northern part of the affected area to environmental protection and biodiversity conservation. It also established a more immediate target – to create protected areas on 20 percent of the area north of the forty-ninth parallel by 2020.

Limited initial consultation with Indigenous groups and other stakeholders, as well as ambiguity of the plan itself, raised concerns about its intent and potential impacts (Canadian Press 2011). The Cree response, articulated in "Cree Vision of Plan Nord" released in February 2011, insisted that they be "fully involved in the definition of the concept and principles that will guide the Plan Nord" and that they receive real and tangible benefits as an outcome of it (Cree Nations of Eeyou Istchee 2011, 12). The willingness of Quebec to accede to these terms was indicated in May 2011 with the announcement of its commitment to create the Assinica National Park Reserve, near the village of Oujé-Bougoumou (MDDELCC 2011a). Grand Chief Matthew Coon Come described the collaboration as "one more important step which demonstrates, in very tangible ways, the extraordinary potential of the new partnership and the new relationship between the Cree Nation and the Government and the people of Quebec when we decide to work together. It is an eloquent expression of what can happen when there is acknowledgement of rights and inclusion of our people and communities." (Coon Come 2011)

A year later the Government of Quebec and the Cree Nation signed the Agreement on Governance in the Eeyou Istchee (James Bay) Territory giving the Crees expanded powers over lands and resources in exchange for Cree support of the Plan (Secrétariat aux affaires autochtones 2012). Plan Nord was shelved by the Parti Québécois government in 2012 but was revived as a scaled-down project in April 2015 (with a $50 billion investment plan). The province's newly elected Liberal government reaffirmed the participation of local and Indigenous communities as a priority for the relaunched Plan Nord "in keeping with the principles of sustainable development and in a manner respectful of their culture and identities" (Secrétariat au Plan Nord 2015, 9).

A new partnership and collaboration agreement between the Grand Council of the Crees and the Quebec government to settle a major logging and road-building dispute highlighted ongoing tensions around jurisdiction over territory and resources. On July 15, 2015, the government and the Crees reached an agreement that included the formal designation of the Broadback River Conservation Area, the reintroduction of woodland caribou, and the creation of a Cree-Innu-Quebec joint forestry working group (Société du Plan Nord du Québec 2015). However, important locations to the south of the Broadback River were not included in the protected area, leaving many Crees feeling disappointed and vulnerable to development threats (East 2015; Cree First Nation of Waswanipi 2016). Disputes over territory were also involved, with Innu Nations claiming that the land covered by the agreement was Innu territory, not Cree (Dougherty 2015). Thus, despite measures to support improved relations between the provincial government and Indigenous people overall, the experience of individual nations or communities, particularly when protected area creation and resource extraction is involved, can be fraught.

Overall, though Quebec has made progress in the expansion of its protected areas, most of these are still under interim protection status with their boundaries and management plans yet to be confirmed. There is also the possibility of a new category of "multi-purpose" protected areas that would allow industrial use such as logging, undermining the effectiveness and integrity of their protection (CPAWS 2015). Finally, whether Quebec has the capacity and commitment to deliver on its promise to protect 50 percent of northern Quebec remains to be seen. In November 2018 the Cree Nation submitted a proposal for 30 percent of its terrestrial territory to be protected from development (Bell 2018). The ability of Quebec to respond appropriately and in a timely fashion to such initiatives will be an important measure of its success in meeting its conservation targets.

Thus, we should be cautious about celebrating the "emerging" conservation paradigm too soon. There is still a long way to go. Protected area policies in Canada and Quebec have their own drivers and dynamics, which are not necessarily the same as those of the international arena. Indeed, the developments in Canada and Quebec illustrate how new conservation thinking can combine with evolving relationships between the state and Indigenous groups, and how these changes can be translated into new policies at national and provincial levels.

WHY SOME NORTHERN INDIGENOUS GROUPS EMBRACE PROTECTED AREAS

Indigenous and other rural groups that depend on land and natural resources have historically been excluded from conservation discourses, in both Canada (Sandlos 2007) and internationally (West and Brechin 1991; Adams and Hutton 2007). Protected area creation has displaced, dispossessed, and marginalized many millions of Indigenous, tribal, and local people across the planet (West, Igoe, and Brockington 2006; Agrawal and Redford 2009). The new conservation and new Indigenous political realities can help explain the apparent paradox of why some Indigenous groups have begun to welcome such areas (Murray and King 2012; Stevens 2014a). A related question is the extent to which this new paradigm is a good fit with the contemporary socio-economic, cultural, and political context of northern Indigenous communities.

As discussed earlier, globalizing discourses of protected area development increasingly acknowledge the contributions of Indigenous peoples and call for policies that safeguard their interests and respect their customary resource practices and tenure systems. Legal recognition of collective tenure rights is understood as essential for positive conservation outcomes; its neglect "constitutes failure to recognize many of the fundamental freedoms and human rights of Indigenous peoples" (Stevens 2010, 193). Like a growing number of international organizations, the World Wildlife Foundation has officially adopted a "rights-based approach" to conservation and development projects that affect Indigenous peoples. It recognizes "that conservation planning and partnerships that do not build on rights are neither equitable, nor are they likely to be sustainable and constructive" (Larsen and Springer 2008, 36).

Indigenous peoples are availing themselves of legal and political opportunities created by this convergence of their aspirations for economic and political autonomy with the long-term health and protection of their traditional territories (Mulrennan 2008). However, as demonstrated by the November 2017 Supreme Court of Canada decision on *Ktunaxa Nation v. British Columbia* – which examined whether a First Nation belief that a proposed development project would drive away a Grizzly Bear Spirit constituted a violation of a religious right under the Canadian Charter of Rights and Freedoms – legal avenues can be long and complicated, with uncertain outcomes (Hopkins-Utter 2017). The establishment of protected areas may represent a more appealing alternative for some Indigenous groups, with the potential to fulfill the twin goals of reinforcing customary institutions of tenure and resource use while also enhancing protection of critical habitat and sustainable use of wildlife species.

If the Canadian protected areas network is to expand markedly, this must occur in the North, which has the country's only remaining large tracts of relatively intact ecosystems. Thus, any opportunities for setting aside significant land and marine areas are directly tied to the traditional territories of Indigenous peoples and depend upon their collaboration. At the same time, increasing resource development pressures in the North, including hydro-electric projects, gas exploration, and various mining operations, pose a serious threat to the integrity of these socio-ecological systems and the Indigenous communities that depend upon them. Many northern Indigenous groups, concerned about the impact of large-scale industrial development, recognize the need to find innovative mechanisms to enhance local systems of authority and to protect special places for future generations (Stadel, Taniton, and Heder 2002). Their engagement with government agencies in the process of protected area development is based on mutual interests and overlapping goals.

The primary framework for engagement between Indigenous groups and the Canadian state consists of treaties, comprehensive land claim agreements, and self-government agreements. Over the past four decades, the land claim settlement process has facilitated the creation of a number of northern national parks and altered the relationship between Indigenous peoples and federal, provincial, and territorial governments. Provisions in land claim agreements for the exercise of traditional harvesting activities, resource revenue sharing, and other economic opportunities, as well as Indigenous participation in park management, have enabled the establishment of these parks (Gertsch et al. 2003; Dearden and Bennett 2016). Less formal arrangements, such as memoranda of understanding, have also been employed when no comprehensive land claim agreement exists (Thomlinson and Crouch 2012).

The fact that northern Indigenous peoples play a more significant role in national park planning and development than do their counterparts in southern Canada can be partly attributed to constitutionally mandated arrangements associated with the settlement of their land claims (Dearden and Bennett 2016). Obviously, there are benefits to be gained from these agreements, but the failure of the Government of Canada to implement the terms of modern land claim agreements impedes broader joint initiatives involving Indigenous people (Natcher, Davis, and Hickey 2005; Spaeder and Feit 2005; Timko and Satterfield 2008; Dearden and Langdon 2009). Co-management "can be a means by which states empower marginalized and disadvantaged groups. But it is also a means by which state control is extended and confirmed. It can restore relations to country and to lands indigenous peoples value, but rarely on terms they determine" (Brockington, Duffy, and Igoe 2008, 11). It is also worth noting that co-management arrangements, as established under land claim agreements, may be incommensurable with ICCAs that don't exclude collaborative governance but require that primary decision making rests with the concerned community (ICCA Consortium 2017).

For northern Indigenous leaders, protected areas can present a counterweight to industrial-scale development projects, including those endorsed by Indigenous communities. As deputy grand chief for the Cree Nation, Rodney Mark expressed this delicate balance in reference to the anticipated creation of the Broadback River Conservation Area: "The Crees consider resource development on their lands with a basic question: does the project add value to the Cree Nation without compromising our core rights and traditions?" (Grand Council of the Crees [Eeyou Istchee] 2014, 2). These dual imperatives tend to arise in response to the social and economic realities of living in remote northern areas, often with poorly developed infrastructure and rapidly growing young populations in urgent need of jobs. In this respect, the opportunity to articulate an ongoing commitment to environmental stewardship through the establishment of a protected area in the traditional territory presents a welcome balance to more controversial development projects or compromised resource extraction approvals elsewhere (see Chapter 3, this volume).

In the Canadian North as a whole, there is a need for integrated planning that is consistent with the conservation and revitalization of its socio-ecological landscapes, as well as the economic well-being of its people. A case in point is the Whitefeather Forest Initiative in northwest Ontario, where planning for protected areas is very much intertwined with decisions about forestry use, control of land by the Pikangikum First Nation, and its ability to direct jobs and other economic benefits toward local people (Davidson-Hunt, Peters, and Burlando 2010; Whitefeather Forest Initiative 2012). According to Iain Davidson-Hunt, Nathan Deutsch, and Andrew Miller (2012), protected areas, such as the Pimachiowin Aki World Heritage Site, are part of a package of benefits and opportunities that can provide much-needed social and economic development, at the risk of opening up Anishinaabe lands to outsiders. Likewise, the provision of economic benefits and development opportunities through impact benefit agreements linked to protected area establishment (see Chapter 3, this volume) can represent an important acknowledgment of the challenges faced by northern Indigenous communities with respect to conservation and development. The Wemindji Cree Nation faces similar challenges and strives for conservation planning that goes hand in hand with the continued viability of the local economy, reconciling conservation, and development objectives (Mulrennan, Mark, and Scott 2012).

Due to the recent wave of new protected area designations in the Canadian North, the experience of such initiatives is immediate and tangible, not removed and abstract. In some cases, first-hand reports from communities that have already taken on the process can be useful in its assessment. The positive experience of the Mistissini Cree in the development of Albanel-Témiscamie-Otish Park, as communicated by Deputy Chief Kathleen Wootton during early meetings of the Wemindji Protected Areas Project, certainly strengthened the resolve of the Wemindji community to pursue a protected area strategy. Similarly, the willingness of government agencies to acknowledge mistakes and current limitations to their role and to commit to working on improving relations with Indigenous peoples also opens up spaces for productive dialogue and creative solutions (see Chapters 11 and 12, this volume).

A convergence of factors is therefore responsible for the new phase of protected area development in Canada, in which Indigenous peoples have become "enthusiastic instigators" of such areas (Mulrennan 2015, 72). This phase is marked by the increasing recognition of Indigenous peoples that parks and protected areas have the potential to enhance local institutions of tenure, knowledge, and practice; to protect traditional lands and resources; and to limit large-scale, externally driven development projects.

PROTECTED AREAS AND CREE CONSERVATION

Many aspects of this emerging paradigm are consistent with Indigenous notions of "taking care of the land," a relationship in which culture and land use are embedded in natural processes and informed by customary rights and responsibilities. There is no separation of use and management, or of user and manager (Berkes 2018). This contrasts with the old paradigm, in which protected areas were set aside purely for conservation purposes, in wilderness areas that were supposedly free of humans and their impacts. Viewed primarily as a national asset, they were governed centrally, using a technocratic approach to management (Sandlos 2007). Table 1.2 diagrams the two approaches.

TABLE 1.2

Conventional and emerging protected area paradigms

	Conventional paradigm	*Emerging paradigm*
Objectives	Focus on setting aside areas for conservation	Recognition of social, economic, and political purposes in addition to biodiversity protection
	Established for scenic preservation and protection of spectacular wildlife	Focus on natural ecosystem functioning
	Focus on wilderness, areas believed to be free of human impacts	Recognition of wilderness areas as culturally important areas
	Focus on protection of natural and landscape assets	Focus on restoration and rehabilitation, as well as protection
Governance	Run by central government	Run by many partners
Involvement of local people	Planned and managed against the impact of local people	Run with, for, and, in some cases, by local people
	Managed for visitors and tourists, with little regard to the local community	Managed to help meet the needs of local communities
Wider context	Developed separately from each other	Planned as part of a national or international system
	Managed as stand-alones	Developed as networks (strictly protected areas buffered and linked by green corridors)
Perceptions	Viewed exclusively as a national asset	Viewed as a community asset
	Viewed as a national concern, with limited regard to international obligations	Guided by international responsibilities, as well as local and national concerns
Management technique	Technocratic approach to management	Planning and management viewed as political exercise, involving consultations and sensitivity
	Reactive management, focused on the short term	Adaptive management, focused on the long term
Finance	Paid for by taxpayers	Paid for by a variety of means
Management skills	Managed by natural scientists or natural resource experts	Managed by people with a range of skills, especially people-related skills
	Management is expert-led	Management values and draws upon local knowledge

Source: Adapted from Adrian Phillips (2003) and Michael Lockwood and Ashish Kothari (2006).

The emerging paradigm, with its inclusion of cultural landscapes and human presence, as well as its recognition of multiple objectives for protected areas (social, economic, political, and conservation), is much friendlier than its precursor to Indigenous notions of conservation. To illustrate, a survey of five ICCAs from around the world indicated a broad range of local objectives: access to livelihood resources, security of land and resource tenure, protection from outside threats, financial benefit from resources or ecosystem functions, rehabilitation of degraded resources, participation in management, empowerment, capacity building, and the reinforcement of cultural identity and cohesiveness (Berkes 2009). Although each case included conservation objectives, local people did not see these as distinct from economic, social, and political objectives (ibid.). Various conservation-development projects from around the world, some of them Indigenous, reveal that these values and objectives generally exist in case-specific combinations, making universal application of a particular framework impossible and culturally inappropriate (Berkes 2013).

Many Indigenous cultures, from the Crees and Anishinaabe of northern Canada to the Maori of New Zealand, reject a notion of conservation that separates people from the environment. In the worldview of these groups, use and protection go together; respect for a resource arises from the use of that resource, as does a sense of responsibility for it. Hence, conservation without use makes little sense because it alienates people from their lands and from their stewardship responsibilities (Berkes 2018). The concept of a protected area that accommodates Indigenous land use and that is seen as a community asset (as well as a national one), to be managed not by government technicians but jointly with the use of Indigenous values and knowledge, closes the gap between Western scientific conservation and Indigenous conservation.

For many Indigenous groups, both person-land and person-person relationships are governed through a common, integrated set of rules. This is true in the Cree lands of the James Bay area (Tanner 1979; Feit 1987; Bearskin et al. 1989; Scott 1996) and elsewhere in the Canadian North. Learning from traditional management is important for broadening conservation objectives and developing a more inclusive, cross-cultural definition of conservation. Crucial here is support for traditional and customary institutions of stewardship, such as the tallyman system of the James Bay Cree, which refers to the Cree person recognized by the community as responsible for the supervision of harvesting activities within each multi-family trapline or hunting territory (Scott 1986).

There is a largely untapped reservoir of local knowledge and practice that may be adapted for new purposes in the dynamic management of ecosystems (see Chapter 2, this volume). The knowledge systems of the Crees show a high degree of sophistication related to an understanding of complex environmental processes, such as disturbance regimes and multiscale management. For example, the Cree goose hunt is informed by constant monitoring of shifts and changes, allowing hunters to live with uncertainty and deal with variability (Peloquin and Berkes 2009; Chapter 8, this volume). Several customary practices of the James Bay Crees – such as rotating use of hunting territories – are thought to be related to the realities of their environment (Berkes 1998; Chapter 9, this volume). These findings are relevant to debates on current international conservation strategies that are based on static reserves but should perhaps be complemented with dynamic reserves in which the boundaries may shift in space and time (Bengtsson et al. 2003; Elmqvist et al. 2004).

The resolution of cross-cultural differences that may arise between Crees and government managers, and the creation of an informed collective vision about protecting important ecosystems and cultural landscapes, can be achieved through sustained dialogue and "a collaborative decision-making process that favors shared ideals about wilderness protection" (Pfister 2002, 27). Collaboration with Crees in protected area management will probably have an important impact on the retention of Indigenous knowledge and the transmission of culture. Place-based education initiatives to involve youth and to cultivate the next generation of stewards should be seen as an essential part of the protected area planning process (Mitchell et al. 2005). However, each protected area has unique characteristics, and each community has its own decision-making processes. Informed support by the representatives of government agencies is crucial to the development of culturally appropriate conservation plans in northern regions. This process can be assisted through the sharing of information and experience, as well as through access to technical support.

CONCLUSION

A paradigm shift on an international scale is well under way regarding protected areas and the role of people in them. This creates opportunities for a fresh consideration of Indigenous approaches to conservation and protected areas, and for a search for policy alternatives to create landscapes and seascapes that are consistent with Indigenous concepts of occupying, using, and taking care of the land and sea. Of particular interest is the broadening of possibilities offered by protected areas to encompass "a mosaic of areas and resource units under different ownerships and regulations, possibly including several government-run protected areas, along with Community Conserved Areas, Co-managed Protected Areas and even private reserves" (Borrini-Feyerabend, Kothari, and Oviedo 2004, 21).

Much is happening in Canada and Quebec to accommodate Indigenous rights and interests in protected area planning and environmental protection in general. ICCAs have received surprisingly limited attention, but this may soon change (Herrmann et al. 2012). The flux in policy and practice indicates a search for alternatives to protected area planning that complement the changing role of Indigenous people in resource and environmental governance. The local stewardship approach is interesting in that it speaks to issues of politics and power, putting conservation in the hands of the people whom it most affects.

In the case of James Bay, the local stewardship approach is also significant in that we have some understanding of how Cree people take care of the land and respond to the particular characteristics of subarctic ecosystems. Learning from traditional management systems and involving Indigenous institutions of stewardship (such as the tallymen who are in charge of trapline areas) in protected area planning is a promising approach. It also helps broaden conservation objectives to include the cultural and economic well-being of local people and to develop a more inclusive, cross-cultural definition of conservation.

Local stewardship alone cannot deal with external threats to protected areas, nor would it be sufficient to deal with the multiple jurisdictions that characterize the contemporary nation-state. Protected areas are assets at all levels (local, regional, provincial, national, and international), and they require the conservation of diversity at all levels of the ecosystem, from the genetic and species levels to the landscape level. The future of protected area management and the new conservation paradigm will therefore depend on the development of multi-level linkages involving numerous partners (Berkes 2007). Ultimately, however, protected areas, such as the Paakumshumwaau-Maatuskaau Biodiversity Reserve and the proposed Tawich (Marine) Conservation Area, need to support an Indigenous way of life and Indigenous resources that are linked to the protection and sustainable use of ancestral lands and waters.

WORKS CITED

Adams, William M., and Jon Hutton. 2007. "People, Parks and Poverty: Political Ecology and Biodiversity Conservation." *Conservation and Society* 5 (2): 147–83.

Agrawal, Arun, and Kent H. Redford. 2009. "Conservation and Displacement: An Overview." *Conservation and Society* 7 (1): 1–10.

Bearskin, Joab, G. Lameboy, R. Matthew, J. Pepabano, A. Pisinaquan, W. Ratt, and D. Rupert. 1989. *Cree Trappers Speak.* Comp. and ed. Fikret Berkes. Chisasibi, QC: Cree Trappers Association's Committee of Chisasibi and the James Bay Cree Cultural Education Centre.

Bell, Susan. 2018. "Cree Nation identifies 30 per cent of its territory in conservation wish list." *CBC News,* 12 December. https://www.cbc.ca/news/canada/north/conservation-cree-quebec-plan-nord-hunt-trap-1.4941383.

Bengtsson, Janne, Per Angelstam, Thomas Elmqvist, Urban Emanuelsson, Carl Folke, Mmargareta Ihse, Fredrik Moberg, and Magnus Nyström. 2003. "Reserves, Resilience and Dynamic Landscapes." *Ambio* 32 (6): 389–96.

Berger, Thomas R. 1977. *Northern Frontier, Northern Homeland.* 2 vols. Toronto: James Lorimer.

Berkes, Fikret. 1998. "Indigenous Knowledge and Resource Management Systems in the Canadian Subarctic." In *Linking Social and Ecological Systems,* ed. Fikret Berkes and Carl Folke, 98–128. Cambridge: Cambridge University Press.

–. 2007. "Community-Based Conservation in a Globalized World." *Proceedings of the National Academy of Sciences* 104: 15188–93.

–. 2009. "Community Conserved Areas: Policy Issues in Historic and Contemporary Context." *Conservation Letters* 2: 19–24.

–. 2013. "Poverty Reduction Isn't Just about Money: Community Perceptions of Conservation Benefits." In *Biodiversity Conservation and Poverty Alleviation,* ed. Dilys Roe, Joanna Elliott, Chris Sandbrook, and Matt Walpole, 270–85. London: Wiley-Blackwell.

–. 2018. *Sacred Ecology.* 4th ed. New York: Routledge.

Berkes, Fikret, Rob Huebert, Helen Fast, Micheline Manseau, and Alan Diduck, eds. 2005. *Breaking Ice: Renewable Resource and Ocean Management in the Canadian North.* Calgary: University of Calgary Press.

Borrini-Feyerabend, Grazia, Ashish Kothari, and Gonzalo Oviedo. 2004. *Indigenous and Local Communities and Protected Areas: Towards Equity and Enhanced*

Conservation. Best Practice Protected Area Guidelines Series No. 11. Gland, Switzerland: IUCN WCPA.

Brandon, Katrina, Kent H. Redford, and Steven E. Sanderson, eds. 1998. *Parks in Peril: People, Politics and Protected Areas.* Washington, DC: Island Press.

Brockington, Dan. 2002. *Fortress Conservation: The Preservation of the Mkomazi Game Reserve, Tanzania.* Oxford: James Currey.

Brockington, Dan, Rosaleen Duffy, and Jim Igoe. 2008. *Nature Unbound: Conservation, Capitalism and the Future of Protected Areas.* London: Earthscan.

Brosius, Peter J. 2004. "Indigenous Peoples and Protected Areas at the World Parks Congress." *Conservation Biology* 18 (3): 609–12.

–. 2006. "What Counts as Local Knowledge in Global Environmental Assessment and Conventions?" In *Bridging Scales and Knowledge Systems: Concepts and Applications in Ecosystem Assessment,* ed. Walter V. Reid, Fikret Berkes, Thomas J. Wilbanks, and Doris Capistrano, 129–44. Washington, DC: Island Press.

Brosius, Peter J., and Sarah L. Hitchner. 2010. "Cultural Diversity and Conservation." *International Social Science Journal* 61: 141–68.

Brown, Jessica, and Brent Mitchell. 2000. "The Stewardship Approach and Its Relevance for Protected Landscapes." *George Wright Forum* 17 (1): 70–79.

Brown-John, C. Lloyd. 2006. "Canada's National Parks Policy: From Bureaucrats to Collaborative Managers." Paper presented at the annual meeting of the Canadian Political Science Association, York University, York, June 1–3. http://www.cpsa -acsp.ca/papers-2006/Brown-John.pdf.

Canadian Council on Ecological Areas. 2017a. *Conservation Areas Reporting and Tracking System (CARTS).* Based on data current as of December 31, 2017. https:// ccea.org/wp-content/uploads/2019/04/CARTS2017ReportEN.pdf.

–. 2017b. *Canada's Protected Areas Network. Conservation Areas Reporting and Tracking System (CARTS) Map.* Based on data current as of December 31, 2017. https://ccea.org/wp-content/uploads/2019/04/CCEA_CANADA_15M_LETTER_ CARTS_IUCN_20171231.pdf.

Canadian Parks Council. 2008. *Aboriginal Peoples and Canada's Parks and Protected Areas: Case Studies.* Ottawa: Government of Canada.

Canadian Press. 2011. "Charest Unveils $80B Plan for Northern Quebec." *CBC News,* May 9. http://www.cbc.ca/news/canada/montreal/story/2011/05/09/Québec -northern-plan-charest.html.

Cittadino, Federica. 2014. "Applying a UNDRIP Lens to the CBD: A More Comprehensive Understanding of Benefit-Sharing." *Indigenous Policy Journal* 24 (4): 18.

Clark, Douglas A., Shaun Fluker, and Lee Risby. 2008. "Deconstructing Ecological Integrity Policy in Canadian National Parks." In *Transforming Parks and Protected Areas: Policy and Governance in a Changing World,* ed. Kevin S. Hanna, Douglas A. Clark, and D. Scott Slocombe, 154–68. New York: Routledge.

Coon Come, M. 2011. "Premier Announces Creation of Assinica National Park Reserve." Press Release, May 17. http://www.mddelcc.gouv.qc.ca/communiques_ en/2011/c20110517-assinica.htm.

CPAWS (Canadian Parks and Wilderness Society). 2015. "Protecting Canada: Is it in our Nature? How Canada can Achieve its International Commitment to Protect our Land and Freshwater." Parks Report. https://cpaws.org/wp-content/uploads/2019/07/CPAWS_Parks_Report_2015-Single_Page.pdf.

–. 2017. "From Laggard to Leader? Canada's Renewed Focus on Protecting Nature Could Deliver Results." CPAWS annual report, July. http://cpaws.org/uploads/CPAWS-Parks-Report-2017.pdf.

Cree First Nation of Waswanipi. 2016. "Broadback: Cree First Nation and Coalition to Urge Quebec to Take a Stand to Protect One of Last Intact Forests." Press release, February 22. http://www.waswanipi.com/en/all-news/471-press-release-broadback -cree-first-nation-and-coalition-to-urge-Québec-to-take-a-stand-to-protect-one -of-last-intact-forests.

Cree Nations of Eeyou Istchee. 2011. "Cree Vision of Plan Nord." https://www.cngov. ca/wp-content/uploads/2018/03/cree-vision-of-plan-nord.pdf. Davidson-Hunt, Iain J., Nathan Deutsch, and Andrew M. Miller. 2012. *Pimachiowin Aki Cultural Landscape Atlas: Land That Gives Life*. Winnipeg: Pimachiowin Aki Corporation.

Davidson-Hunt, Iain J., P. Peters, and C. Burlando. 2010. "Beekahncheekahmeeng Ahneesheenahbay Ohtahkeem (Pikangikum Cultural Landscape)." In *Indigenous People and Conservation,* ed. Kristen Walker Painemilla, Anthony B. Rylands, Alisa Woofter and Cassie Hughes, 137–44. Arlington, VA: Conservation International.

Dearden, Philip, and Nathan Bennett. 2016. "The Role of Aboriginal Peoples in Protected Areas." In *Parks and Protected Areas in Canada,* 4th ed., ed. Philip Dearden, Rick Rollins, and Mark Needham, 357–90. Toronto: Oxford University Press.

Dearden, Philip, and Steve Langdon. 2009. "Aboriginal Peoples and National Parks." In *Parks and Protected Areas in Canada,* 3rd ed., ed. Philip Dearden and Rick Rollins, 373–402. Toronto: Oxford University Press.

Dougherty, Kevin. 2015. "Cree-Quebec Forestry Agreement Is Unacceptable, Innu Say." *Montreal Gazette,* July 13. http://montrealgazette.com/news/local-news/cree -Québec-forestry-agreement-is-unacceptable-innu-say.

Dowie, Mark. 2011. *Conservation Refugees: The Hundred-Year Conflict between Global Conservation and Native Peoples.* Cambridge, Massachusetts: MIT Press.

Dudley, Nigel, ed. 2008. *Guidelines for Applying Protected Area Management Categories.* Gland, Switzerland: IUCN.

Dudley, Nigel, Jeffrey D. Parrish, Kent H. Redford, and Sue Stolton. 2010. "The Revised IUCN Protected Areas Management Categories: The Debate and Ways Forward." *Oryx* 44 (4): 485–90.

East, Jeremy. 2015. "Terms Reached in Baril-Moses Dispute and Broadback Protection." *The Nation,* July 24.

Elmqvist, Thomas, Fikret Berkes, Carl Folke, Per Angelstam, Anne-Sophie Crépin and Jari Niemelä. 2004. "The Dynamics of Ecosystems, Biodiversity Management and Social Institutions at High Northern Latitudes." *Ambio* 33 (6): 350–55.

Enkerlin-Hoeflich, Ernesto C., et al. 2015. "IUCN/WCPA Protected Areas Program: Making Space for People and Biodiversity in the Anthropocene." In *Earth Stewardship, Linking Ecology and Ethics in Theory and Practice*, ed. Ricardo Rozzi, F. Stuart Chapin III, J. Baird Callicott, S.T.A. Pickett, Mary E. Power, Juan J. Armesto, and Roy H. May Jr., 339–50. Dordrecht: Springer International.

Environment Canada. 2015. *Canadian Protected Areas: Status Report 2006–2011*. Environment Canada, Gatineau, QC. http://publications.gc.ca/collections/collection_2016/eccc/En81-9-2011-eng.pdf.

Feit, Harvey A. 1987. "Waswanipi Cree Management of Land and Wildlife: Cree Cultural Ecology Revisited." In *Native People, Native Lands: Canadian Indians, Inuit and Métis*, ed. B. Cox, 75–91. Ottawa: Carleton University Press.

Foster, Janet, and Lorne Hammond. 1998. *Working for Wildlife: The Beginning of Preservation in Canada*. 2nd ed. Toronto: University of Toronto Press.

Gelfand, J. 2017. "Report 1 – Progress on Reducing Greenhouse Gases – Environment and Climate Change Canada." Fall reports of the Commissioner of the Environment and Sustainable Development to the Parliament of Canada. http://www.oag-bvg.gc.ca/internet/English/parl_cesd_201710_01_e_42489.html.

Gertsch, Frances, Graham Dodds, Micheline Manseau, and Joadamee Amagoalik. 2003. "Recent Experiences in Cooperative Planning and Management for Canada's Northernmost National Park: Quttinitpaaq National Park on Ellesmere Island." Paper presented at the Fifth International Science and Management of Protected Areas Association Conference, Victoria, BC, May 11–16. http://www.sampaa.org/PDF/ch2/2.7.pdf *(site now discontinued)*.

Gladu, Jean Paul, Doug Brubacher, Brad Cundiff, Anna Baggio, Anne Bell, and Tim Gray. 2003. "Honouring the Promise: Aboriginal Values in Protected Areas in Canada." National Aboriginal Forestry Association and Wildlands League. http://wildlandsleague.org/attachments/Honouring%20the%20Promise.pdf.

Grand Council of the Crees (Eeyou Istchee). 2014. "Crees Share Vision of Land Stewardship and Conservation at World Parks Congress in Sydney Australia." Press release, November 14. http://www.eeyouconservation.com/crees-share-vision-of-land-stewardship-and-conservation-at-world-parks-congress-in-sydney.html.

Herrmann, Thora Martina, Michael A.D. Ferguson, Gleb Raygorodetsky, and Monica E. Mulrennan. 2012. "Recognition and Support of ICCAs in Canada." In *Recognising and Supporting Territories and Areas Conserved by Indigenous Peoples and Local Communities: Global Overview and National Case Studies*. CBD Secretariat Technical Series No. 64, ed. Ashish Kothari, with Colleen Corrigan, Harry Jonas, Aurélie Neumann, and Holly Shrumm, 160. Montreal: Secretariat of the Convention on Biological Diversity, ICCA Consortium, Kalpavriksh, and Natural Justice.

Hopkins-Utter, Shane R. 2017. "Case Comment: Ktunaxa Nation v. British Columbia (Forests, Lands and Natural Resource Operations)." Shane R. Hopkins-Utter Law Corporation. https://canliiconnects.org/fr/commentaries/46976.

ICCA Consortium. 2010. "Bio-Cultural Diversity Conserved by Indigenous Peoples and Local Communities – Examples and Analysis." Companion document to IUCN/CEESP Briefing Note No. 10.

–. 2017. ICCA Consortium website. http://www.iccaconsortium.org.

Igoe, Jim, and Dan Brockington. 2007. "Neoliberal Conservation: A Brief Introduction." *Conservation and Society* 5 (4): 432–49.

IUCN. 1994. *Guidelines for Protected Area Management Categories.* Gland, Switzerland: IUCN.

–. 2005. *Benefits beyond Boundaries: Proceedings of the Vth IUCN World Parks Congress.* Gland, Switzerland: IUCN.

–. 2014. "The Promise of Sydney Vision." IUCN World Parks Congress, Sydney. http://worldparkscongress.org/about/promise_of_sydney_vision.html.

Jessen, Sabine, Lance Morgan, and Juan Bezaury-Creel. 2016. "Dare to Be Deep: SeaStates Report on North America's Marine Protected Areas (MPAs)." Ottawa, Canadian Parks and Wilderness Society and Marine Conservation Institute. http://cpaws.org/uploads/CPAWS-Oceans-Report-2016.pdf.

Langdon, Steve, Rob Prosper, and Nathalie Gagnon. 2010. "Two Paths One Direction: Parks Canada and Aboriginal Peoples Working Together." *George Wright Forum* 27 (2): 222–33.

Larsen, Peter Bille, and Jenny Springer. 2008. "Mainstreaming WWF Principles on Indigenous Peoples and Conservation in Project and Programme Management." Gland, Switzerland, WWF.

Locke, Harvey, and Philip Dearden. 2005. "Rethinking Protected Area Categories and the New Paradigm." *Environmental Conservation* 32: 1–10.

Lockwood, Michael. 2006. "Global Protected Area Framework." In *Managing Protected Areas: A Global Guide,* ed. Michael Lockwood, Graeme L. Worboys, and Ashish Kothari, 73–100. London: Earthscan.

Lockwood, Michael, and Ashish Kothari. 2006. "Social Context." In *Managing Protected Areas: A Global Guide,* ed. Michael Lockwood, Graeme L. Worboys, and Ashish Kothari, 41–72. London: Earthscan.

Loo, Tina. 2006. *States of Nature: Conserving Canada's Wildlife in the Twentieth Century.* Vancouver: UBC Press.

McNamee, Kevin. 2010. "Filling in the Gaps: Establishing New National Parks." *George Wright Forum* 27 (2): 142–50.

MDDELCC. 2007. "Parc National des Pingualuit. Québec Premier Jean Charest Inaugurates First Norther Park." Press release, November 30. http://www.environnement.gouv.qc.ca/communiques_en/2007/c20071130-pingualuit.htm.

–. 2011a. "Premier Announces Creation of Assinica National Park Reserve." Press release, May 17. http://www.mddelcc.gouv.qc.ca/communiques_en/2011/c20110517-assinica.htm.

–. 2011b. "We Take Growth Seriously: Strategic Guidelines for Québec Protected Areas 2011–2015 Period." http://www.environnement.gouv.qc.ca/biodiversite/aires_protegees/orientations-strateg2011-15-en.pdf.

–. 2017. "Protected Areas in Québec." http://www.mddelcc.gouv.qc.ca/biodiversite/aires_protegees/aires_Québec-en.htm.

MDDEP. 2000. "Cadre d'orientation en vue d'une stratégie sur les aires protégées: Québec vise une superficie en aires protégées de 8% d'ici 2005." Press release, July 10. http://www.mddep.gouv.qc.ca/communiques/2000/c000710a.htm.

–. 2003a. "Québec's Strategy for Protected Areas: Québec Announces the Protection of 3220 Km2 of Territory in Northwestern Québec." Press release, March 3. http://www.environnement.gouv.qc.ca/communiques_en/2003/c20030304-area.htm.

–. 2003b. "Stratégie québécoise sur les aires protégées: La majestueuse rivière Moisie: un joyau patrimonial dorénavant protégé." Press release, February 8. http://www.mddep.gouv.qc.ca/infuseur/communique.asp?no=318.

–. 2005. "Québec's Strategy on Protected Areas: The Government Confirms Its Intention of Teaming Up with the Cree Nation of Mistissini to Create the First Inhabited Park in the Boreal Forest: Albanel-Temiscamie-Otish Park." Press release, November 9. http://www.environnement.gouv.qc.ca/communiques_en/2005/c20051109-audiencesATO.htm.

–. 2009. "Protected Areas in Québec: A Lifelong Heritage." http://www.mddep.gouv.qc.ca/biodiversite/aires_protegees/articles/090329/synthese-en.pdf.

Miller, Thaddeus R., Ben A. Minteer, and Leon-C. Malan. 2011. "The New Conservation Debate: The View from Practical Ethics." *Biological Conservation* 144 (3): 948–57.

Mitchell, Nora, Jacquelyn Tuxill, Guy Swinnerton, Susan Buggey, and Jessica Brown. 2005. "Collaborative Management of Protected Landscapes: Experience in Canada and the United States of America." In *The Protected Landscape Approach: Linking Nature, Culture and Community,* ed. Jessica Brown, Nora Mitchell, and Michael Beresford, 191–204. Gland, Switzerland: IUCN.

Morrison, James. 1997. "Conservation Parks in Cree Territories." Discussion paper prepared for the Cree Regional Authority, Direction du plein air et des parcs (Ministère de l'Environnement et de la Faune du Québec), World Wildlife Fund of Canada, February 28.

Mortimer-Sandilands, Catriona. 2009. "The Cultural Politics of Ecological Integrity: Nature and Nation in Canada's National Parks, 1885–2000." *International Journal of Canadian Studies* 39–40: 161–89.

Mulrennan, Monica E. 2008. "Reaffirming 'Community' in the Context of Community-Based Conservation." In *Renegotiating Community: Interdisciplinary Perspectives, Global Contexts,* ed. Diane Brydon and William D. Coleman, 66–82. Vancouver: UBC Press.

–. 2015. "Aboriginal Peoples in Relation to Resource and Environmental Management." In *Resource and Environmental Management in Canada: Addressing Conflict and Uncertainty,* 5th ed., ed. Bruce Mitchell, 56–79. Toronto: Oxford University Press.

Mulrennan, Monica E., Rodney Mark, and Colin H. Scott. 2012. "Revamping Community-Based Conservation through Participatory Research." *Canadian Geographer* 56 (2): 243–59.

Murray, Grant, and Leslie King. 2012. "First Nations Values in Protected Area Governance: Tla-o-qui-aht Tribal Parks and Pacific Rim National Park Reserve." *Human Ecology* 40 (3): 385–95.

Nash, Roderick Frazier. 2001. *Wilderness and the American Mind*. New Haven: Yale University Press. Originally published 1965.

Natcher, David C., Susan Davis, and Clifford G. Hickey. 2005. "Co-management: Managing Relationships, Not Resources." *Human Organization* 64 (3): 240–50.

National Parks Act, S.C. 1930, c. 33.

Neufeld, David. 2008. "Indigenous Peoples and Protected Heritage Areas: Acknowledging Cultural Pluralism." In *Transforming Parks and Protected Areas: Policy and Governance in a Changing World*, ed. Kevin S. Hanna, Douglas A. Clark, and D. Scott Slocombe, 181–99. New York: Routledge.

Neumann, Roderick P. 1998. *Imposing Wilderness: Struggles over Livelihood and Nature Preservation in Africa*. Berkeley: University of California Press.

Northwest Territories Lands Act, S.N.W.T. 2014, c. 13.

NWT-PAS (Northwest Territories Protected Areas Strategy Advisory Committee). 1999. "Northwest Territories Protected Areas Strategy: A Balanced Approach to Establishing Protected Areas in the Northwest Territories." September 27. Unpublished report.

Oelschlaeger, Max. 1991. *The Idea of Wilderness*. New Haven: Yale University Press.

Otto, Jonathan, et al. 2013. *Natural Connections: Perspectives in Community-Based Conservation*. Washington, DC: Island Press.

Parks Canada. 1964. "Parks Canada Policy Statement." Ottawa, Department of Environment.

–. 1994. "Guiding Principles and Operational Policies." Ottawa, Department of Canadian Heritage.

–. 2000. *Unimpaired for Future Generations? Protecting Ecological Integrity with Canada's National Parks*. Vol. 2, *Setting a New Direction for Canada's National Parks*. Ottawa: Government of Canada.

–. 2017a. "Federal and Provincial Governments Create National Advisory Panel on Canada's Biodiversity Conservation Initiative." Press release, June 8. http://www.newswire.ca/news-releases/federal-and-provincial-governments-create-national-advisory-panel-on-canadas-biodiversity-conservation-initiative-627230281.html.

–. 2017b. "Working Together with Indigenous Peoples." https://www.pc.gc.ca/en/docs/pc/rpts/elnhc-scnhp/2016/coll-work.

Peepre, Juri, and Philip Dearden. 2002. "The Role of Aboriginal Peoples." In *Parks and Protected Areas in Canada: Planning and Management*, 2nd ed, ed. Philip Dearden and Rick Rollins, 323–53. Toronto: Oxford University Press. Originally published 1993.

Peloquin, Claude, and Fikret Berkes. 2009. "Local Knowledge, Subsistence Harvests, and Social-Ecological Complexity in James Bay." *Human Ecology* 37: 533–45.

Pfister, Robert E. 2002. "Collaboration across Cultural Boundaries to Protect Wild Places: The British Columbia Experience." In *Wilderness in the Circumpolar North: Searching for Compatibility in Ecological, Traditional, and Ecotourism Values,* comp. Alan E. Watson, Lilian Alessa, and Janet Sproull, 27–35. Ogden, UT: United States Department of Agriculture, Forest Service, Rocky Mountain Research Station.

Phillips, Adrian. 2003. "Turning Ideas on Their Head: The New Paradigm for Protected Areas." *George Wright Forum* 20 (2): 8–32.

Redford, Kent H., John G. Robinson, and William M. Adams. 2006. "Parks as Shibboleths." *Conservation Biology* 20 (1): 1–2.

Sahanatien, Vicki. 2007. "Land Claims as a Mechanism for Wilderness Protection in the Canadian Arctic." In *Science and Stewardship to Protect and Sustain Wilderness Values,* comp. Alan Watson, Janet Sproull, and Liese Dean, 199–203. Fort Collins, CO: United States Department of Agriculture, Forest Service, Rocky Mountain Research Station.

Sandlos, John. 2007. *Hunters at the Margin: Native People and Wildlife Conservation in the Northwest Territories.* Vancouver: UBC Press.

–. 2014. "National Parks in the Canadian North." In *Indigenous Peoples, National Parks, and Protected Areas: A New Paradigm Linking Conservation, Culture, and Rights,* ed. Stan Stevens, 133–49. Tucson: University of Arizona Press.

Scott, Colin H. 1986. "Hunting Territories, Hunting Bosses and Communal Production among Coastal James Bay Cree." *Anthropologica* 28 (1–2): 163–73.

–. 1996. "Science for the West, Myth for the Rest? The Case of James Bay Cree Knowledge Construction." In *Naked Science: Anthropological Inquiry into Boundaries, Power and Knowledge,* ed. Laura Nader, 69–86. London: Routledge.

Scott, Colin H., and Monica E. Mulrennan. 2010. "Reconfiguring Mare Nullius: Torres Strait Islanders, Indigenous Sea Rights and the Divergence of Domestic and International Norms." In *Indigenous Peoples and Autonomy: Insights for a Global Age,* ed. Mario Blaser, Ravi de Costa, Deborah McGregor, and William D. Coleman, 148–76. Vancouver: UBC Press.

Secrétariat au Plan Nord. 2015. *The Plan Nord: Toward 2035, 2015–2020 Action Plan.* Gouvernement du Québec. http://plannord.gouv.qc.ca/wp-content/uploads/2015/04/Long_PN_EN.pdf.

Secrétariat aux affaires autochtones. 2012. "Signing of an Agreement between the Government of Québec and the Cree Nation. Creation of the Regional Government of Eeyou Istchee James Bay." Press release, July 24. https://www.autochtones.gouv.qc.ca/centre_de_presse/communiques/2012/2012-07-24-en.htm.

Shadie, Peter. 2006a. "Convention on Biological Diversity Programme of Work on Protected Areas." In *Managing Protected Areas: A Global Guide,* ed. Michael Lockwood, Graeme L. Worboys, and Ashish Kothari, 704–6. London: Earthscan.

–. 2006b. "The IUCN and Protected Areas." In *Managing Protected Areas: A Global Guide,* ed. Michael Lockwood, Graeme L. Worboys, and Ashish Kothari, 78. London: Earthscan.

Shafer, C.L. 2015. "Cautionary Thoughts on IUCN Protected Area Management Categories V–VI." *Global Ecology and Conservation* 3: 331–48.

Shultis, John, and Susan Heffner. 2016. "Hegemonic and Emerging Concepts of Conservation: A Critical Examination of Barriers to Incorporating Indigenous

Perspectives in Protected Area Conservation Policies and Practice." *Journal of Sustainable Tourism* 24: 8–9, 1227–42.

Société de la faune et des parcs du Québec. 2000. "Provisional Master Plan: Parc des Pingaluit." http://www.fapaq.gouv.qc.ca/en/consultation/pingualuit/plan_ping_a. pdf *(site now discontinued)*.

Société du Plan Nord du Québec. 2015. "The Québec Government Signs a New Agreement with the Cree." Press release, July 13. https://www.autochtones.gouv. qc.ca/centre_de_presse/communiques/2015/2015-07-13-en.asp.

Spaeder, Joseph A., and Harvey J. Feit, eds. 2005. "Co-management and Indigenous Communities: Barriers and Bridges to Decentralized Resource Management." *Anthropologica* 47 (2): 147–54.

Spence, Mark D. 1999. *Dispossessing the Wilderness: Indian Removal and the Making of the National Parks*. New York: Oxford University Press.

Stadel, Angela, Raymond Taniton, and Heidi Heder. 2002. "The Northwest Territories Protected Areas Strategy: How Community Values Are Shaping the Protection of Wild Spaces and Heritage Places." In *Wilderness in the Circumpolar North: Searching for Compatibility in Ecological, Traditional, and Ecotourism Values,* comp. Alan E. Watson, Lilian Alessa, and Janet Sproull, 20–26. Ogden, UT: United States Department of Agriculture, Forest Service, Rocky Mountain Research Station.

Standing Committee on Environment and Sustainable Development. 2016. 42nd Parliament, 1st session, Number 027. "Evidence: Tuesday, October 4, 2016." http://www.ourcommons.ca/DocumentViewer/en/42-1/ENVI/meeting-27/ evidence.

Stevens, Stan. 2010. "Implementing the UN Declaration on the Rights of Indigenous Peoples and International Human Rights Law through the Recognition of ICCAs." *Policy Matters* 17: 181–94.

–, ed. 2014a. *Indigenous Peoples, National Parks, and Protected Areas: A New Paradigm Linking Conservation, Culture, and Rights.* Tucson: University of Arizona Press.

–. 2014b. "A New Protected Area Paradigm." In *Indigenous Peoples, National Parks, and Protected Areas: A New Paradigm Linking Conservation, Culture, and Rights,* ed. Stan Stevens, 47–83. Tucson: University of Arizona Press.

Stevens, Stan, Neema Pathak Broome, and Tilman Jaeger. 2016. *Recognising and Respecting ICCAs Overlapped by Protected Areas: A Report for the ICCA Consortium.* https://www.iccaconsortium.org/wp-content/uploads/2016/11/publication -Recognising-and-Respecting-ICCAs-Overlapped-by-PAs-Stevens-et-al-2016-en. pdf.

Stronghill, Jessica, Murray B. Rutherford, and Wolfgang Haider. 2015. "Conservancies in Coastal British Columbia: A New Approach to Protected Areas in the Traditional Territories of First Nations." *Conservation and Society* 13 (1): 39–50.

Tanner, Adrian. 1979. *Bringing Home Animals: Religious Ideology and Mode of Production of the Mistassini Cree Hunter*. London: Hurst.

Tauli-Corpuz, Victoria. 2016. "Report of the Special Rapporteur of the Human Rights Council on the Rights of Indigenous Peoples." United Nations. http://unsr.vtaul-icorpuz.org/site/images/docs/annual/2016-annual-ga-a-71-229-en.pdf.

Terborgh, John. 1999. *Requiem for Nature*. Washington, DC: Island Press.

Thomlinson, Eugene, and Geoffrey Crouch. 2012. "Aboriginal Peoples, Parks Canada, and Protected Spaces: A Case Study in Co-management at Gwaii Haanas National Park Reserve." *Annals of Leisure Research* 15 (1): 69–86.

Timko, Joleen A., and Terre Satterfield. 2008. "Seeking Social Equity in National Parks: Experiments with Evaluation in Canada and South Africa." *Conservation and Society* 6 (3): 238–54.

Wells, Michael P., and Thomas O. McShane. 2004. "Integrating Protected Area Management with Local Needs and Aspirations." *Ambio* 33 (8): 513–19.

West, Paige, James Igoe, and Dan Brockington. 2006. "Parks and Peoples: The Social Impact of Protected Areas." *Annual Review of Anthropology* 35: 251–77.

West, Patrick C., and Steven R. Brechin, eds. 1991. *Resident Peoples and National Parks: Social Dilemmas and Strategies in International Conservation*. Tucson: University of Arizona Press.

Western, David, and Michael Wright. 1994. "The Background to Community-Based Conservation." In *Natural Connections: Perspectives in Community-Based Conservation*, ed. D. Western and R.M. Wright, 1–14. Washington, DC: Island Press.

Whitefeather Forest Initiative. 2012. "Our Stewardship Vision." http://www.white featherforest.ca/stewardship/our-stewardship-vision.

2

The Politics of Traditional Ecological Knowledge in Environmental Protection

WREN NASR and COLIN H. SCOTT

For Eeyou of Eeyou Istchee, the journey for full Eeyou
governance begins and ends with and within the historical
and traditional authorities of self-governing power – the
people of the land.

— Ted Moses (2002a, 7)

The epigraph from former grand chief Ted Moses expresses the Crees' demand for self-governance of their ancestral lands on their own cultural terms. This sensibility is paralleled by widespread support in academic and policy discourse for the idea that traditional ecological knowledge (TEK) should participate alongside conventional science in projects of environmental protection and development in Indigenous homelands. There is little agreement on how to arrange this, but it will have major consequences for Indigenous goals of cultural continuity, land and resource protection and management, and affirmations of sovereignty.

In this chapter, we address implications of the fact that resource development and environmental protection projects are spaces of representation and knowledge production marked by power imbalances. Within these spaces, hegemonic thought and action wielded by often non-Indigenous authoritative actors effect a distillation of Indigenous knowledge into what is considered by them to be relevant TEK. This isolates objective data and empirical practices from their larger social context, including precisely those social relations and processes upon which, locally, the instrumental effectiveness and political authority of Indigenous knowledge rest (Nadasdy 2005; Nasr and Scott 2010). This appropriation of TEK risks throwing out of court important domains of cultural context, such that the ontological and epistemological character of the knowledge is repositioned and transformed. The risks are multi-form: the distortion of subjectivity, the denial of agency, and the exclusion of Crees from full participation in civil society.

For clarity, we distinguish between Indigenous knowledge (not solely ecological and not only traditional) and TEK. The former is the knowledge of people who are Indigenous to a place, born of their relationships within and with that place (Berkes 2008). The latter has become a category at the interface and may fruitfully be analyzed as a product of the interaction of Indigenous and Western orders of knowledge. From the perspective of those who selectively appropriate it, aspects of Indigenous knowledge writ large lie beyond the scope of objective data and empirical practice. Indeed, certain commonplaces distinguish between Indigenous knowledge and science. The former is more oriented to the sacred and the figurative, is more "holistic," yet more "concretely" immersed in specific places. The latter is more secular, literal, specialized, abstract, and portable. The scientifically "inconvenient" (irrelevant? potentially destabilizing?) dimensions of Indigenous knowledge, as viewed against these simplistic contrasts, may be quietly passed over by scientific managers who are sympathetic to the idea of Indigenous knowledge, or actively denounced by those who are not.

In development discourse, Indigenous knowledge is often constructed as similar to that of the West (at least in the case of its objective, empirical practice bits); in environmental protection discourse, Indigenous knowledge may also be constructed as different (the sacred and figurative, ethics-reinforcing bits). The characterization of the Indigenous Other as either similar to or an inversion of the West produces a sociological abstraction that selectively reveals and conceals aspects of community life and worldview, with a number of implications. Both constructions involve problematic assumptions about the marginal status of "Native customs and traditions" vis-à-vis the *realpolitik* of development and conservation – a superficial multiculturalism, not a robust political pluralism.

To be sure, the Indigenous claim to justice, correcting the historical relationship between colonizers and colonized, has been insistent. There are peculiarities in the relations of domination involving Indigenous peoples that demand, at least, special techniques of recognition and dissimulation through the selective incorporation of difference by constitutional provisions, treaties, and comprehensive claims settlements. These have required some not insubstantial concessions by the state; however, asymmetrical distributions of power remain. Some such concessions involve new relations of knowledge in contexts of co-management between Indigenous people and state institutions. More broadly, the current call for justice by Indigenous people necessarily leads to a questioning of the normative structures underlying current and past modes of recognition of Indigenous participation in development and conservation. As Marisol de la Cadena and Orin Starn (2007, 7) note, "reckoning with indigeneity demands recognizing it as a relational field of governance, subjectivities, and knowledges that involve us all – indigenous and non-indigenous – in the making and remaking of its structures of power and imagination."

The Cree people of Eeyou Istchee have been leaders in rejecting limited framings of their rights and in asserting themselves as fully modern (but different) subjects, with interests in controlling their own economic destiny. Regional strategies are being articulated by the Cree Trappers Association that identify the institution of family hunting territories and its attendant values and knowledge as the basis for environmental development and protection (see Cree Trappers Association 2009). Similarly, we argue that policy seeking participation and representation from Indigenous people should focus on strengthening local institutions of tenure, management, and knowledge. Engagement at this level inspires mutual respect and partnership on terms that are real for Indigenous communities in the present social and historical context.

If, as we suggest, the participation of Indigenous people in environmental governance cannot be equated with cherry picking their knowledge in the abstract, the challenge is to specify the conditions in which space can be secured for the free play of Indigenous socio-ecological relations and ontologies. These conditions should be generative of knowledge that is authoritative in its own cultural terms and in relations with others. How this may be ensured in the instance of protected area creation remains a major practical preoccupation of our partnership as researchers with Cree communities and leaderships. Thus, the business of protected area creation and its accommodation of Cree knowledge and resource management arrangements are part of the larger challenge of Cree empowerment with respect to homeland and territory.

POLITICAL EVACUATION OF INDIGENOUS KNOWLEDGE

The goals of documenting Indigenous communities' traditional knowledge to inform projects of environmental protection and development are to increase our understanding of the environment and to empower the people who hold this knowledge (Henkel and Stirrat 2001; Nadasdy 2003). Outcomes, however, depend on the definition given to the knowledge to be collected. Social researchers often define it generously: according to Peter Usher (2000, 183), "TEK can be classified as knowledge about the environment, knowledge about the use of the environment, values about the environment, and the knowledge system itself." For geographer Deborah McGregor (2004, 78), TEK "encompasses such aspects as spiritual experience and relationships with the land ... Rather than being just the knowledge of how to live, it is the actual living of that life." However, in the specific instances of its definition and collection, this polythetic bundle of facts and values tends to be pruned down to stand-alone empirical content and specific land-use practices. This kind of purified TEK – often taken to represent the totality of relevant knowledge – is subsequently used by state actors to index local participation in governance procedures, by industry actors to support eligibility for extractive development-related compensation, or by conservation NGO actors to contribute to the project of biodiversity conservation. A focus on the positivistic dimensions of Indigenous knowledge (its empirical manifestations and extant forms) disconnects it from its local political and ethical authority. This process frames the political and ethical dimensions of Indigenous knowledge as externalities to identifying and collecting what has been narrowly defined as "knowledge" commensurate with the mandate at hand.

Defining and collecting TEK entails a series of displacements and transformations that may be viewed as a translation (Callon 1986). Translation, intimately bound up in creating and maintaining power relationships by delineating the nature and limits of visible and relevant objects, renders unruly objects commensurable with external requirements and institutional arrangements in the process of their delineation. In his study on the sociology of translation, Michel Callon (1986, 224) writes that "translation is the mechanism by which the social and natural worlds progressively take form. The result is a situation in which certain entities control others. Understanding what sociologists generally call power relationships means describing the way in which actors are defined, associated and simultaneously obliged to remain faithful to their alliances."

The "traditional ecological knowledge" that is produced through these processes of translation and formalization could be understood as expressing philosophical traditions underlying the modernist epistemologies that inform the political and cultural institutions of dominant Euro-Canadian society. These traditions, here understood in the sense of naturalized ideologies, when applied to the identification and collection of what is seen as relevant Indigenous knowledge, operate to occlude elements of Indigenous territorial presence that are potentially incommensurable with and inconvenient for dominant modes of governance. The ostensible universals of "tradition" and "knowledge," as framed in the institutional languages of technical expertise and assigned to the definition and collection of TEK, represent an extension of the consensual foundation (the "traditions") of bureaucratic rationality into the realm of Indigenous political subjectivities.[1] The resulting TEK expresses and delineates, within the parameters of modernist discourse, the agency of its holders, defining the limits of what they say and what they want, why they act as they do, and how they associate with others and with the land.

DEVELOPMENT

When allowed to define TEK on their own terms, resource development companies have understood it as an empirical catalogue of activities performed on the land that can be represented in their entirety through land-use mapping (Nasr and Scott 2010). The nomothetic, quantitative approach of land-use mapping, when unaccompanied by sufficient interpretative engagement, can recognize empirical data only, reframing the relationships between people and the land as mappable features and activities while neglecting a vast structuring and productive context (Drummond 2001). Here, the form and function of TEK are significantly shaped by the pragmatic requirements of economic development and the ontological assumptions of modern knowledge institutions. The result is the creation of a malleable and free-floating datum that may be inserted into situations and contexts far removed from the social frame in which it originated (see Ferguson 1994). Prone to respond more to the exigencies of powerful actors at the national or international scale than to the advancement of specific local interests, empirical land-use data, once divorced from any consideration of their historical and cultural context, may be applied to almost any purpose, reinforcing the projects and jurisdictional roles of their collectors.

The facility with which empirical land-use data may be integrated into environmental impact assessments, and their enthusiastic embrace as a form of local participation in development, is testament to their commensurability with capitalist transformations of the land (see Hale 2002).[2] Complex socio-ecological alterities and histories of dispossession are bypassed in this reduction to mappable empirical activities (rendered amenable to compensation), while at the same time, this invocation of TEK can "in effect shift the burden of proof, of generation of evidence, onto the shoulders of Aboriginal communities" (Alan Penn, pers. comm., October 28, 2017). Elements of Indigenous life that might constitute obstacles to extractive transformations of the land, such as socio-ecologies rooted in distinctive ontologies and epistemologies, complex histories of misrecognition or abrogation of rights, and the social costs of these latter, remain unacknowledged. However, as we elaborate below, these elements and the larger political and economic claims that they animate constitute in no small measure the development aspirations of Indigenous knowledge holders (Noble 2007; Povinelli 2001).

CONSERVATION

A principal organizing representation in environmental conservation efforts that involve Indigenous people is the notion that their cultures are sustainable and that their environmental behaviour is governed by an overarching ethic. Links between Indigenous cultural diversity and biodiversity have stimulated "global encounters between conservationists and indigenous peoples and the generation and 'discovery' of new kinds of knowledge, of community-based conservation and indigenous environmental wisdom"

(Igoe 2005, 378). The sustainable practices of Indigenous cultures are a major justification for their presence in what remains an enterprise committed to the ideal of wilderness as refuge from human industry. Pursuant to this default ideal, and its notion of protected areas as spaces "where nature might be left alone to flourish by its own pristine devices" (Cronon 1996, 82), the acceptability of human presence in protected areas hinges on the maintenance of cultural forms (generally understood as sets of values and land-use practices) that conform to the modern definition of environmental sustainability.

The difficulty of meeting these standards varies from community to community, but generally speaking, the bar is high. Sets of values and land-use practices that are recognized as traditional become enrolled in a modern-traditional dichotomy that sets up an evaluative framework for determining their integrity. As anthropologist Brian Noble (2007, 342) observes in relation to the United Nations Working Group on Intellectual Property, "by means of translation, TK is first polarized against modern knowledge and then enrolled into the latter's domain." The result is a paradoxical (modern) notion of tradition that is continually purged of modern influences, while being partially or fully eclipsed in the process.

When Indigenous people respond to the injunction to demonstrate sustainability in order to participate in the project of biodiversity protection, the values and practices that justify their inclusion become selectively enrolled in the tradition-modernity dichotomy. Thus enrolled, these forms are subject to a standard of traditionality that is defined by the absence of the effects and practices associated with modernity. Paradoxically, the increased cultural integration resulting from the intertwining of the local and the global scales that inevitably occurs when protected areas are created on Indigenous lands could be seen to impoverish Indigenous conservation values (for an elaboration of the many ways this "Catch-22" operates, see Holt 2005). This has led to a renewed call for the exclusion of communities from the project of biodiversity protection by some conservation biologists, especially those adherent to a "protectionist paradigm" (as described in Wilshusen et al. 2002).

A further paradox of the expectation of traditional cultural forms (absent modernity) is the requirement that they must lack self-consciousness or intentionality. If a cultural form, value orientation, or land-use practice is seen to be self-consciously held in an instrumental way (for political gain, for example), it becomes vulnerable to disqualification from being authentically traditional, as though habitus could be devoid of agency. And yet, for some scholars, intention is a key criterion that determines the presence or absence of environmental conservation in a given socio-ecological system (Krech 1999). Traditional forms then must be simultaneously unintentional and intentional, and political neutrality must be communicated in a context in which tradition, and Indigeneity itself, is instrumental to establishing rights and protecting political interests.

Rhetoric in which modernity is opposed to tradition, finally, insinuates that traditional cultures will naturally and inevitably be displaced by modernity ("progress"), glossing over the power differentials and alternative possibilities at play. Reading the history of colonial governments and resource-rich Indigenous homelands as the "modernization of the wild" obscures an alternative narrative, in which Indigenous peoples are engaged in political struggle with resource-hungry institutions (Tsing 2005, 255), with outcomes to be determined.

HEGEMONY AND RESISTANCE IN CONSERVATION AND DEVELOPMENT

Our analysis draws some inspiration, but also diverges, from that of Tania Li (2001, 2005), who argues that cultural differences vested in "specific identities and sets of practices" on defined territories prove a problematic foundation for claims to community empowerment and justice (2005, 448). She observes that "respect for cultural difference associated with nature conservation has become hegemonic" (Li 2001, 670) and that recognition on this basis "runs the risk of replicating old patterns of discrimination in new, environmental garb" (Li 2005, 448). In Li's (2001, 671) assessment, "land and resource rights made contingent upon stewardship are a pale version of the rights other citizens effectively enjoy. So are rights linked to demonstrated 'difference' ... Difference both enables claims to be made, and limits those claims by locating them within particular fields of power." In addition, it sets out normative standards for environmental knowledge and practice that shape people's behaviour on the ground, insofar as access to rights may require a potentially disempowering adherence to externally defined norms.

Whereas Li's Indonesian analysis sheds light on the hegemonic potential of discourses that join biodiversity conservation to Indigenous stewardship, there are evident difficulties in transferring it holus-bolus to the northern Quebec context and perhaps to other contexts as well. Li (2005, 448) is critical of a regime that would "segment the social and physical terrain and allocate rights and obligations on a differential basis," yet without Aboriginal title and treaty rights, which are surely anchored in "specific identities and sets of practices," Cree people would be subject to far greater economic and political inequality. Critiques of the devices of modernity are insufficient in and of themselves for devising genuine alternatives to neo-colonial discourses, often because they stop short of understanding how Indigenous societies are creatively and strategically adapting to and engaging with the dominant polity (Smith 2000). For example, David Scott (1995, 215) asserts that accessing pre-modern political rationalities is impossible in the post-colonial context. The old, pre-modern world has been "systematically displaced by a new one such that the old would now only be imaginable along paths that belong to new, always-already trans-formed, sets of co-ordinates, concepts, and assumptions." This stance suggests that Indigenous people are completely circumscribed by the modernist discourses under analysis. Persistent portrayals of Indigenous societies as submitting in subtle ways to neo-colonial concepts may also encourage despair or acquiescence. To understand how Indigenous people are resisting, accommodating, and otherwise engaging with these concepts, we need to avoid any totalizing vision and to consider the specific possibilities of particular contexts.

In the context of the sovereign claim of the state, external adjudication of rights is not fully escapable. But Crees' reliance on strategies of Aboriginal entitlement and cultural distinctiveness (including their knowledge and practices of environmental stewardship) has supported a relatively power-ful negotiating position and a measure of shared sovereignty through "treaty federalism" (White 2002). The ability of Crees to resist the hegemonic implications of TEK, and perhaps turn TEK to counter-hegemonic advan-tage, depends on the broader network of political and legal positions and resources in which the politics of TEK are engaged. There is no question that Crees must contend with the notion of TEK, as employed by state authorities, corporations, and conservationists. The crux of hegemonic/

counter-hegemonic dynamics is the tension between the bureaucratic integration of TEK in environmental governance and Wemindji Crees' practice and understanding of their own knowledge.

In the following section, we sketch the political and ethical dimensions of knowledge that are excluded by a selective focus on land-based empirical practices and cultural forms. Recovery of the excluded dimensions of traditional knowledge requires an account of the social fields of historical interactions between differently situated actors. From this position, we can see how the historical misrecognition and suppression of Cree political sovereignty and modes of relationship to the land have inflected subsistence hunting as a force for decolonization.

POLITICAL AND ETHICAL DIMENSIONS OF CREE KNOWLEDGE

Connected with the rise of Indigenous identity politics in a post-colonial globalized era, engagement with concepts of Indigenous knowledge is part of a larger strategy of expressing and protecting Cree difference in relation to the rest of Canada to obtain remedial justice. On the one hand, regional Cree identity emerged from transformations precipitated by colonization and industrial development, including the transition from nomadic bush-based life to settlement in permanent villages and the flooding of vast tracts of productive land by Hydro-Québec. At the same time, this identity was informed by the rise of international Indigenism and its articulation of Indigenous alterity in relation to settler societies. Like other Indigenous peoples, Crees seek the "affirmation of their collective rights, recognition of their sovereignty, and emancipation through the exercise of power" (Niezen 2003, 18). Because it explicitly references culturally distinctive relationships to the land, the knowledge of hunters extends this claim of difference into the realm of environmental protection and development, asserting Indigenous presence on and authority over contested territory. The Crees' distinctive modes of relationship to their territory are inextricably bound up in assertions of sovereignty (Jenson and Papillon 2000).

Can Indigenous knowledge, situated within broader struggles for Indigenous rights, be a means of embedding cultural difference in a modern politics of identity? Thus situated, can the political and ethical dimensions of Indigenous knowledge that are externalized in conventional appropriations of TEK be recouped? We find it helpful to think of the externalized elements of TEK, at least those that belong to the realm of politics and ethics, as phantasmagorias (in the sense used by Fischer 2003, 143) that come back to haunt neo-colonial governmentality. In other words, the qualities that are occluded in the process of its definition may become formative of the terrain to which TEK is appropriated. Beyond a collection of purified facts and sets of behaviours, Cree traditional knowledge and hunting practices are nodes of enduring identity that centralize and communicate political and ethical qualities of collective memory, suffering, and aspiration, whose salience seems to intensify in reaction to their exclusion within modern regimes of recognition.

KNOWLEDGE AND AUTHORITY AMONG WEMINDJI CREE HUNTERS

The Wemindji Cree hunters among whom we conducted our fieldwork are reminded every day that their authority on their hunting territories is contested. This occurs in their interactions with the transformed landscapes of development, in their cohabitation with increasing numbers of southern sport hunters and fishers, and through their changing roles in a community that is attempting to balance traditional forms of land use with economic development. The exercise of traditional knowledge is now steeped in awareness of the potential for attrition or annihilation of the hunting way of life. In Cree hunters' stories of interactions with agents of "modern rationality," this awareness manifests as a vigorous assertion of the capacity to understand and act on their environments in the face of modes of relationship with their territory that undermine or challenge this capacity.

The voices of people who lost significant portions of their family hunting territories to flooding illustrate the sense of loss and mourning that accompanies interrupted land use. Far from converting these individuals to the new way of experiencing the land, the dams and reservoirs have become potent symbols of danger and disempowerment. One of these individuals remembered going on a helicopter tour before the flooding to survey the gravesites (of parents, grandparents, and great-grandparents) that would soon be submerged. He remarked that he suffered a lot due to the flooding, both emotionally and materially. During an interview through a translator, this man's father expressed his personal sense of loss with an economy of words that was quite typical:

He thinks about when he used to trap, before the flooding. Today his son can't do it. He doesn't like that. You can't really do anything on the reservoir. It's dead. There's no beaver, no otter. The fish are contaminated. You can't get as much food from the reservoir as you can from the lakes. The food from the lakes that were drowned went far away. The beaver and the fish. Half of his territory is flooded. (Nasr field notes, August 14, 2006)

Crees put their own stories into play against modernist narratives about the forces transforming their land, challenging their use of it, and resituating their affective relationships with its other inhabitants. In their stories, lands, waters, animals, and other humans are embedded in relationships that take place within a coherent social and moral universe invoked in the telling. The universe of interaction is permeated with pressures to change and an awareness of externally imposed limits. These colour the exercise and invocation of locally held knowledge and ethics as markers for Cree identity. The stories of Cree hunters reject modes of relationship that would negate their capacity to meaningfully engage and invoke the power of the land, a negation that re-creates the need and amplifies the importance of reasserting their authority. The following passage illustrates one such interaction involving the erasure and reassertion of authority.

One trapper saw some southerners fishing on a lake where he has his camp. These people come every summer to fish. Normally, he leaves them alone because they don't bother him. One time, he was beckoned over by these people on this lake. One man asked him if he had a gun, and what kind of gun it was. The trapper said, "yes," and showed them his gun. Then the sport fishers asked him if he would shoot a bear that had been coming around their camp at night. The trapper looked around him at their campsite and saw lots of garbage lying around everywhere. The trapper hung around for a couple hours but didn't see the bear. The sport fisher said, "It only comes at night, you have to come back at night." The trapper said, "Well, I don't want to shoot the bear, because it's the wrong time of year, and it might have cubs." The sport fisher said, "Well, actually, there are two cubs with the bear." The trapper said, "I'm not going to shoot those bears. Clean your camp." And then he left. (Nasr field notes, August 11, 2006)

In this story, the southern sport fishers wanted the bear to be killed because they saw it as a nuisance and found it frightening. They stand outside the Cree sphere of meaningful relationships with the bear, an animal that, for Crees, exerts significant power over hunting success in general (see Colin Scott 2006). This episode represents a Cree trapper's effort to maintain a relationship with the bear that is threatened with negation by the sport fishers. It underlines the differences in ethical conduct and valuation of the land between the Crees and the southerners. Too, unclean campsites are a sensitive issue in Cree stewardship, inviting as they do inappropriate relations with animals.

From their own moorings in Indigenous knowledge and institutions, Cree hunters may also establish patterns of relationships with newcomers on their family territories that enable them to protect their lands and continue to manage their resources. We have witnessed, for example, Cree hunting camps at the edge of hydro-electric transmission corridors, taking advantage of hydro access roads not only for personal access, but also to monitor the comings and goings of recreational hunters from the South (Colin Scott 2017). This also enables Cree territory leaders to form relationships with a few responsible sport hunters who visit every year. These hunters benefit from Cree knowledge of the locale to fill their moose tag; in return, the visibility and social ties of these guests, through the etiquette of the sport hunting fraternity, help to both educate and limit access to others who are strangers to Cree traditional owners. Harvey Feit (2004a), similarly, observes that such strategies pervade Cree practice vis-à-vis a broad spectrum of newcomers and that they are deeply rooted in onto-logical premises of reciprocity and mutual respect.

If we emphasize here the political dimensions of Indigenous know-ledge over its substantive content, it is to foreground the point that from Indigenous perspectives, the purpose of TEK's engagement in develop-ment and conservation policy is not just preservation of subsistence hunting practices for their own sake. Rather, subsistence hunting has come to embody claims to political rights, resources, and cultural survival in the face of forces negating Cree presence and authority on the traditional territories. Our experience with Wemindji hunters confirms the enduring emotive dimensions of subsistence hunting and the continuing vitality of the frames of reference associated with it. Indeed, hunting knowledge as embodied experience has become charged with new relevance and is asserted by Indigenous people as a privileged frame of reference – even as it has become an object of political discourse and ethnic identification. Naomi Adelson (2000) shows how the process of cultural assertion and political mobilization that has characterized Cree society in recent decades has inflected notions of health (*miyupimaatisiiun* – literally, "being alive well") by taking up residence in the bodies and experiences of members of Cree society. In a similar way, subsistence hunting and associated knowledge, when practised by individual hunters, become permeated with meaning through their complex interface with identity reformulation, the negotiation of power, and resistance to the elision of Cree authority on the land. Practising traditional knowledge, along with enacting the Cree notion of health, seems to be linked to strategies of cultural assertion and the exercise of power within a particular dynamic involving the Crees, the government, and individual land users (see Adelson 2000).

Cultural continuity is commonly a local value and goal in its own right, but that continuity from pre-colonial times to the present is not the stereotypical reproduction of cultural forms; these have always been in constitutive relationship with a spectrum of dynamic forces. These forces include the reality and continual historical consciousness of a relationship of domination and the persistent need for redress. In common with those of other Indigenous peoples, Cree goals include cultural preservation/perpetuation, authority over land and resource protection and management, and affirmation of collective sovereignty (Niezen 2003). The context is one of power relations, and the political importance of TEK as an avenue of Indigenous cultural politics should be understood, following Sidsel Saugestad (2001), in a relational rather than essentialist way. For Justin Kenrick and Jerome Lewis (Asch et al. 2004, 263), "a relational understanding of the term ['Indigenous'] focuses on the fundamental issues of power and dispossession that those calling themselves 'indigenous' are concerned to address ... 'Indigenous rights' describes a strategy for resisting dispossession that employs a language understood by those wielding power." Similarly, the politics of TEK in development and conservation arenas involve contested processes of translation that operate within particular power dynamics.

This matrix of (political) relations suffusing Indigenous knowledge locally is precisely what goes unrecognized in its incorporation as TEK, which tends to turn on the existence of Indigenous cultural forms that are distinctive from, yet functionally reconcilable with, Euro-Canadian norms and institutions. It is promised by state actors, corporations, or conservation NGOs that "empowerment" (within acceptable limits) will somehow occur through selective participation of Indigenous values and knowledge. This begs the question of how these processes can contribute to the recognition of local relations of power and authority when associated rights and considerations are indexed to delimited phenomena of compensable and empirically demonstrable hunting resources and practices.

We suggest that scaling up from selective accommodation and appropriation of TEK to a more comprehensive recognition of evolving land tenure institutions in policy and law would enable Cree perspectives to be heard and enacted, less fettered by the political majority's regimes of recognition, discourses of cultural authenticity, and institutional routines. Cree land tenure practices and institutions afford an "organic" rather than a "totalizing" critique (see Smith 2000) of modes of recognition and inclusion by external actors, foregrounding past and present Cree strategies for achieving accommodation on terms partly of their own making. As we discuss in the following section, land tenure institutions resist analysis as shadows or mimetic echoes of colonial domination. Rather, they are loci of emergent difference – integral to a "territory of difference" in Arturo Escobar's (2008) terms – arising from a long history of struggle against external circumscription and political domination. They reveal distinctively Cree ontological commitments to relation making vis-à-vis non-Cree administrations and to the continued stewardship of the traditional territory. They thus potentially provide a solid basis for developing transformative strategies and outcomes regarding co-governance policy.

CREE INSTITUTIONS AS THE BASIS FOR POLICY

Cree governance has evolved dramatically over the past three decades, mainly in response to fundamental changes in the political, social, and economic landscape of Eeyou Istchee, the Cree homeland. This evolution is customary and natural, as political power is universal and inherent in human nature. The meaning and practice of Eeyou governance have evolved and have been redefined by Eeyouch (Cree people) on the basis of traditional law and customs, and the intent and spirit of the James Bay and

Northern Québec Agreement. Eeyouch are presently using their governments and other Eeyou authorities to meet goals and needs such as Eeyou Nation administration, land management and administration, housing, economic development, financial administration, human resources development, traditional pursuits, cultural and language development, policing, administration of justice, education, social services, health, delivery and administration of services and programs, community development, environmental protection, wildlife management, and political representation to conduct government-to-government as well as nation-to-nation relations (Moses 2002a, 6).[3]

As former grand chief Ted Moses, who negotiated the recent Paix des Braves (Anonymous 2002),[4] states, Cree governance has evolved to meet the changing needs of the Cree people and the increasing complexity required for the administration of a host of services. To the Crees, change does not diminish the authenticity or legitimacy of political claims embodied in modern Cree governance, because according to Moses, "political power is universal and inherent in human nature" (Moses 2002a, 6). In light of the preceding discussion, it is important to underscore the explicit connection that Moses is making here between the evolution of governance, cultural authenticity, and political power. The wide range of services delivered by Crees for Crees, as well as the effectiveness of the grand council in negotiating on a nation-to-nation basis with the provincial and federal governments, testifies to the vitality of Cree power in current configurations of governance. This power is evident in the Crees' flexibility in negotiating state institutional requirements for social and political recognition. Cree society has become increasingly complex and bureaucratized over the years, as political mobilization and claims to Indigenous rights and authority (including the negotiations of land claims and treaties) required the formation of regional identities and institutions (Salisbury 1986; Jenson and Papillon 2000). These institutions have played a vital role in advancing current levels of recognition of Cree autonomy by provincial and federal governments, enabling nation-to-nation negotiations and the formation of an administrative structure that can challenge and shape the status quo of development and Aboriginal rights on Cree territory.

For the purposes of subsistence land use and stewardship, the institutions of the tallyman and the trapline are crucial to Cree governance. Each trapline, or family hunting territory, is overseen by an *uchimaau* (steward and hunting leader, known as a tallyman).[5] *Uchimaauch,* whose role tends to be passed down through the male line of the family, are respected holders of ecological and cultural knowledge, responsible for the health of the land and the maintenance of harmonious relationships between humans and other-than-humans (Colin Scott 1988). There is long-standing debate about the antiquity of Algonkian family hunting territories, most recently resurrected by anthropologists Shepard Krech (1999) and Harvey Feit (2004b, 2007), with the former discounting and the latter affirming their historical role in conservation (Hames 2007, 182). In our view, Feit provides compelling evidence that the hunting territory system, regardless of its origins, has long served to effectively conserve and manage beaver populations. The system remains the institutional setting for the induction and socialization of young Crees into bush-based knowledge, values, and subsistence techniques. For centuries, it has underwritten Cree commercial and political transactions with the Hudson's Bay Company and the Canadian state, has thus been an enduring marker of Cree identity, and remains a key locus for a personal and collective sense of well-being. As Peter Awashish (2003, 166) explains,

> Eeyou Istchee consists of nine communal lands and about three hundred Indoh-hoh Istchee (Eeyou hunting territories, or "Cree traplines"). As the land that Eeyouch have used and occupied for millennia, Eeyou Istchee is essential for the *meeyou pimaat-tahseewin* or holistic well being, of Eeyouch. Eeyou Istchee comprises the foundation of Eeyou governance, culture, identity, history, spirituality, and traditional way of life. The unique and special relationship between Eeyouch and their homeland is a fundamental part of the nature of being Eeyou.

The trapline system has been the focus of negotiated inclusion in co-management regimes since about 1925. Family hunting territories were first registered and mapped by the federal government in the 1940s, after the numbers of fur bearing animals crashed disastrously in the region during the 1920s and 1930s. A rise in fur prices, a massive influx of white trappers to the more southerly James Bay Cree territories, and the subsequent breakdown of the Cree land tenure system occasioned the crash.

This hardship was compounded by a trough in the caribou population cycle and widespread tuberculosis epidemics.

The move to rehabilitate the beaver through the implementation of beaver preserves marked the extensive administrative intervention of federal government agents into James Bay Cree society. Preserves were created at various sites in the James Bay region between the 1930s and the 1950s, with local Cree trappers participating in their maintenance and management. Indian Affairs agents worked with the Crees to map and register the family hunting territories.

With the land tenure system again functioning effectively, thanks to the exclusion of white trappers from the region and official recognition of Cree tallymen's authority, the beaver population had fully recovered from its catastrophic slump by the years after the Second World War. Quotas declared by the government and in effect from the 1930s to the 1970s were respected by the Crees. Though nominally regulated by government agents, the quotas were based on information supplied by Cree tallymen. However, "with their extensive knowledge of the resource populations, the Cree did not feel bound to follow the advice of government agents, which was based on simply following the trends in the number of lodges. Cree decisions were based on far more extensive knowledge" (Feit 1995, 198).

From the perspective of the Crees, the government was recognizing their authority through formalizing their land tenure institutions and the role of the tallyman.[6] Tallymen have responsibility for, and authority over, all activities that affect the land and animals within the boundaries of a given trapline. They dispense and revoke permission for hunting, fishing, and trapping, and are respected for their knowledge and guidance in the proper management of land-based activities and maintenance of relationships with animal and other non-human inhabitants.[7] By restricting non-Native trapping and helping to set up beaver reserves, government agents created favourable conditions for both the beaver to recover and the Cree system of land tenure to regain its effectiveness. At the same time, the episode marked an extension of state administration into Cree society and territory: "The Cree were still exercising extensive control and autonomy in their hunting culture, but they were now doing so as part of the Canadian polity" (Feit 1995, 199).

The 2002 Paix des Braves with Quebec is one instance of the reinforcement of customary tenure in land-use policy. On the most local scale, family hunting territories became basic frames of reference and regulation in the new forest co-management regimes established under the agreement (described in Colin Scott 2005). On the regional scale, Quebec confirmed that the James Bay Crees were a nation, which had a right to negotiate directly with the provincial government. For the first time, the province explicitly acknowledged that the Cree Nation possessed the authority to have a say and an ongoing stake in development on Eeyou Istchee. As Ted Moses (2002b) stated, "our new agreement recognizes that the Crees are entitled to share meaningfully with Québec in the benefits from natural resources that accrue within the entire traditional territory of the Cree Nation." Brian Craik (2004, 176) notes that what was distinctly new here was "the possibility of the Crees becoming involved in the development of their territory, rather than being compensated to step aside."

DISCUSSION AND CONCLUSIONS

A challenge for our collective research, described more fully in Chapter 1 of this volume, is to situate the development of protected areas in a manner that is politically achievable in negotiations with Quebec and Canada, while materially reinforcing the value and security of relations of Cree tenure. In the process of collaborative research between Cree hunters and the various disciplines represented on our team, the assessments of local experts and biologists have been mutually affirmative in our exploration of priorities for protection. In this regard, a real and pragmatic intersection of conservation interests emanates from the distinct ontological and institutional standpoints of hunters and scientists, which we have sought to advance in negotiations with responsible state authorities. Establishing a network of protected areas in Wemindji territory is a work in progress, and conclusive assessment would be premature, but our Cree collaborators understand protected areas, on appropriate terms yet to be negotiated, as spaces that could potentially reinforce, not subjugate, the complex of Cree institutions, knowledge, values, and practices comprising life on, and stewardship of, the land.

In support, the present chapter has addressed pitfalls and possibilities in the dance of hegemonic and counter-hegemonic discourses that make up the politics of TEK. We argue for "scaling up" from a narrow focus on TEK to the broader problem of empowering Indigenous knowledge, values, and the Cree institutions in which they are embedded. These are autonomous domains, with their own history of power relations, which must be the starting point for Indigenous people's engagement in environmental protection and development on their territories. The TEK focus too often separates (empirical) knowledge artifacts from their generative social practices and institutions, narrowing the latter to stereotypical subsistence behaviours and empirical observations that can then be validated or delegitimized according to modernist criteria regarding governance and development. The selectivity involved in these manoeuvres purges this knowledge and behaviours of ontological significance, ethical orientation, and political vitality. Further, Indigenous subjects are exposed to the double jeopardy of dismissal should they deviate from imposed standards of authenticity and sustainability, when in reality, "interrogating the 'integrity' of IK [Indigenous knowledge] is an unproductive project, because it implies an immutability that is neither possible nor desirable" (Butler 2006, 121).

Hence, we have advocated a broadly framed institutional account of the politics of knowledge, with a focus on family hunting territories and leadership. Contemporary Cree polity involves an internal dynamic, a sometimes contested and negotiated division of responsibility and authority between the system of hunting territories and uchimaauch, the local community councils, and the regional grand council. Each draws strength from its involvement with the others, and within the whole, the values of territorial stewardship are universally acknowledged. The special role of uchimaauch and other occupational hunters in this political order is to speak on behalf of the land, relations that are seen as fundamental to Cree power and identity, not just internal to Cree society. Ted Moses (2002a, 7) notes, "Eeyou governance is ... the practice and exercise of stewardship, guardianship and custodianship of Eeyou Istchee." Whereas the values of "stewardship, guardianship, and custodianship" are claimed by regional politicians and disseminated in macro-political arenas, they are embodied locally in the family hunting territory, the authority of the uchimaau, and knowledge of the land.

The Cree hunting territory remains a vitally important institution to the present day, the product of a history of environmental and political relations, at once internal and external. Acknowledging and building upon this institution for decision making in conservation, resource management, and development have the pragmatic advantage of relying on an existent socio-ecological reality, in relational terms that are locally grounded and culturally intelligible to resident participants.[8] It is an institution, moreover, with various layers of legal recognition in the James Bay and Northern Québec Agreement of 1975 and the Paix des Braves of 2002. There is the potential for further reinforcement and elaboration of the role of family hunting territories and tallymen in the current wave of parks and protected areas creation in Eeyou Istchee. Cree agency, from this institutional base, can viably negotiate the confluence of goals in conservation, community empowerment, social justice, and cultural continuity – always in dialogue with institutions of community and regional council governance that have emerged from band and inter-band relations (Colin Scott 2018). The engagement and social reproduction of Indigenous knowledge in this institutional and practical context are sufficiently robust to resist the abstraction and appropriation of TEK by external authority and to negotiate more equitable conditions for knowledge exchange in collaborative decision making. In such a context, sustained by its ontological and experiential moorings and the institutional relations of its social reproduction, the trajectory of Indigenous knowledge remains part and parcel of self-determined cultural continuity.

NOTES

1 Ronald Niezen (2003, 96) describes the universalizing directionality of the human rights movement in a similar way.

2 Charles Hale (2002, 485) argues that the cultural project of neo-liberalism entails "pro-active recognition of a minimal package of cultural rights, and an equally vigorous rejection of the rest."

3 Eeyou Istchee, also spelled Iyiyuu Ischii (or Aschii), refers to the Cree homeland. "Eeyou," the word for Cree people, also applies to the more general notions of "(the being of) a person" or the "totality of life." "Istchee" means "the world" or "the ground of being" (Colin Scott 2006, 61–62).

4 The Crees of Quebec and the Government of Quebec signed a "New Relationship Agreement" on February 7, 2002. This agreement was dubbed the "Paix des Braves" in the early days of its unveiling and we refer to is as such throughout the chapter.

5 So-called because for a time beginning in the 1930s, the government paid these individuals an annual honorarium to tally the number of beaver lodges on their territories.

6 For a discussion of this process and its impact on the pre-existing system of land tenure, see Harvey Feit (1995). See Colin Scott (1989) for Cree understandings of relationships with "the Whiteman State" at this period.

7 For an extensive discussion of the Cree system of land tenure and the role of the tallyman, see Colin Scott (1988).

8 Arun Agrawal and Clark C. Gibson (1999) argue the merits of grounding decentralized, participatory conservation in locally operative institutional arrangements.

WORKS CITED

Adelson, Naomi. 2000. *"Being Alive Well": Health and the Politics of Cree Well-Being.* Toronto: University of Toronto Press.

Agrawal, Arun, and Clark C. Gibson. 1999. "Enchantment and Disenchantment: The Role of Community in Natural Resource Conservation." *World Development* 27 (4): 629–49.

Anonymous. 2002. "Agreement concerning a New Relationship between Le Gouvernement du Québec and the Crees of Québec." Nemaska, QC, Cree Regional Authority/Le Gouvernement du Québec.

Asch, Michael, Colin Samson, Dieter Heinen, Justin Kenrick, Jerome Lewis, Sidsel Saugestad, Terry Turner, and Adam Kuper. 2004. "On the Return of the Native." *Current Anthropology* 45 (2): 261–63.

Awashish, Peter. 2003. "From Board to Nation Governance." In *Reconfiguring Aboriginal-State Relations,* ed. M. Murphy, 165–83. Montreal and Kingston: McGill-Queen's University Press for the Institute of Intergovernmental Relations.

Berkes, Fikret. 2008. *Sacred Ecology.* New York: Routledge.

Butler, Caroline. 2006. "Historicizing Indigenous Knowledge." In *Traditional Ecological Knowledge and Natural Resource Management,* ed. C.R. Menzies and C. Butler, 107–26. Lincoln: University of Nebraska Press.

Callon, Michel. 1986. "Some Elements of a Sociology of Translation: Domestication of the Scallops and the Fishermen of St. Brieuc Bay." In *Power, Action and Belief: A New Sociology of Knowledge?* ed. J. Law, 196–233. London: Routledge.

Craik, Brian. 2004. "The Importance of Working Together: Exclusions, Conflicts and Participation." In *In the Way of Development: Indigenous Peoples, Life Projects and Globalization,* ed. Mario Blaser, Harvey A. Feit, and Glenn McRae, 166–86. London: Zed Books.

Cree Trappers Association. 2009. "Eeyou Indoh-hoh Weeshou-Wehwun: Traditional Eeyou Hunting Law." Cree Trappers Association, Eastmain. QC.

Cronon, William. 1996. "The Trouble with Wilderness; or, Getting Back to the Wrong Nature." In *Uncommon Ground: Rethinking the Human Place in Nature,* ed. William Cronon, 69–90. New York: Norton.

de la Cadena, Marisol, and Orin Starn. 2007. "Introduction." In *Indigenous Experience Today,* ed. M. de la Cadena and O. Starn, 1–32. Oxford: Berg.

Drummond, Susan G. 2001. "Writing Legal Histories on Nunavik." In *Aboriginal Autonomy and Development,* ed. Colin H. Scott, 41–62. Vancouver: UBC Press.

Escobar, Arturo. 2008. *Territories of Difference: Place, Movements, Life, Redes.* Durham: Duke University Press.

Feit, Harvey A. 1995. "Hunting and the Quest for Power, the James Bay Cree and Whitemen in the Twentieth Century." In *Native Peoples: The Canadian Experience,* 2nd ed., ed. R.B. Morrison and C.R. Wilson, 181–223. Toronto: McClelland and Stewart.

–. 2004a. "James Bay Crees' Life Projects and Politics: Histories of Place, Animal Partners and Enduring Relationships." In *In the Way of Development: Indigenous Peoples, Life Projects, and Globalization,* ed. Mario Blaser, Harvey A. Feit, and Glenn McRae, 92–110. London: Zed Books.

–. 2004b. "Les territoires de chasse algonquiens avant leur 'découverte'? Études et histoires sur les tenures, les incendies de forêt et la sociabilité de la chasse." *Recherche Amérindiennes au Québec* 34 (5): 5–21.

–. 2007. "Myths of the Ecological Whitemen." In *Native Americans and the Environment: Perspectives on the Ecological Indian,* ed. D.R.L.M. Harkin, 52–94. Lincoln: University of Nebraska Press.

Ferguson, James. 1994. *The Anti-politics Machine: "Development," Depoliticization and Bureaucratic Power in Lesotho.* Minneapolis: University of Minnesota Press. Originally published 1990.

Fischer, Michael M.J. 2003. *Emergent Forms of Life and the Anthropological Voice.* Durham: Duke University Press.

Hale, Charles R. 2002. "Does Multiculturalism Menace? Governance, Cultural Rights and the Politics of Identity in Guatemala." *Journal of Latin American Studies* 34: 485–524.

Hames, Raymond. 2007. "The Ecologically Noble Savage Debate." *Annual Review of Anthropology* 36: 177–90.

Henkel, Heiko, and Roderick Stirrat. 2001. "Participation as Spiritual Duty: Empowerment as Secular Subjection." In *Participation: The New Tyranny?* ed. B. Cooke and U. Kothari, 168–84. London: Zed Books.

Holt, Flora Lu. 2005. "The Catch-22 of Conservation: Indigenous Peoples, Biologists and Cultural Change." *Human Ecology* 33: 199–215.

Igoe, James. 2005. "Global Indigenism and Spaceship Earth: Convergence, Space, and Re-entry Fiction." *Globalizations* 2 (3): 377–90.

Jenson, Jane, and Martin Papillon. 2000. "Challenging the Citizenship Regime: The James Bay Cree and Transnational Action." *Politics and Society* 28 (2): 245–64.

Krech, Shepard, III. 1999. *The Ecological Indian: Myth and History.* New York: W.W. Norton.

Li, Tania M. 2001. "Masyarakat Adat, Difference, and the Limits of Recognition in Indonesia's Forest Zone." *Modern Asian Studies* 35 (3): 645–76.

–. 2005. "Engaging Simplifications: Community-Based Natural Resource Management, Market Processes, and State Agendas in Upland Southeast Asia." In *Com-*

munities and Conservation: Histories and Politics of Community-Based Natural Resource Management, ed. J.P. Brosius, A.L. Tsing, and C. Zerner, 427–57. Walnut Creek, CA: AltaMira Press.

McGregor, Deborah. 2004. "Traditional Ecological Knowledge and Sustainable Development: Towards Coexistence." In *In the Way of Development: Indigenous Peoples, Life Projects, and Globalization*, ed. Mario Blaser, Harvey A. Feit, and Glenn McRae, 72–91. London: Zed Books.

Moses, Ted. 2002a. "Eeyou Governance beyond the Indian Act and the James Bay and Northern Québec Agreement." Address delivered at the Pacific Business and Law Institute Conference, "Beyond the Indian Act," Ottawa, April 17–18.

–. 2002b. "Grand Chief Celebrates Agreement in Europe: Full Text of Moses Speech in London." November 26. http://www.ottertooth.com/Reports/Rupert/News/rupert-news3.htm.

Nadasdy, Paul. 2003. *Hunters and Bureaucrats: Power, Knowledge, and Aboriginal-State Relations in the Southwest Yukon*. Vancouver: UBC Press.

–. 2005. "The Anti-politics of TEK: The Institutionalization of Co-management Discourse and Practice." *Anthropologica* 47 (2): 215–32.

Nasr, Wren, and Colin H. Scott. 2010. "The Politics of Indigenous Knowledge in Environmental Assessment: James Bay Crees and Hydro-Electric Projects." In *Cultural Autonomy: Frictions and Connections*, ed. Petra Rethmann, Imre Szeman, and Will Coleman, 132–55. Vancouver: UBC Press.

Niezen, Ronald. 2003. *The Origins of Indigenism: Human Rights and the Politics of Identity*. Berkeley: University of California Press.

Noble, Brian. 2007. "Justice, Transaction, Translation: Blackfoot Tipi Transfers and WIPO's Search for the Facts of Traditional Knowledge Exchange." *American Anthropologist* 109 (2): 338–49.

Povinelli, Elizabeth A. 2001. "Radical Worlds: The Anthropology of Incommensurability and Inconceivability." *Annual Review of Anthropology* 30: 319–34.

Salisbury, Richard F. 1986. *A Homeland for the Cree: Regional Development in James Bay, 1971–1981*. Montreal and Kingston: McGill-Queen's University Press.

Saugestad, Sidsel. 2001. "Contested Images: 'First Peoples' or 'Marginalized Minorities' in Southern Africa." In *Africa's Indigenous Peoples: "First Peoples" or "Marginalized Minorities"?* ed. A. Barnard and J. Kenrick, 299–322. Edinburgh: Centre of African Studies.

Scott, Colin H. 1988. "Property, Practice and Aboriginal Rights among Québec Cree Hunters." In *Hunters and Gatherers*. Vol. 2, *Property, Power and Ideology*, ed. Tim Ingold, David Riches, and James Woodburn, 35–51. New York: Berg.

–. 1989. "Ideology of Reciprocity between the James Bay Cree and the Whiteman State." In *Outwitting the State*, ed. Peter Skalnik, 81–108. New Brunswick, NJ: Transaction.

–. 2005. "Co-management and the Politics of Aboriginal Consent to Resource Development: The Agreement concerning a New Relationship between Le Gouvernement du Québec and the Crees of Québec (2002)." In *Canada: The State of the Federation, 2003; Re-configuring Aboriginal-State Relations: An Examination of Federal Reform and Aboriginal-State Relations,* ed. Michael Murphy, 133–63. Montreal and Kingston: McGill-Queen's University Press.

–. 2006. "Spirit and Practical Knowledge in the Person of the Bear among Wemindji Cree Hunters." *Ethnos* 71 (1): 51–66.

–. 2017. "The Endurance of Relational Ontology: Encounters between Eeyouch and Sport Hunters." In *Entangled Territorialities: Negotiating Indigenous Lands in Australia and Canada,* ed. Françoise Dussart and Sylvie Poirier, 51–69. Toronto: University of Toronto Press.

–. 2018. "Family Territories, Community Territories: Balancing Rights and Responsibilities through Time." In *Who Shares the Land? Algonquian Territoriality and Land Governance. Anthropologica* 60 (1): 90–105.

Scott, David. 1995. "Colonial Governmentality." *Social Text* 43: 191–220.

Smith, G.H. 2000. "Protecting and Respecting Indigenous Knowledge." In *Reclaiming Indigenous Voice and Vision,* ed. M. Battiste, 209–24. Vancouver: UBC Press.

Tsing, Anna L. 2005. *Friction: An Ethnography of Global Connection.* Princeton: Princeton University Press.

Usher, Peter J. 2000. "Traditional Environmental Knowledge in Environmental Assessment and Management." *Arctic* 53 (2): 183–93.

White, Graham. 2002. "Treaty Federalism in Northern Canada: Aboriginal-Government Land Claims Boards." *Publius: The Journal of Federalism* 32 (3): 89–114.

Wilshusen, Peter R., Steven R. Brechin, Crytal L. Fortwangler, and Patrick C. West. 2002. "Reinventing a Square Wheel: Critique of a Resurgent 'Protection Paradigm' in International Biodiversity Conservation." *Society and Natural Resources* 15: 17–40.

3

A Balancing Act
Mining and Protected Areas on Wemindji Territory

UGO LAPOINTE and COLIN H. SCOTT

> We are required to keep the Land and all of our environment in
> a healthy and clean state, for ourselves, our future generations,
> and all living things that share our Territory with us. It is part of
> our obligation as Elders, Grandparents, Parents and Community
> Leaders to provide a good example in this way, and to teach
> the Youth.
>
> — Cree Nation of Wemindji, n.d., 2

A mining project, like other externally initiated industrial development on Indigenous lands, brings a host of community objectives into sharp focus. These include protecting the environment, retaining cultural autonomy and identity, maintaining traditional livelihoods, and pursuing economic development through market participation. Achieving these goals, on terms negotiated with the state and with corporate developers, is a complicated exercise. Implicit in the power asymmetries of such negotiation are sometimes painful trade-offs. As a result, decisions made by Indigenous communities about development and environmental protection may be perceived by outsiders, and indeed may be locally experienced, as conflicting and contradictory.

Beyond the community, there may be little appreciation for the multiple visions, goals, and pressures that are at play locally. Mining companies, government ministers, environmentalists, and interested or disinterested consumers of the media may employ simplistic and opposing images to assess the politics of Indigenous community involvement in resource decision making and protected area creation. One such image portrays Indigenous communities as fully entangled in modernity, on a slippery slope into cultural assimilation. An opposing image is that of the environmental Indian and cultural traditionalist, standing against the menace of industrial development. Too often, Indigenous politics are cynically understood as driven by the goals and interests of modernity, while simultaneously proclaiming an allegiance to traditional values and identities, as a means of generating sympathy and leverage.

Indigenous communities in the Canadian North are increasingly aware of the politics of such imagery, but one would be deeply mistaken to think that they are resigned to the hegemony of a mainstream modernity or that the politics of Indigenous cultural difference are mere smoke and mirrors. A more nuanced account is both possible and necessary if we are to grasp realities on the ground, and in the imaginations, of Indigenous people. Such an account must embrace a range of actualities and possibilities, from determined resistance to mutual accommodation in relations with mining companies and non-Indigenous government entities. We cannot understand why accommodation may be achievable in one instance but not in another if we fail to assess local visions, projects, realities, interests, and anxieties.

This chapter focuses on the responses of the Cree Nation of Wemindji to mining and its relationship to protected area creation. In the early 2000s, prospecting claims proliferated throughout the hunting territories (also referred to as traplines) of Wemindji, complicating the goal of establishing a network of protected areas that is appropriate to the socio-ecological order of Cree hunting and environmental stewardship. Fortunately, the portion of Wemindji territory in which the Paakumshumwaau-Maatuskaau Biodiversity Reserve lies was marked by relatively few mining claims, although there were important exceptions, including a large claim encircling Paakumshumwaau (Old Factory Lake) at the headwaters of Paakumshumwaashtikw (Old Factory River). Strong community opposition to advanced exploration on a claim in the biodiversity reserve contrasted with considerable community support for negotiating a "collaboration agreement" with Goldcorp, the developer of the rich Éléonore gold deposit that lies about forty or fifty kilometres east of Paakumshumwaau and outside the biodiversity reserve.

The *nituuhuu uchimaauch,* or tallymen, who were responsible for the family hunting territories in the biodiversity reserve unanimously supported negotiating with Goldcorp. By contrast, the tallyman for the family territory at the Éléonore deposit gave qualified support. An important part of the story is the manner in which Indigenous institutions of tenure and management authority shape agendas at the level of community leadership. Further, a project such as Goldcorp's activates the relationship between the Cree Nation of Wemindji and the Grand Council of the Crees (Eeyou Istchee), which involves mutual as well as competing interests.[1] The positions and political resources of tallymen, community councils, and the regional council are all conditioned, environmentally, by the impacts of hydro-electric development in eastern James Bay, and institutionally, by land categories, environmental review procedures, and various other provisions of the James Bay and Northern Québec Agreement (Quebec 1975) and the Agreement Concerning a New Relationship between the Government of Québec and the Crees of Québec, hereafter referred to as the Paix des Braves (Quebec 2002). In turn, these environmental and institutional elements interact with distinct bodies of Quebec legislation for mineral extraction and environmental protection that present both obstacles and opportunities for local projects.

Hence, this story reaches from the very local – the personal and culturally authorized commitments and responsibilities of individual tallymen – to the archetypically global, the circumstances of international gold markets. It is based on the authors' engagements at various institutional nodes and junctures. We spoke with tallymen and other community members, community councillors, representatives of prospecting and mining companies, local and regional Cree leaders, public servants of the provincial government, and members of non-governmental organizations. We were participant-observers in several meetings that brought various combinations of these actors into dialogue with one another.

PERCEPTIONS OF AND RESPONSES TO MINING DEVELOPMENT

Mineral exploration activities have increased in both number and intensity on Wemindji territory since the early 2000s (see Table 3.1). Favourable

TABLE 3.1

Mineral exploration trends in Wemindji territory, Eeyou Istchee, and Quebec (1985–2008)

	1985	1990	1995	2000	2001	2002	2003	2004	2005	2006	2007	2008
Number of active projects[a]												
Wemindji	<5	<5	5	12	24	10	9	8	20	42	45	40
Eeyou Istchee[b]	5	10	12	42	53	55	47	43	66	93	117	109
Quebec	–	–	–	472	462	420	312	345	347	357	486	463
Expenses in exploration work (M$)[c]												
Wemindji	–	–	–	2	6	3	4	5	12	35	44	39
Eeyou Istchee[b]	<5	<5	<5	5–10	7–10	18	29	28	39	77	113	106
Quebec	135	145	132	90	110	115	150	227	205	295	476	450[d]

a Numbers of active exploration projects are conservative and do not include dormant projects and claims.
b This refers to Eeyou Istchee north of the fiftieth parallel. Waswanipi and Oujé-Bougoumou territories, both subject to significant mining activity since the early twentieth century, are not included.
c All in current dollars: figures for Wemindji and Eeyou Istchee are estimations (+/- 20 percent) based on a pro rata of active exploration projects in the region versus those for the overall province of Quebec.
d Provisory estimate.
Sources: AMQ (2006–08); MER (1986, 1991, 1996); MRNF (2001–08).

geology, supportive government policies with incentives to develop northern projects, and an expanding network of roads and other transportation infrastructure have all prompted companies and individuals to prospect in the area, particularly since the late 1990s, as stated in the 2001 annual report of the Ministère des Ressources naturelles et de la Faune (MRNF 2001–08). Gold, diamonds, copper, iron, uranium, and lithium are among the main commodities sought. A significant exploration rush swept the western part of Wemindji territory in 2001, when till sampling by Majescor Resources found minerals indicative of potential diamond deposits (ibid.). This rush, however, was short-lived and did not produce any significant diamond findings. On the other hand, the 2004 discovery of the Éléonore gold deposit by Virginia Gold Mines (a property later bought by Goldcorp) was significant. It attracted dozens of new mining companies and prospectors to stake claims and conduct exploration work in the area.

Mounting Development Pressures

The discovery of the Éléonore gold deposit contributed to a five-fold increase in the number of active mining exploration projects on Wemindji territory between 2004 and 2008. In 2008, sixty-eight companies and prospectors held mining claims in eighteen of Wemindji's twenty traplines (CMEB 2009). Cumulatively, these claims covered about one-fifth of Wemindji territory (Figure 3.1). From 2005 to 2008, the most recent period for which we have data, an estimated $130 million was spent in exploration work in the Wemindji area, more than in the previous twenty years combined (Table 3.1). Proportionally, Wemindji accounts for nearly 10 percent of all mineral exploration work conducted in Quebec during those four years, and over 30 percent in Eeyou Istchee (Table 3.1). The Éléonore deposit finally began production in 2014 (Goldcorp 2013).

Although several mineral deposits have been found in various parts of Eeyou Istchee since the 1950s, particularly in the southeastern portion, Éléonore (with its 7 million ounces of gold worth over $7 billion) was the first significant high-value example. The Troilus mine near Mistissini (1997–2009, worth about $3 billion), the Renard diamond deposit north of Mistissini (about $4 billion), the Nemaska lithium deposit (about $4 billion), and the Blackrock iron-vanadium-titanium deposit near Oujé-Bougoumou (unknown value) are perhaps the only comparable discoveries in Eeyou Istchee's recent history. Yet, they generally represent a lesser value in current dollar terms (except perhaps for Blackrock), and their discoveries did not generate the same rush of exploration (MRNF 2010, 2013; InfoMine 2013). The discovery of gold at Éléonore and the sustained rush that followed probably marked the intensification and acceleration of mineral exploration in Wemindji territory and, more broadly, in Eeyou Istchee.

Although Éléonore presented a new scale of mining development in the region, many Wemindji community members perceived it as a continuation of major development projects that had affected Wemindji territory and contributed to rapid social and ecological changes since the 1980s. These include the La Grande hydro-electric complex, with its associated large-scale damming, flooding, and networks of roads and transmission lines, as well as the construction, during the 1990s, of an access road to connect Wemindji village to both southern cities and eastern parts of the territory. Whereas many Crees welcomed the positive aspects of these

FIGURE 3.1

Paakumshumwaau-Maatuskaau Biodiversity Reserve as initially proposed, showing regional development pressures

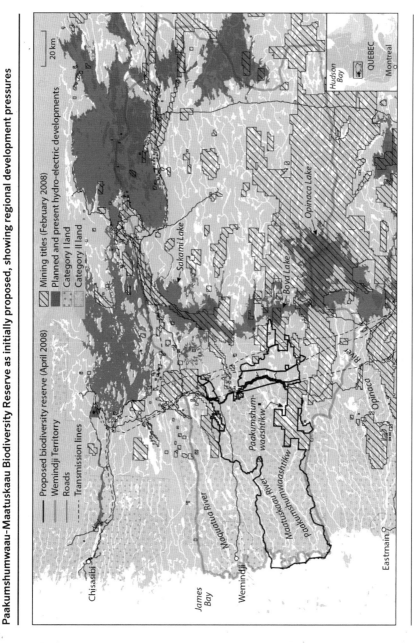

Source: Adapted from Christopher Wellen et al. (2008).

changes (such as greater access to health and education services, as well as broader economic opportunities), they also dealt with significant and complex socio-cultural impacts, including those associated with the loss of major tracts of lands and waters. The hunters, tallymen, and families who depended on these lands felt such changes most strongly. The La Grande project of mass flooding and large-scale landscape modification had lasting significance for Wemindji, and it continues to shape community responses whenever significant developments are proposed that may similarly affect the land and community.

Mineral exploration and exploitation projects and their associated roads, transmission lines, air traffic, and ground works are no exception here. In August 2007, the Cree Nation of Wemindji organized a community conference on mining issues, during which several tallymen stated that the mining development projects were simply a continuation of earlier pressures. The following comments are typical:

Development of EM1 [the Eastmain-1-A and Rupert Diversion Project was a multi-billion dollar hydroelectrical project involving the diversion of the Rupert River into the Eastmain River], and now the mining exploration and exploitation that is taking place in Eeyou Istchee ... I am concerned about the cumulative impacts of all this development, in particular about the fish, which is already affected by current and past developments. (tallyman 23, August 14, 2007)

I am concerned that the road will make the territory more accessible; more people will use the land. (tallyman 11, August 13, 2007)

I am concerned about the potential impacts on environment related to this Éléonore project, especially for the potential contamination of the reservoir and downstream waters. (tallyman 20, August 13, 2007)[2]

Given the impact of previous development projects on Wemindji territory, the fears voiced by these tallymen are reasonable. Portions of the territory were heavily affected by hydro developments during the 1970s and 1980s, with the largest rivers and lakes dammed or flooded – among them the Opinaca River, Sakami River, Sakami Lake, and Boyd Lake. The loss of good hunting, fishing, and trapping grounds, and the methyl-mercury contamination of fish caused by flooding for hydro-electricity are still vividly remembered and induce caution toward future developments. The hydro projects also brought roads, transmission lines, and airstrips, which facilitated access to land and resources for both Crees and non-Crees. Roads and other transportation infrastructure aided the recent expansion of mineral exploration in the territory, since they significantly reduced the transportation and operation costs of developers. Due to its more northern location, and unlike Cree communities in southern Eeyou Istchee, Wemindji has not had to deal with the consequences of large-scale commercial logging. Nevertheless, past and current hydro developments,[3] expanding transportation infrastructure, and the regular influx of large numbers of sport hunters all contribute to a perception of ongoing and increasing pressures on land and resources, to which recent mining activities are a significant addition.

At the same time, in the case of the Éléonore project, the fact that the Opinaca reservoir and downstream portions of the Boyd Lake–Sakami Lake diversion corridor have already been degraded by hydro-electric development is relevant to how the community responds to mining and the location of new protected areas. This is not to say that contamination from mining would be acceptable, but that local calculation of risks, losses, and benefits is different for areas that have already been damaged.

Notwithstanding serious concerns about the possible direct environmental impacts of mining, and about its contribution to cumulative environmental change in Wemindji territory, the community has a history of proactive entrepreneurship in response to business opportunities that have accompanied developments since the 1970s. Accordingly, Wemindji was greatly interested in revenue-sharing potential from a mine at Éléonore, as well as in training, jobs, and service contracts in a sector of the economy that could become a regional fixture.

Thus, whereas many community members recognize the intrusive nature of mineral exploration and development, as well as the significant changes they may bring to the land and community life, ideas differ about how to respond to this progressively significant reality. Some wish to resist it while others are ready to embrace it. The latter group see mining as a positive opportunity for a community in which population growth is rapid, unemployment rates are high, and dependency on the labour market economy is increasing. For many of them, mining offers a potential opportunity to remedy locally perceived socio-economic fragility. However, this will depend on a political and regulatory system that has failed to adequately engage Crees in the possibilities and benefits of mining development in Eeyou Istchee.

Voice of a Community Leader

August 2006 saw the first public presentation of Goldcorp's Éléonore project in Wemindji. The following comment, made by a community leader the day before, perhaps best expresses the diverging and complex perspectives regarding the recent wave of mineral exploration and development in the territory:

> Yes, mining is very intrusive ... and everything seems to happen very fast ... But I feel we can have some control [over the Éléonore project]. I always thought that there would be a mine. So let's be open to this occasion, let's plan something that will focus on our youth, on education ... We are not victims. We have to embrace it for the long term.
>
> I don't agree with the victimized discourse I sometimes hear when I speak to other people about development or mining on the land ... I prefer asking myself "how could we embrace this kind of development without considering ourselves victims of outside circumstances?" We don't have an industry in the community, and we don't have many employment opportunities; the band office and the school board are the main good-paying outlets. So, mining development could be good for some people who wish not to work for the band or the board ... We have to be proactive and positive instead of staying put. (pers. comm., August 30, 2006)

This view seemed to be widely shared in Wemindji, although, as mentioned earlier and as the leader also pointed out, perspectives were not homogeneous. Some community members emphasized the potential socio-economic opportunities and benefits of the project, whereas others, generally more concerned about consequences on the land, stressed the lack of information and control over mineral exploration and development on the territory.

These differences indicate variability in preferred balances of priorities, rather than diametrically opposed positions. That is, most people who called for community involvement in mining also advocated continuity of the hunting way of life; they themselves were hunters and they also supported measures to preserve ecologically important zones, such as the creation of protected areas.

The community leader's comments also highlight broader socio-economic and regulatory issues, which must be considered to better understand Wemindji's feelings regarding lack of control and lack of choice when confronted with relatively few other economic opportunities. These sentiments are well founded, since, under the regulatory regime that governs mining in Quebec and Eeyou Istchee, Wemindji is not well positioned to address certain community concerns or to fully orient mining according to its needs, interests, and aspirations. The difficult socio-economic realities of Wemindji also mean that it cannot easily dismiss the potential economic benefits of mining.

SOCIO-ECONOMIC OPPORTUNITIES

Many Wemindji residents justified taking an open approach to mining development by citing the greater youth employment and improved economic opportunities that would accompany the Éléonore project. Some socio-economic statistics are revealing here. Wemindji's unemployment rate has been approximately 20 percent in recent years, with a rate above 40 percent for fifteen- to twenty-four-year-olds (CHRD 2005; Mark 2008). In this community of 1,300 people, whose population is growing rapidly, up to twenty new jobs would need to be created on a yearly basis over the next ten years just to maintain current employment figures. The public sector is the primary employer in Wemindji and already accounts for nearly 70 percent of local jobs. The land can support only a certain number of occupational hunters, so expansion in the subsistence economy is also constrained (Notzke 1999; Gnarowski 2002), although some increased participation may be feasible according to a local administrator at the Cree Trappers Association (pers. comm., August 5, 2009). Tourism, mining, wind energy, and small-scale hydro-electric projects are alternative strategies under consideration by the local development corporation. In this context, many Wemindji Crees saw mining the Éléonore deposit as a significant, rare, and positive source of both employment and training.

During our various interviews, public presentations, and consultative meetings in Wemindji about the Éléonore project, people commonly referred to its potential social and economic benefits. This is understandable: the project could create up to 800 jobs during its construction phase (about two years) and 400 once mining began, for a minimum of fifteen years (Goldcorp 2007, 2011). The Éléonore exploration and development camp had already provided 60 to 130 jobs annually since 2005, of which up to 30 or 40 percent were held by Crees, not to mention the parallel opportunities for Crees in contracting and servicing jobs in transportation, construction, supply stores, and hospitality, laundry, janitorial, outfitting, and administrative services (Goldcorp 2011). The additional exploration projects on Wemindji territory also represented a significant source of employment, which could lead to even more work if minerals were discovered. Given the socio-economic realities of Wemindji and the difficulty of finding employment there, these facts are not negligible.

However, certain challenges and well-documented issues are associated with mining employment in, or near, Indigenous communities, particularly when a mine starts operating. At that point, a "glass ceiling" can limit the roles played by Indigenous employees, language and cultural issues can come into play, and negative social and family impacts can arise. Whether Wemindji could benefit fully from the extraction phase of the Éléonore project remained an open question for some time. Experience elsewhere was not encouraging, as reported by Alan Penn (2007) for the case of the Troilus mine, which lay near the Cree community of Mistissini and was in operation from 1997 to 2009. Maintaining Cree employment levels at 15 percent proved difficult, and they never surpassed 25 percent. Penn (ibid.) identifies a series of language, cultural, educational/training, and social issues that constituted ongoing challenges, both on and off the mining site, and that diminished community members' capacity to access some benefits of Troilus. These issues included the lack of specific experience and technical skills to hold a wide array of jobs, the difficulty of providing adequate and timely training for such a variety of positions, and the extra barriers created by the differing languages and cultural backgrounds of the Crees and the largely French-speaking labour force (ibid.). Adapting to a work schedule away from home also proved difficult for many Cree workers and their families, who experienced, to various degrees, psychological and health effects of such separation (ibid.).

These difficulties should not be overstated, nor should they be under-estimated, and they need to be carefully examined and addressed during the early stages of planning for a mine (O'Faircheallaigh 2004, 2006; Penn 2007). In this regard, Wemindji undertook a number of initiatives in col-laboration with Goldcorp and regional authorities, including building a Cree training centre in Wemindji to teach skills that transcend the mining industry, financing scholarships for a large array of training programs, and participating in regional efforts with the Cree School Board, Cree Human Resources Development, and the Cree Regional Authority to offer access-ible programs that are specific to the mining industry (German 2011, 2012, 2013). The general hope was that "a significant portion of the mine's em-ployees will be from the Cree community, especially women and youth" (Goldcorp 2011, 4). The Éléonore mine began production in 2014, with Crees comprising "approximately 30 percent of the workforce present on site" (Goldcorp 2014, 15). In the years since, many young people from Wemindji, including several women, have found employment at the mine. Many work in catering and janitorial services, but there are also highly skilled miners, drivers, and machine operators. In early 2019 there were approximately 225 Crees employed on site, and while none of them are at manager level, these jobs represent a significant source of local employment. Beyond the mine, a small number of spin-off business ventures have also been created, such as the 100 percent Cree owned Wemindji Laundry Inc. (Goldcorp 2014).

REGULATORY CONSTRAINTS AND OPPORTUNITIES

Under the James Bay and Northern Québec Agreement (JBNQA) and the provincial mining regime, mineral rights in Eeyou Istchee are held by the Quebec government and are made accessible to companies and prospect-ors through the free mining system (Barton 1993; Lapointe 2009, 2010; MRNF 2009). This system, which is also called the free entry system, en-ables individual or corporate mining entrepreneurs to freely access, acquire, explore, and mine publicly owned mineral resources, providing that they meet basic conditions. By definition, a free mining system imposes few constraints on the rights of developers to access and mine mineral resources (Barton 1993; Bankes 2004; Hoogeveen 2008; Laforce, Lapointe, and Lebuis 2009; Lapointe 2009, 2010).[4] Although sections 5, 22, and 24 of the JBNQA protect certain Cree rights to lands and resources, they are eclipsed to some extent by rights granted to mining entrepreneurs under the free mining system. For instance, companies and individual prospectors can freely acquire mining claims in most of Eeyou Istchee, and nothing in the provincial Mining Act compels them to provide notice to, consult with, or obtain consent from the affected Crees (Quebec 2013b, chapter I.2.3). Once acquired, mining claims confer full ownership of the subsurface mineral rights on their holders and can be extended for years or decades, as long as the holders keep investing in exploration work to the minimum levels prescribed in regulations. Depending on the size of the claim and how long the current owner has possessed it, this usually amounts to a few thousand dollars a year (ibid.). Those who hold mining claims can also access the land and conduct most types of exploration work there, from low-impact prospecting, to mechanized trenching and drilling, to helicopter and plane surveying, without any obligation for prior notice, environmental review, or local consultation (ibid.; Quebec 2013a). Only when exploration projects reach their more advanced stages – usually after several years – can the Crees formally voice their concerns and suggestions through the environmental and social impact assessment and review procedure set forth in section 22 of the JBNQA (Quebec 1975). The procedure also takes into account the hunting, fishing, and trapping regime of the JBNQA.

Exceptions occur on Category I lands, defined under the JBNQA as lands in and around Cree communities that are reserved exclusively for their use; formal Cree consent is required before mining titles are granted or any type of mining work is conducted (Quebec 1975, paragraph 5.1.10). On Category II lands, defined as lands on which Crees have exclusive hunting, fishing, and trapping rights, mineral exploration can occur if measures are taken "to avoid unreasonable conflict with harvesting activities" (ibid., paragraph 5.2.5). The JBNQA allows for mining development on Category II lands, "provided such lands are replaced" with lands "of similar characteristics" or, by agreement, the Crees are financially "compensated" (ibid., paragraph 5.2.3). Category I and II lands represent a small proportion of Eeyou Istchee (about 1.6 percent and 18.6 percent, respectively). The bulk of the territory remains in Category III, where Crees have more limited hunting and harvesting rights, and where, except for advanced exploration, most mining activities occur without an environmental review or formal consultative processes. The Mining Act of December 2013 did not change this fact, although it did introduce a requirement to consult with "native communities" (Quebec 2013b, chapter I.2.3).

The hunters, tallymen, and families who still depend on the land are among the first to be affected by these regulatory limitations and the sense of intrusion they generate. In a group meeting held with local tallymen on August 13, 2007, as part of the community conference mentioned above, it rapidly became apparent that lack of proper information and consultation was a main source of anxiety for tallymen and families whose traplines were subject to mining exploration.[5] Most tallymen said that they were poorly informed about mining activities on their traplines. They were concerned about the real and potential impacts on the land, waters, and animals, particularly the cumulative impacts of past and present hydro developments and the construction of new roads, trails, and transportation infrastructure. Two senior hunters said that they had already observed changes in the behaviour of beaver and sturgeon populations because of growth in mineral exploration and development. The potential impacts and conflicts that could ensue from the increased presence of non-Cree hunters using new access roads was also a source of concern. Many tallymen said that they periodically received information letters from exploration companies, which they described as "announcements" of the types of work, camp, and transportation to be carried out on the land, rather than genuine invitations for open dialogue and consultation. The letters often arrived days or even weeks after the work on the ground had been completed. Following are a few representative comments:

> I am concerned about the lack of consultations; they are supposed to consult, but in practice they do this through written letters. Tallymen have no control over how prospectors operate. Tallymen should benefit somehow and remain key players in territorial management. They should be advised earlier, way in advance. Tallymen can be of help because they

know the land. Maybe [prospectors and companies] have the wrong impression; maybe they don't realize that they need to consult the tallymen. (tallyman 22, August 13, 2007)

I received a letter about claims on our trapline and how far along they were in their exploration work, but never anything before they got started. (senior hunter 27, August 13, 2007)

Same situation, I didn't get a letter from the company until the exploration was finished, three weeks after the guys were finished prospecting. (tallyman 13, August 13, 2007)

Although these statements contain criticism, we emphasize that the hunters and tallymen did not reject all mining activity on Wemindji territory (although some would refuse it for their own family territories). Nor did they expect to exert full control over such development. Instead, the comments demonstrate the tallymen's awareness that they were poorly informed and consulted, and that the local socio-cultural values, knowledge, and institutions they represented were inadequately considered prior to the commencement of mineral exploration and development. As customary stewards, and under the traditional land tenure and resource-harvesting system, tallymen expect to be informed and consulted about activities on the land. Through this consultation, they share their knowledge of the land to help orient family members or outsiders in avoiding or minimizing undesired impacts. Consultation also expresses respect for the authority of tallymen, who have direct responsibility for the moral and reciprocal relationships between humans and other life on their territories (Feit 1992; Scott 1996; Berkes 2008). For tallymen, damage to the land brings a sense of loss, anguish, and failure for not having nurtured the health of the land and fulfilled the responsibilities inherited from their elders and ancestors (Scott 1996). Development by negotiation and mutual consent is the required approach.

For Cree hunters, rules of reciprocity are the order of life, and tallymen take a positive view of kin and community benefitting from what the land has to share, which may include exploitation of mineral resources within respectful environmental limits.[6] Under the authority and knowledge of the tallymen, the traditional land tenure and harvesting system has served

Cree society well for generations, promoting sustainable, adaptive, and resilient land and resource management (Feit 1992). These institutions are also central to the community's socio-cultural identity and organization today, and they continue to play key roles in the maintenance and renewal of land-based livelihoods and associated ecological, cultural, and spiritual benefits (Scott 1988; Feit 2004).

The environmental and social protection regime of the JBNQA (section 22) is intimately tied to its hunting, fishing, and trapping regime (section 24), and its mission is to protect the Crees' relationship to their environment (Peters 1999). For instance, it establishes an impact assessment and review procedure "to minimize the environmental and social impact of development when negative" and to protect "the Cree people, their economies and the wildlife resources upon which they depend," with respect to "developmental activity affecting the Territory" (Quebec 1975, paragraphs 22.2.2 and 22.2.4). Section 22 also recognizes "special status and involvement for the Cree people" in assessing and reviewing proposed development projects (Quebec 1975, paragraph 22.2.2), notably through joint environmental review arrangements with Quebec and Canada (ibid., paragraphs 22.5.6 and 22.6.1). Development projects that are subject to section 22 are usually first reviewed by the joint evaluating committee, which then recommends to the responsible minister (or "administrator") whether a project should be further assessed by the joint review committee (ibid., paragraphs 22.5.13 and 22.5.14). In practice, however, because the legislative status of mineral exploration and projects is not clear, very few mining activities have been assessed and reviewed under section 22 of the JBNQA. For example, the Quebec environmental public registry indicated that the co-evaluating committee reviewed fewer than fifteen exploration projects between 2000 and 2008 (MDDEP 2009).[7] Given that several hundred of these projects took place in Eeyou Istchee during the same period, this represents a very small proportion (Table 3.1).

Impacts from mining are typically greatest during the production phase, but they can also be significant during exploration, depending on the equipment and machinery used; the scale, frequency, and intensity of the work; the season; and the social, cultural, and ecological sensitivity of the affected area. Initial prospecting methods, involving hand tools and the occasional use of all-terrain vehicles and helicopters, may be of small consequence, but exploration involving frequent low-level flights and heavy machinery for drilling and trenching (the removal of soils and organic matter above bedrock) may be more problematic, particularly when numerous projects focus intensively on a "hotspot," as was the case for the Éléonore gold deposit. Impacts worsen when projects reach advanced exploration and development stages, with heavy machinery carrying out extensive drilling and trenching. This stage often produces dozens to hundreds of drilling holes to sample the subsurface rock, as well as the removal of soils and organic material – which the industry calls "overburden" – from several hectares. To achieve this, and to move machinery and workers from place to place, trails, temporary bush roads, and airlifts must often be constructed.

When advanced exploration detects lucrative mineral deposits, mining itself is usually the next step. Its principal impacts include the permanent disposal of large quantities of mining wastes and tailings on the surface, which may be more or less toxic, depending on the nature and composition of the mineral deposit. Extracting and transforming deposits also involves heavy machinery, blasting equipment, and relatively large quantities of water, energy, and chemical products.

Under section 22 of the JBNQA, only extractive "mining operations" are automatically subject to an impact assessment and review procedure (Quebec 1975, section 22, schedule 1). Exploration projects are not automatically subject to this provision, but neither are they explicitly exempt from it (ibid., section 22, schedule 2). Thus, they fall into a grey zone, and their legal status is not clear under the JBNQA. On the one hand, section 22 specifies that "*preliminary* investigation, research ... and technical survey works prior to any project, work or structure" shall be "*exempt*" from an impact assessment and review procedure (ibid., emphasis added). On the other hand, the provincial Environment Quality Act specifies that "no person may undertake or carry out any project which *is not automatically exempt* from the assessment and review procedure" without first obtaining an authorization from the responsible minister, in consultation with the environmental co-evaluating committee in which the Crees are represented (Quebec 2013a, sections 154–57, emphasis added). Given this second provision, one could argue that all mineral exploration projects should be assessed and reviewed under section 22 since they are "not automatically exempt" from it.[8] As mentioned above, however, very few exploration projects have been assessed and reviewed in Eeyou Istchee under section 22. This suggests that mining developers and the Quebec administrator have routinely interpreted mineral exploration projects as "preliminary" investigative and research activities. As such, they are exempt from the obligation to subscribe to section 22's impact assessment and review procedure.

An examination of the few exploration projects that were assessed and reviewed under section 22 also reveals that most were at advanced exploration or development stages and that their instigators were seeking authorization as required under the general Quebec environmental regulations. In other words, where mining exploration and projects in Eeyou Istchee are concerned, section 22 of the JBNQA appears to provide little social and environmental protection beyond the general Quebec regulations. This situation arguably undermines the general philosophy and objectives of the JBNQA to provide extra protection and considerations of Cree societal needs and interests. Furthermore, this is exacerbated by reported irregularities and deficiencies in the application of provincial regulations in the mining sector (Péloquin 2008; Quebec Auditor General 2009, 2013). In 2007, for example, Quebec's northwestern regional environmental office employed the equivalent of five to six permanent staffers. How effectively could this small number of people inspect and control dozens of active and past mine sites? How well could they monitor the several hundred mineral exploration projects dispersed throughout the Abitibi-Temiscamingue region, Eeyou Istchee, and Nunavik – an area twice the size of France – without adequate travel and financial resources (former regional employee of the Ministère du Développement durable, de l'Environnement et des Parcs, Ugo Lapointe, pers. comm., July 1, 2008)?

The Paix des Braves (Quebec 2002) between the Crees and the Quebec government did not bring about significant changes to the mining regulatory regime, although it encouraged, in principle, more collaborative approaches between the Crees, Quebec, and mining developers, and it guaranteed forms of revenue sharing for resources extracted in Eeyou Istchee (ibid., chapters 5 and 7). It also established the Cree Mineral Exploration Board, a joint Cree-Quebec committee with the mandate to promote greater Cree involvement in mineral exploration. In contrast to the forestry regime (ibid., chapter 3), however, the Paix des Braves does not establish detailed protective measures and institutional mechanisms regarding the mining sector (see Scott 2005; Chapter 11, this volume, for a discussion of the forestry regime). In fact, mining issues have not yet been central to any agreement negotiated between the Crees and governments, perhaps because hydro and forestry developments seemed more pressing when the JBNQA (1974–75) and the Paix des Braves (2001–02) were negotiated. As Table 3.1 shows, mining exploration had barely begun to intensify in Eeyou Istchee when the Paix des Braves was concluded in 2002 and investment in exploration and development started to climb steeply in 2004–05.

This omission and the increasing presence of mining activities in Eeyou Istchee may help explain why the Grand Council of the Crees and the Cree Regional Authority felt the need to establish a Cree Nation Mining Policy in 2010 (GCC-EI 2010). The policy lists guiding principles for mining in Eeyou Istchee, such as the needs to respect Cree rights and interests, ensure environmental protection, protect traditional activities, provide benefits to affected communities, and meet social acceptability through collaboration agreements. The Crees demand to be "active partners" in mining developments, not just "passive bystanders" (Bosum 2012). The policy also attempts to provide for a more co-ordinated and coherent approach between the Cree communities and regional authorities in relation to mining in Eeyou Istchee (ibid.). Nevertheless, compliance with its principles is voluntary, as the policy is not enforced by law or a government-to-government binding agreement.

Substantially reforming the mining regulatory regime would demand a significant political investment by the Grand Council of the Crees, as a defender of rights recognized under the JBNQA and subsequent agreements. It might also require increased resistance to mining projects, pending the achievement of significant reform. No single Cree community could bring such a campaign to fruition and would thus be obliged to work in solidarity with other Cree communities, perhaps also with outside partners. To some extent, the path to resolution might lie in the environmental protection provisions of the JBNQA – *if* review bodies accepted that impacts at exploration stages were sufficiently problematic to trigger review. The review process would then become a vehicle for more rigorous scrutiny of the whole process, from early exploration, through advanced exploration, and into the operational phase of a mine, with appropriate conditions attached when a mine is given the green light to proceed. The environmental review process, of course, can also be a forum for mounting opposition to mining activity that is perceived to be harmful. And it could contribute to greater leverage by Cree communities and the grand council in negotiating improved environmental guarantees through "impact benefits" or collaboration agreements.

Partly in reaction to the increased mining pressures and the lack of control over mineral exploration and mining, the Cree Nation of Wemindji supports the idea of a network of protected areas to safeguard vital ecological and cultural places. Our experience at Wemindji tells us that for the large majority of residents, mining may be acceptable in some parts of their territory, but it is not acceptable in others. Protected areas offer a means of addressing the latter.

BALANCING MINING AND PROTECTED AREAS AND DIALOGUING WITH DEVELOPERS

It is against the general background of intensifying mineral exploration that Wemindji is developing new strategies to balance the socio-economic benefits that may accrue from controlled industrial mining with the need to protect the land for the continuing, and evolving, Cree way of life. Ecological conditions that are essential to hunting, fishing, and trapping remain a high political priority for the community. Reaching consensus about which portions of the territory to target for protection, and which portions to negotiate mining developments on, hinges on complex factors and various local calculations of risk, loss, and benefit.

Mining and Protected Areas

In March 2006, the Wemindji Council submitted its proposal for the creation of Paakumshumwaau-Maatuskaau Biodiversity Reserve to the Ministere du Développement durable, de l'Environnement et des Parcs du Québec (MDDEP). Afterward, and much to its dismay, it learned that a company named Azimut Exploration had just registered an extensive mining claim at Paakumshumwaau (Old Factory Lake). This lake, which lies near the headwaters of Paakumshumwaashtikw (Old Factory River), is the largest lake on Wemindji territory that has not been modified by hydro-electric development. As the community's main riverine artery to the interior from historic trading establishments at the Old Factory post, Paakumshumwaashtikw holds special significance in Cree cultural memory. Furthermore, its watershed is an ecologically rich part of Wemindji territory, notable for its population of yellow sturgeon, a species known to Cree hunters as especially vulnerable to environmental disturbance (knowledge borne out by the loss of sturgeon populations in the Boyd Lake–Sakami Lake hydro-electric diversion corridor).

The Wemindji Council decided that it had no choice but to oppose the claim outright. Thus, Chief Rodney Mark wrote to Azimut Exploration on February 28, 2007, outlining his community's objections:

> As you may be aware, large portions of Wemindji's family hunting territories have been heavily modified by hydro-electric development since the late 1970s, particularly those affected by the damming of the La Grande Rivière and the diversion of the Eastmain and Opinaca rivers through the Sakami and Boyd lakes corridor. More recently, our community territory has been subject to increasing mining activities, particularly since the discovery of the Éléonore gold deposit in 2004. While the Cree Nation of Wemindji participates actively in developing the mineral potential in parts of its community territory ... it is also important for our community members to keep areas of high historical, cultural and/or ecological value available for different forms of land-uses. In this regard, Wemindji community members reaffirmed in General Assembly last August 2006 their wish to see no mining activities in the area and to support the establishment of the Paakumshumwaau-Maatuskaau Biodiversity Reserve in its entirety.

MDDEP administrators were supportive of Wemindji's proposal for the biodiversity reserve, but they needed the co-operation of the Ministère des Ressources naturelles et de la Faune (MRNF), which was responsible for mining. MRNF was resistant, on the grounds that implementing the reserve would pre-empt mining exploration in a significant portion of Wemindji territory. It suggested that the reserve be reduced to encompass only the sections of the Paakumshumwaau and Maatuskaau watersheds that lay near the coast. Wemindji took the position that whatever the mineral potential might be, the watersheds were so culturally and ecologically important that they should be protected in full. No further mining development should be pursued. These positions, the subject of an initial round of correspondence, were discussed by representatives from Wemindji, MDDEP, and MRNF at a pair of meetings, the first in Montreal in February 2007. MRNF proposed that the biodiversity reserve exclude inland portions of the watershed where the potential for mineral extraction was highest, until an adequate geological inventory had been completed. The rationale did not persuade the Wemindji representatives. At this meeting, the president of the Wemindji Cree Trappers Association, Edward Georgekish, reiterated the community position: "The land is affected by rapid development and changes, hydro-developments, mining activities, etc. Our people are not anti-development; we have been supportive of some development, but there is a time when we need to draw the line" (Edward Georgekish, pers. comm., February 2007).

The second meeting, in April 2007, was held at Wemindji and was open to the public. Community leaders, including Chief Rodney Mark, most Wemindji tallymen, several elders, and various other residents attended. High-ranking MRNF officials, as well as MDDEP representatives, flew in from Quebec City. Wemindji was determined to demonstrate to MRNF that it was serious in its wish to protect the Paakumshumwaau-Maatuskaau watersheds in their entirety. Prior to the meeting, MRNF had expressed some doubts about the unity of the community on the issue. It reiterated its position that majority inland portions of the watersheds should not be protected pending further exploration. If MRNF expected to hear differences of local opinion, it encountered only unanimity; speaker after speaker voiced opposition to mining in the watersheds. By the end of the day, the senior MRNF official stated that Wemindji had made its position clear.

In August 2007, MDDEP informed Wemindji that it would collaborate with MRNF toward achieving biodiversity reserve status for the Paakumshumwaau-Maatuskaau watersheds. However, existing mining titles, most importantly, the large claim encircling Paakumshumwaau, would not be included in the reserve. In reply, Wemindji urged MRNF and MDDEP to work to ensure that these titles would not be renewed in the future, that the areas they covered would be integrated into the reserve once the titles expired, and that strict environmental conditions would be applied to any exploration work performed in the meantime.

Rights to the Paakumshumwaau claim held by Azimut Exploration (and optioned for development to a major industry player, Iamgold) were protected by the provincial Mining Act so long as the claim continued to be developed. But Wemindji held what turned out to be a trump card. As it reminded both MRNF and the gold companies, most of the Old Factory watershed (including portions of four Wemindji family hunting territories) was Category II land, as was a majority of the Azimut/Iamgold "property" at Paakumshumwaau. Of course, the JBNQA did not preclude Quebec from taking Category II land for development, but its stipulation that such land must be replaced with land of similar quality from Category III was a serious obstacle for the would-be developers. As the Wemindji Council pointed out, no watershed and no major lake of comparable ecological value and integrity remained in Wemindji's Category III lands. Paakumshumwaau (the lake), nearby McNab Lake, Paakumshumwaashtikw (the river), and the estuary were a unique and irreplaceable system, from both ecological and cultural points of view. Other lake and river systems of comparable size in Wemindji territory had already been heavily degraded by hydro-electric development. Although Wemindji could have accepted cash compensation for the lost Category II lands, in lieu of replacement, the JBNQA made no provision for imposing such an option against a community's wishes. Wemindji was adamant that it had no intention of negotiating for cash compensation. Hence, there was a high probability, if not certainty, that the mining companies and MRNF were headed for litigation if they persisted in developing the Paakumshumwaau claim.

A waiting game ensued, but eventually the companies abandoned the claim, and in March 2010, the director of the Services des aires protegées informed the Wemindji Council that the area, totalling some 488 square kilometres, would be protected as part of the Paakumshumwaau-Maatuskaau Biodiversity Reserve.

Negotiating Collaborative Agreements with Mining Developers

Concurrently, the council was negotiating a collaboration agreement with Goldcorp for development of the Éléonore deposit, which the company had acquired from Virginia Gold Mines in April 2006. The road to this agreement was not straightforward. As mentioned, the Quebec Mining Act provides broad powers and autonomy to mining companies in terms of access to lands and mineral resources, whereas the JBNQA and Paix des Braves do not specifically require companies to negotiate consensual agreements with affected communities prior to developing a mine. Nor do they give mineral ownership rights to the Crees. A multinational mining company has far greater technical and financial resources at its disposal than does a Cree community, a fact that can hinder the achievement of a balanced negotiation process. Nevertheless, Wemindji took a proactive approach, using various legal, political, and institutional means whenever it could to improve its negotiating position. One pivotal moment occurred at the outset, when Goldcorp tried to segment its project by fast tracking the construction of an airstrip and a sixty-two-kilometre road to the Éléonore site. Not satisfied, Wemindji immediately and successfully petitioned the Grand Council of the Crees to adopt, in July 2006, a unanimous resolution threatening to seek an injunction against a segmented impact review process that would address Goldcorp's road and airstrip separately from the mine itself. The Crees had successfully defended a similar position during the early 1990s in the case of *Cree Regional Authority and Namagoose v. Raymond Robinson*, to block Hydro-Québec's road construction for the Great Whale hydro-electric project (for details, see Posluns 1993, 192–96). In connection with the Goldcorp project, Wemindji was particularly worried by the possibility that the road and airstrip might be built before any comprehensive impact review could be conducted for the mine and before Goldcorp engaged in substantive negotiations and commitments to protect Cree rights and interests. This incident sent a strong and clear message: Goldcorp could not bypass Cree rights and interests without compromising the Éléonore project as a whole.

Rigorous discussions between Goldcorp and Wemindji immediately ensued, on a variety of matters related to environmental review, construction and control of the access road to the mine, and economic participation and benefits for Crees. Shortly afterward, Wemindji took Goldcorp by surprise when it proposed that it become the main proponent to construct the airstrip and road. Wemindji argued that this would demonstrate its good faith in negotiating with Goldcorp (which claimed to need the road and airstrip as early as possible), while at the same time securing greater protection of Cree rights and interests in the planning, construction, and accessing of the road and airstrip. Wemindji also asserted that since the road and airstrip would eventually fulfill a broader purpose than just servicing the Éléonore site, it made sense that it would be the main proponent. Goldcorp agreed to financing the road and airstrip on those terms, and in September 2006, Wemindji and Opinaca Mines (Goldcorp's subsidiary for the project) signed a letter of intent to reach an eventual collaboration agreement, with a schedule of monthly negotiating meetings. Both parties also signed a service agreement, which confirmed Wemindji as the primary Cree entity responsible for recruiting Cree labour contracting for Opinaca Mines until the main collaboration agreement was signed. On the Cree side of the negotiations, there were some challenges, while Wemindji and the Grand Council of the Crees worked out a mutually acceptable balance of authority, responsibilities, and potential benefits in negotiations with Goldcorp, as well as an accepted approach to environmental review for the project.

Meanwhile, Goldcorp communicated with community residents through presentations at local meetings. In addition, Wemindji organized a community conference on mining in August 2007, with the assistance of the Wemindji Protected Areas team, to which several outside experts were invited to address environmental, economic, and social dimensions that should be borne in mind while negotiating a collaboration agreement. Community members generated a list of about forty questions for circulation and reply by conference participants. They expressed numerous and various concerns regarding the potential environmental, social, and cultural impacts of the mine, but they also appreciated the socio-economic benefits that it could bring to Wemindji. The fact that the Éléonore mine would be underground (not a large open pit) and located in an area that had already been degraded by hydro-electric projects may also have been relevant in developing community consensus. In addition, Goldcorp agreed to produce dry tailings (instead of larger volumes of wet tailings), part of which would be returned underground to reduce environmental risks and footprint on the surface. The tallyman for the family territory at Éléonore also gave his support to negotiations with Goldcorp.

The Opinagow Collaboration Agreement was reached early in February 2011, with Opinaca Mines, the Grand Council of the Crees (Eeyou Istchee), the Cree Regional Authority, and the Cree Nation of Wemindji as signatories. Its main features entailed:

- a joint collaboration committee for overall oversight, exchange, and co-ordination;
- dispute resolution procedures;
- training programs (including cross-cultural awareness);
- a committee responsible for establishing annual Cree employment targets;
- a joint business opportunities committee responsible for identifying, on a priority basis, opportunities for Cree enterprises during construction and operational phases of the mine;
- tendering procedures;
- contributions by Goldcorp to promote local business development and educational and cultural initiatives;
- measures for harmonization of mining activity with traditional pursuits on family territories affected by the mine;
- environmental management standards and procedures;
- a joint environment committee; and
- financial provisions for payments to the Cree parties, including revenue sharing and fixed annual payments. Specifics of the agreement were confidential.

In 2014, when our research was conducted, mine construction was advanced, but Éléonore was not yet in full production. According to Rodney Mark, implementation of the collaboration agreement had been a mixed success. Cree employees from Wemindji and other communities in the region numbered in the hundreds. However, some of the intentions in the collaboration agreement had not found concrete expression. In February 2012, a Montreal daily newspaper reported that under Quebec environmental laws, Opinaca Mines had been "fined $400,000 for carrying out work without proper authorizations, releasing a contaminant into the environment and improperly storing hazardous waste" (Beaudin 2012). At the mine itself, Cree personnel had interpreted certain incidents in workforce relations as racist. Goldcorp had delegated to SNC-Lavalin the awarding of contracts for mine construction, and the major contracts had gone to companies based in the Abitibi mining district several hundred kilometres to the south, rather than to Cree companies. The honeymoon was over, and constructive collaboration between Wemindji and Goldcorp would now depend on corrective action and sustained vigilance.

CONCLUSION

The commitments of hunting territory stewards, extended families, and the Wemindji Council involve dual objectives: on the one hand, to care for the land and to safeguard the ecological conditions for valued land-based traditional activities, and on the other hand, to realize opportunities for occupational diversity in the contemporary economy. Individuals may disagree about how heavily each objective should bear on a particular mining prospect, but there is general consensus that both are important. This has enabled substantial community consensus on the potential benefits of a collaboration agreement with Goldcorp, as well as the need to categorically oppose mining development in the Paakumshumwaau-Maatuskaau watersheds.

The contrast between community opposition to the Azimut claim at Paakumshumwaau and the willingness to negotiate regarding the Éléonore site is somewhat iconic in relation to the challenges that lie ahead. Protecting the environment while also satisfying the aspirations of the mining industry on Wemindji territory will not always be possible. The existing pattern of mining claims is a major impediment to the development of a network of protected areas that is effective in ecological and cultural terms. Several factors influence the possibility for give and take, as was the case at Paakumshumwaau. These included the commitment of tallymen to maintain the area in pristine condition; the cultural and historical significance of the area to the Wemindji community; Wemindji's capacity to legally block a development on Category II lands; and the fact that the Azimut claim, though promising at some level, was not necessarily more interesting than other claims that the company could pursue with less political cost and investor uncertainty. Various factors likewise shaped outcomes for the Éléonore project: a proven claim of high potential market value; the willingness of the local tallyman to support initial discussions with the company; the deployment of an early injunction threat, which forced Goldcorp to a more collaborative and rigorous negotiation process, even if the Crees had more limited say in regard to a project on Category III lands; and the fact that Éléonore was a relatively small underground mine in an area that had already been damaged by hydro-electric developments. Local tallymen and Wemindji Council leaders have not abandoned their responsibility for the environment, notwithstanding existing damage. But they are prepared to selectively accommodate some mining development and to seek benefits from it.

As the community makes its way, collective values, economic needs, and the sober assessment of political resources produce a tango of resistance and accommodation that confounds pro-development versus anti-development stereotypes. The choreography is often adaptive and resilient. It is possible to exaggerate community consensus regarding how to respond to a specific project. At the same time, it would be inaccurate and counterproductive to deny that there are real bases for reaching practical consensus. As Wemindji envisions and enacts its larger project for a collective future on its home territory, everyone knows there is a balance to be struck.

NOTES

We would like to express our gratitude to the Cree Nation of Wemindji and to all the participants who generously and patiently offered their time and comments to complete the research and writing of this chapter.

1 Any one of the ten Cree communities represented by the regional Grand Council of the Crees is stronger in political solidarity with the other communities through the regional council. Furthermore, the council has responsibilities and prerogatives throughout Eeyou Istchee in regard to development of all kinds, pursuant to the James Bay and Northern Québec Agreement and subsequent agreements with the Governments of Quebec and Canada. At the same time, it is not automatically clear what proportion of benefits should flow to the community whose territory is most directly affected by a mine or other development and what proportion should flow to the Cree Nation as a whole. Local and regional leadership may differ on these matters, which will require negotiation.

2 Thanks to Edward Georgekish for translating these comments during the community conference and the parallel meeting with tallymen. The names of quoted tallymen are not revealed because permission was not sought to use them.

3 Wemindji is now subject to the impact of further hydro development, though less dramatic than previous ones, and following Cree consent under the Paix des Braves in 2002, the newly diverted waters of the Rupert River flow through the Boyd Lake–Sakami Lake diversion corridor that crosses Wemindji territory.

4 For a more detailed description of the free mining system and its various implications in Quebec and Canada, see Nigel Bankes (2004); Barry Barton (1993); Dawn

Hoogeveen (2008); Ugo Lapointe (2009, 2010); Myriam Laforce, Ugo Lapointe, and Véronique Lebuis (2009). Barton (1993, 115) describes free mining as comprising three sets of rights: "(i) a right of free access to lands in which the minerals are in public ownership, (ii) a right to take possession of them and acquire title by one's own act of staking a claim, and (iii) a right to proceed to develop and mine the minerals discovered." The Quebec government describes free mining in practically the same terms:

> Le régime minier inscrit dans la loi actuelle repose sur un principe de base fondamental le «free mining». Ce principe, qui est bien connu des gens du secteur minier, détermine les règles d'attribution des droits miniers. Il signifie: [i] que l'accès à la ressource est ouvert à tous, sans égard aux moyens du demandeur; [ii] que le premier arrivé obtient un droit exclusif de rechercher les substances minérales qui font partie du domaine public; [iii] que ce premier arrivé à l'assurance d'obtenir le droit d'exploiter la ressource minérale découverte dans la mesure où il s'est acquitté de ses obligations, c'est-à-dire essentiellement qu'il a réalisé des travaux d'exploration. (MRNF 2009)

5 Twelve of the twenty Wemindji tallymen attended this meeting, which was jointly called by the local Cree Trappers Association (CTA) and the co-ordinating committee of the conference, to which we were parties. The meeting concentrated largely on issues connected to the Eléonore deposit but also in relation to Wemindji territory more broadly. Edward Georgekish, CTA president at the time, collected, synthesized, and presented the tallymen's comments and concerns two days later at the conference session entitled "Environmental and Land Protection."

6 Although more research would be useful to document how Cree society metaphorically and symbolically represents the mineral world in relation to land, community, and individuality, some community members and senior hunters described mineral resources as "growing" out of the land over time, or referred to them in "dreams" and "visions" from long ago as a potential sign of wealth and benefits for the broader community.

7 Some projects have been reviewed more than once.

8 Projects that are exempt from section 22's assessment and review procedure include fossil-fuel-fired power-generating systems below three thousand kilowatts, borrow pits for highway maintenance purposes, small wood cuttings for personal and community use, and municipal streets and sidewalks (Quebec 1975, section 22, schedule 2).

WORKS CITED

AMQ (Association minière du Québec). 2006–08. "Rapports Annuels 2006 à 2008." Association minière du Québec.

Bankes, Nigel. 2004. "The Case for the Abolition of Free Entry Mining Regimes." *Journal of Land Resources and Environment* 24: 317–22.

Barton, Barry J. 1993. *Canadian Law of Mining.* Calgary: Canadian Institute of Resources Law.

Beaudin, Monique. 2012. "Quebec's Environment Law Given More Bite." *Montreal Gazette,* February 2. https://montrealgazette.com/news/local-news/quebecs -environment-law-given-more-bite.

Berkes, Fikret. 2008. *Sacred Ecology.* 2nd ed. London: Routledge.

Bosum, Abel. 2012. "The Need for a Cree Nation Mining Policy." Paper presented at "Building the North: A Mining Conference," Quebec City, May 31.

CHRD (Cree Human Resources and Development). 2005. "Cree Labour Market Survey 2003." Cree Regional Authority, Cree Human Resources and Development.

CMEB (Cree Mineral Exploration Board). 2009. "Eeyou Istchee Mineral Exploration Activities for the Year 2009." Cree Mineral Exploration Board.

Cree Nation of Wemindji. n.d. "Mission and Vision Statements." https://www. wemindji.ca/assets/the-vision-we--hold-for-our-people-and-community.pdf.

Feit, Harvey A. 1992. "Waswanipi Cree Management of Land and Wildlife: Cree Ethno-ecology Revisited." In *Native People, Native Lands: Canadian Indians, Inuit and Métis,* ed. B. Cox, 75–91. Ottawa: Carleton University Press.

–. 2004. "James Bay Crees' Life Projects and Politics: Histories of Place, Animal Partners and Enduring Relationships." In *In the Way of Development: Indigenous Peoples, Life Projects and Globalization,* ed. Mario Blaser, Harvey Feit, and Glenn McRae, 92–110. London: Zed Books and the Canadian International Development Research Center.

GCC-EI (Grand Council of the Crees – Eeyou Istchee). 2010. "Cree Nation Mining Policy (Policy 2010–7)." Grand Council of the Crees – Eeyou Istchee.

German, Amy. 2011. "This Is Our Moment." *The Nation,* September 23. http://www. nationnews.ca/this-is-our-moment.

–. 2012. "Ready the Workforce." *The Nation,* June 1. http://www.nationnews.ca/ ready-the-workforce.

–. 2013. "CHRD Taking Human Resources to the People." *The Nation,* September 9. http://www.nationnews.ca/chrd-taking-human-resources-to-the-people.

Gnarowski, Michael, ed. 2002. *I Dream of Yesterday and Tomorrow: A Celebration of the James Bay Crees.* Kemptville, ON: Golden Dog Press.

Goldcorp. 2007. "Creating the Future: Opinaca Mines and the Cree Nation of Wemindji." Paper presented at "Mining near Opinaca and the Éléonore Deposit: Opportunities and Challenges for Our Community, Wemindji Community Conference on Mining," August 14–15.

–. 2011. "Signed, Sealed & Delivered." *Above Ground: Our World of Community Responsibility* 3: 2–7. https://issuu.com/goldcorpinc/docs/above_ground_vol03.

–. 2013. "Annual Information Form for the Financial Year Ended December 31 2012." Vancouver, Goldcorp.

-. 2014. "Opinagow Collaboration Agreement: Looking Back, Moving Forward." https://s22.q4cdn.com/653477107/files/doc_downloads/portfolio_docs/eleonore/ booklet-summit-2014-en-low-res.pdf.

Hoogeveen, Dawn. 2008. "What Is at Stake? Diamonds, Mineral Regulation, and the Law of Free-Entry in the Northwest-Territories." Master's thesis, Simon Fraser University.

InfoMine. 2013. "Charts and Data for the Mining Industry, Commodity Prices." http://www.infomine.com.

Laforce, Myriam, Ugo Lapointe, and Véronique Lebuis. 2009. "Mining Sector Regulation in Québec and Canada: Is a Redefinition of Asymmetrical Relations Possible?" *Studies in Political Economy* 84: 47–78.

Lapointe, Ugo. 2009. "Origins of Mining Regimes in Canada and the Legacy of the Free Mining System." Paper presented at "Rethinking Extractive Industry: Regulation, Dispossession, and Emerging Claims," Centre for Research on Latin America and the Caribbean and the Extractive Industries Research Group, Toronto, March 5–7.

–. 2010. "L'héritage du système du free mining au Québec et au Canada." *Recherches amérindiennes au Québec* 40 (3): 5–25.

Mark, Rodney. 2008. "Wemindji Community Perspective." Québec Exploration 2008. Quebec City, Ministère des Ressources naturelles et de la Faune et Association d'exploration minière du Québec, November 26.

MDDEP (Ministère du Développement durable, de l'Environnement et des Parcs). 2009. "Projets soumis à l'évaluation environnementale et sociale au sud du 55e parallèle depuis le 1er janvier 2000." Gouvernement du Québec. http://www.mddep.gouv.qc.ca/evaluations/projet-sud.htm.

MER (Ministère de l'Énergie et des Ressources). 1986. *Rapports des représentants régionaux: 1985.* Quebec City: Publications du gouvernement du Québec.

–. 1991. *Rapports des géologues résidents sur l'activité minière régionale: 1990.* Quebec City: Publications du gouvernement du Québec.

–. 1996. *Rapports des géologues résidents sur l'activité minière régionale: 1995.* Quebec City: Publications du gouvernement du Québec.

MRNF (Ministère des Ressources naturelles et de la Faune). 2001–08. *Rapports annuels sur les activités minières de 2000 à 2008.* Quebec City: Publications du gouvernement du Québec.

–. 2009. "Free Mining." Gouvernement du Québec. http://www.mrnf.gouv.qc.ca/presse/discours-detail.jsp?id=1053.

–. 2010. "Exploitation minière." In *Rapport sur les activités minières au Québec 2009 (DV2010–01)*, 55–74. Quebec City: Gouvernement du Québec.

–. 2013. "Exploitation minière." In *Rapport sur les activités minières au Québec 2012 (DV2013–01)*, 28–84. Quebec City: Gouvernement du Québec.

Notzke, Claudia. 1999. "Indigenous Tourism Development in the Arctic." *Annals of Tourism Research* 26 (1): 55–76.

O'Faircheallaigh, Ciaran. 2004. "Evaluating Agreements between Indigenous Peoples and Resource Developers." In *Honour among Nations? Treaties and Agreements with Indigenous People*, ed. Marcia Langton, Maureen Tehan, Lisa Palmer, and Kathryn Shain, 303–28. Carlton, Victoria: Melbourne University Press.

–. 2006. *Environmental Agreements in Canada: Aboriginal Participation, EIA Follow-Up, and Environmental Management of Major Projects.* Calgary: Canadian Institute of Resources Law.

Péloquin, Claude. 2008. Unpublished Research Notes. Montreal, James Bay Advisory Committee on the Environment.

Penn, Alan. 2007. "Unpublished Notes: Some Lessons Learned from the Troilus Mine and the Inmet-Mistissini Troilus Agreement." Paper presented at "Mining near Opinaca and the Éléonore Deposit: Opportunities and Challenges for Our Community, Wemindji Community Conference on Mining," August 14–15.

Peters, Evelyn J. 1999. "Native People and the Environmental Regime in the James Bay and Northern Québec Agreement." *Arctic* 52 (4): 395–410.

Posluns, Michael. 1993. *Voices from the Odeyak.* Toronto: NC Press.

Quebec. 1975. *The James Bay and Northern Québec Agreement.* Quebec City: Editeur officiel du Québec.

–. 2002. *The Agreement Concerning a New Relationship between Le Gouvernement du Québec and the Crees of Québec – Paix des Braves.* Quebec City: Editeur officiel du Québec.

–. 2013a. *Environment Quality Act,* c. Q-2. http://legisquebec.gouv.qc.ca/en/showdoc/cs/Q-2/20140101.

–. 2013b. *Mining Act,* c. M-13.1. http://legisquebec.gouv.qc.ca/en/ShowDoc/cs/M-13.1.

Quebec Auditor General. 2009. "Interventions gouvernementales dans le secteur minier." In *Rapport du Vérificateur général du Québec à l'Assemblée nationale pour l'année 2008–2009* (Décembre 2009). Vol. 2, 11–6. Quebec City: Vérificateur général du Québec. http://www.vgq.qc.ca/fr/fr_publications/fr_Rapports-CAP/fr_23-CAP-2009-11-rapport.pdf.

–. 2013. "Suivi d'une vérification de l'optimisation des ressources Interventions gouvernementales dans le secteur minier." In *Rapport du Vérificateur général du Québec à l'Assemblée nationale pour l'année 2012–2013,* 26. Quebec City: Vérificateur général du Québec. http://www.vgq.qc.ca/fr/fr_publications/fr_rapport-annuel/fr_2012-2013-CDD/fr_Rapport2012-2013-CDD.pdf.

Scott, Colin. 1988. "Property, Practice and Aboriginal Rights among Québec Cree Hunters." In *Hunters and Gatherers.* Vol. 2, *Property, Power and Ideology,* ed. James Woodburn, Tim Ingold, and David Riches, 35–51. London: Berg.

–. 1996. "Science for the West, Myth for the Rest? The Case of James Bay Cree Knowledge Construction." In *Naked Science: Anthropological Inquiry into Boundaries, Power, and Knowledge,* ed. Laura Nader, 69–86. London: Routledge.

–. 2005. "Co-management and the Politics of Aboriginal Consent to Resource Development: The Agreement concerning a New Relationship between Le Gouvernement du Québec and the Crees of Québec (2002)." In *Canada: The State of the Federation, 2003: Reconfiguring Aboriginal-State Relations,* ed. Michael Murphy, 133–63. Montreal and Kingston: McGill-Queen's University Press.

Wellen, Christopher, Katheryne Budd, Ugo Lapointe, and Colin Scott. 2008. *Paakumshumwaau Biodiversity Reserve and Development Pressures* [map]. McGill University School of Environment.

4

Collecting Scientific Knowledge
A Historical Perspective
on Eastern James Bay Research

KATHERINE SCOTT

It only served the developers, not the community. But now, research in the community must be always also for the community.

— Rodney Mark (pers. comm., October 2009)

The collaborative and community-based protected areas project at the heart of this volume invites reflection not only on how Indigenous people and academic researchers work together today, but also on the working relationships of those who came before us. This chapter traces the history of that research, acknowledging our debt both to earlier scientists whose publications we reference and to their Cree hosts and guides, who were largely excluded from the published accounts, but whose knowledge of the land and expertise was essential to expeditions.[1] Rooted in colonial and post-colonial aspirations for natural resource acquisition, research in James Bay rarely considered its impacts upon or benefits to Cree communities until recent decades. The story of James Bay research offers insights into relationships between Crees and academic researchers, as well as perspectives on changes in scientific approaches and in broader ongoing Cree, Québécois, Canadian, and international relations.

James Bay was long seen as a "blank space" on the map of Canadian research (Hare 1952; Carlson 2004), although state-run expeditions had set out periodically to name, map, and report upon its details. Reports, maps, and articles were published but relatively few academic analyses appeared until the 1970s. Travel was challenging and costly until road construction began late in the twentieth century, and research funding was often hard to secure in the absence of profitable resource extraction or an expanding settler population. When more up-to-date environmental information was required for development planning during the post-war era, and later when the hydro-electric projects were announced, earlier studies were found to have inadequately addressed or anticipated the requirements of new projects. At each stage, more studies and inventories were commissioned. With the realization of major development schemes, the region became one of the most researched in northern Canada. Government agencies, industry, universities, and the Crees themselves have undertaken research to examine various aspects of the James Bay region, in pursuit of objectives that have differed over time with shifting economic, environmental, political, and scientific contexts, in turn determined by contemporary local/regional, national, and global events and agendas. These studies provide glimpses of contemporary politics, scientific practices, funding agency priorities and objectives, and changing research relationships, all of which helped to shape future research directions.

Reports and articles also reflect the imprecise boundaries of the area. The idea of a "James Bay region" is nearly as ephemeral as notions of where "the North" is located. Definitions and boundaries depend on who you are and where you stand. Cree hunters, university scientists, federal government agents, or provincial politicians might each find very different reasons to care where the lines are drawn. My focus is on Eeyou Istchee, the east coast watersheds of James Bay and southernmost Hudson Bay. This is the territory of the ten Cree Nations of Eeyou Istchee, extending from Whapmagoostui on Hudson Bay to Washaw Sibi south of James Bay. All of Eeyou Istchee comes under the James Bay and Northern Québec Agreement (JBNQA).

Environmental historians, political scientists, and others draw attention to the political and economic entanglements that have been as inescapable for James Bay science as for any other. They describe the role that northern science has played in government projects calculated to control nature, territory, and all inhabitants.[2] Political geographer Bruce Willems-Braun (1997, 5) refers to "knowledge wielded in the interest of empire." Stephen

Bocking (2007, 887) describes northern science as the "sharp edge of southern intervention ... Scientists have played numerous roles in northern decisions, justifying its exploitation and sometimes its protection, with their authority and influence shaped not just by the knowledge they contributed, but by ideas about how the North relates – physically, politically, and cognitively – to Canada and the world." Writing about British Columbian geological mapping surveys in the 1860s, Bruce Braun (2000, 12) relies on Michel Foucault's concept of governmentality to explain how state practices and scientific surveys constructed "nature" in support of the modern state's primary commitment to manage "people and things" in the interests of general prosperity.[3] Management for prosperity requires detailed information about both territory and population. Scientific expertise is called upon to supply the data, but methods and analyses are specific to particular moments in time. In *Seeing Like a State,* James Scott (1998, 2) describes how states drew on scientific management of forestry and agriculture to make them "more legible – and hence manipulable." Matthew Farish and Whitney Lackenbauer (2009, 520) build on Scott's themes to describe how research helped turn the North into "an object of investigation and improvement."

Research designed by distant governments, agencies, and corporations to acquire greater control of the territory and its resources eventually generated distrust and harsh criticism, particularly from the Indigenous people who were studied. Marlene Brant Castellano (2004, 98) points out that "because the purposes and meanings associated with its practice by academics and government agents were usually alien to the people themselves, the outcomes were as often as not, misguided and harmful." In response, social scientists and then other scientists began to re-examine their practices and to engage in more collaborative and mutually beneficial projects that privileged traditional knowledge and addressed local concerns. Regional and local governments in many areas of the North have imposed licensing regulations and other means to exert some control over research in their communities. Guidelines for the conduct of ethical research were created by and with university researchers. In order to work with the communities of Eeyou Istchee today, it is essential for researchers to build equitable partnerships that are based in respectful relations, long-term commitments, and participatory processes (Mulrennan, Mark, and

Scott 2012; see also Stocek and Mark 2009). An exploration of the complex interplay of government, corporate, academic, and First Nations' interests is guided here within frameworks offered by Indigenous critiques of research, political ecologists, and environmental historians.

A review of the literature on James Bay from many disciplines laid the groundwork for this chapter.[4] I then spoke with Wemindji Eeyouch about their interactions with researchers who had visited during the last fifty years, and I interviewed researchers from various disciplines who had devoted time to James Bay studies. These discussions, from which brief quotes are used, provided invaluable background material and insights. The goal here has been an overview rather than a comprehensive review of all the published material. I apologize for the numerous but inevitable oversights and omissions. To organize the unwieldy amount of data, I identified four time periods that roughly corresponded to major social changes and changing research paradigms: exploration and collecting, 1867–1950, with brief reference also to earlier "pre-scientific" expeditions; mapping and measuring, 1950–70; developing hydro-electric power, 1970s; and the changing face of research after the advent of the JBNQA. The phases overlap, boundaries between the periods blur, and the partitions are arbitrary in spite of the logic that seemed so obviously to distinguish them at the start.[5] I employ these temporal divisions simply as signposts for tracing the paths of research over some 150 years.

EXPLORING AND COLLECTING (1867–1950)

European exploration of James and Hudson Bays, which began with the search for a northwest passage, continued through the nineteenth century. The Hudson's Bay Company (HBC) received its charter for Rupert's Land in 1670 and began building a monopoly on the highly profitable trade in furs (Francis and Morantz 1983; Morantz 1983). Crees, who supplied the furs, were not consulted when their homelands changed hands and their livelihoods were entangled with those of the HBC. Fur trading posts, established around James Bay and managed by company traders, provided destinations and opportunities for restocking supplies to explorer-scientists who were travelling to even more remote sites. Cree guides who were willing to meet the challenges of cross-cultural communication were usually hired at these posts.

From the earliest days, explorers recorded observations of the territory in journals, maps, and sketches, and many returned to England with collections of plant, animal, and/or mineral specimens. Journals kept by early HBC employees detailed weather, interactions with local people and visitors, numbers of furbearing animals harvested, and other information (Morantz 2002). Traders' letters reveal that they too collected specimens for naturalists in England who, with other European taxonomists, were intent on naming and classifying the natural world (Lindsay 1993). These collections were valuable introductions to new flora and fauna. Later researchers, such as Edward Preble (1902), complained that archaic and poor preservation and labelling techniques had produced collections that were inadequate for modern science. This argument was used repeatedly to justify extended research trips and to restock collections.

Geology, Exploration, and Early Scientific Research
Canada became a leader in the geological sciences during the second half of the nineteenth century, when the Geological Survey of Canada (GSC) was established to map and assess the mineral potential of the Province of Canada (Zaslow 1975). GSC maps and abstracts, collections, and inventories were rooted in "an economic and political language of possibility" (Braun 2000, 24) as the foundation of a prosperous national future. Under the scrutiny of the GSC, James Bay lands were made more legible, and the groundwork for future administrative policy-making and decisions was laid by geological explorers.

Toward the end of the nineteenth century, geologists Robert Bell and Albert Peter Low (known as A.P.), both later directors of the GSC, led major expeditions to James Bay. Maps and reports produced on these trips were essential to later travel and scientific inquiry.[6] With the help of Cree guides, Bell mapped the southern rivers and made significant botanical collections (Bell 1881, 1887; Vodden 1992). He also travelled both coasts of the bay, assessing diverse possibilities for economic development. At about the same time, Low began exploratory canoe journeys on long sections of the Rupert, Eastmain, La Grande, and other rivers (Low 1888). Low's engaging reports for the GSC detail geography, weather, wildlife, timber, and other vegetation, as well as mineral potential. In government reports, journal articles, and letters, both men wrote of the people whom they met and the knowledge they acquired from their Cree guides and other informants.

Economic constraints in the years leading up to and during the First World War forced the GSC to limit its research scope to the more "practical" or, rather, more immediately profitable (Zaslow 1975). This meant that coal and iron ore discoveries began to drive GSC northern research, along with surveys for a northern rail route to link the Saskatchewan and Alberta wheat fields to Port Nelson on the bay and to European markets (Fleming et al. 1874; Cormie 1917). Bell and others, including Robert Flaherty, a geologist later famous for his film *Nanook of the North,* were hired to carry out both of these initiatives (Flaherty 1918a, 1918b). Commercial fishery potentials were assessed by a Canadian Naval Services expedition in the same decade (Melvill, Comeau, and Lower 1915). Research supported by American universities and museums in the region highlighted the need to assert Canadian sovereignty in the North.[7]

Cree and Inuit Involvement in Early James Bay Research

Explorer-scientists were almost entirely dependent on the expertise, assistance, and language skills of the Cree guides whom they hired at HBC posts for logistical support. The expert knowledge of these guides was acknowledged in many early publications. Low and Preble clearly credit specific information to their guides and to "Indians in these parts," though mentions are brief and the names of individuals are rarely given (Low 1888, 1896; Preble 1902). Flaherty's (1918a, 1918b) articles include generous descriptions of his guides,[8] their survival and navigational skills (including cartography), and lively depictions of other people he met. His writings reveal a keen appreciation of Cree and Inuit guides and seem to indicate his enjoyment of the cross-cultural intellectual exchange (Flaherty 1918b, 1924). For others, such as E.M. Kindle (1925), who described his travels on the rivers of western James Bay, relationships with local people were clearly uncomfortable. Anthropologist Alanson Skinner, who travelled with guides to Moose Factory and Eastmain in 1908, lamented the "roughness of the country, the exigencies of the weather, and the scarcity of food," as well as difficult communications (Skinner 1912, 7; Preston 1975). He never returned to James Bay.

Cree assessments of fieldworkers and their research were never recorded (Morantz 2002). I asked people in Wemindji about interactions they remembered with scientists over the past fifty years. Marion Stewart (pers. comm., October 2009) recalled that some of her relatives had worked for visiting scientists in the 1950s and said her impression was that her father, for example, "always did everything he could to help them get their work done." In the late 1940s, ethnobotanist Jacques Rousseau interviewed Siméon Raphael, who guided for A.P. Low in the late 1880s (Rousseau 1948; Caron 1965; Morantz 2002). Rousseau (1948, 103) included a few generous words from Raphael about Low's good character, providing just a glimpse of the view through Cree eyes. The limited number of local stories about Cree reactions to early fieldwork indicates perhaps that most people saw visiting scientists as little more than transient disturbances, largely irrelevant to more pressing everyday concerns.

James Bay Anthropology

Anthropological interest in James Bay began in the early twentieth century, when the discipline was relatively new, and prior to the active intervention of federal administrative agencies in the lives of James Bay Crees. According to Richard Preston's 1975 overview of James Bay ethnographic work, the first anthropologist in the area was Alanson Skinner, whose single visit was museum- and artifact-based. Frank Speck's fieldwork for the University of Pennsylvania began at about the same time. In contrast, Speck's attention to the region was sustained over thirty years. He collected stories from Mistissini people at Lac St-Jean and studied Naskapi hunting practices and spirituality (see Speck 1915, 1923, 1935). Regina Flannery (1935, 1939, 1995), an anthropologist from the Catholic University of America, described women's lives on the southern James Bay coast. Interest in Cree religious views took her supervisor, Father John Cooper, to both the east and west coasts between 1923 and 1946 (Cooper 1929, 1933, 1938). During this time, Preston (1975, 270) observes, James Bay anthropology shifted from short-term, broad-based "reconnaissance field work" to longer stays at single sites and "problem-oriented research." These anthropologists' research programs, methods of inquiry, and representation of people and territory reflect general focal shifts in that discipline from the collection of material culture to more theory-based analyses.

Ecology and Wildlife Management

Developments in the emerging field of ecology during the period 1900–50 also echo broad disciplinary transformations. Stephen Bocking (2007) traces debates over the fluctuations of northern wildlife populations through the 1940s and 1950s, linking them to changing scientific theory and practice. These debates, he argues, followed the advent of new ideas about ecosystems, ecosystem fragility, and the management of animal populations. The idea that northern wildlife populations might need "management" or could be managed at all was itself a relatively new one.

In the 1930s, scientists began to research cyclical changes in animal populations, with an eye on greater economic stability for the fur trade.[9] HBC records, which confirmed that animal populations did indeed fluctuate dramatically, were mined for data. In addition, questionnaires rather than scientists were sent to the HBC posts for completion by Crees and other trappers (Chitty and Elton 1937). According to Bocking (2007), a decade later, the scientific value of trappers' knowledge and the validity of data collected by the HBC were questioned by a new generation of ecologists. Bocking attributes this rejection of local knowledge to a general shift in scientific practice that favoured laboratory-based experiments and more particularly to a decline in the market value of furs.

Population management, rather than observation and classification, became the focus of study for biologists in the North. By the 1960s, interest was growing rapidly in the control and maintenance of stable populations of larger game animals, such as caribou (Sandlos 2007). John Sandlos (2007) argues that game management projects in Canada's Northwest Territories were conceived from afar, with little attention to the particulars of local conditions. In consequence, they could be, and often were, disastrous for both Indigenous people and wildlife.

In eastern James Bay, the rapid decline of beaver populations in the 1930s was calamitous for the Crees, who depended on them for food and for the fur trade. The re-establishment of the beaver demonstrated the shift in scientific and state thinking about game management, but in this case Cree knowledge of beaver biology was also important. The story is a complex one that has been well documented elsewhere (see Morantz 2002; Feit 2007; Carlson 2008). Although, as these authors and others have noted, the idea of "population management" runs counter to Cree notions of animals as fellow members of a community that includes the land and all its inhabitants, Crees actively participated in the restoration efforts (Berkes 1986; Morantz 2002; Feit 2007). In other instances, such as caribou management, expert Indigenous knowledge of the land and animal behaviour ceased to be important to studies of northern fauna and was not incorporated into it for several decades.

MAPPING AND MEASURING (1950–70)

Post-war Developments

New technologies contributed to increasing southern intervention in the North after the Second World War, interventions with a primary goal of defending the nation-state's territorial sovereignty. Aerial reconnaissance expertise developed during the war made accurate mapping and resource assessment in remote areas far more viable than that performed by earlier land-based surveys (Lackenbauer and Farish 2007). Construction in 1954 of the Mid Canada Radar Line (MCRL) along the fifty-fifth parallel began the first of a series of briefly active, US-initiated, Cold War installations, the only lasting legacy of which is a string of contaminated sites on several coasts.[10] Mapping the James Bay region was completed in the mid-1950s by the Royal Canadian Air Force and the Department of Mines and Technical Surveys.[11]

Mapping and the military presence in the North paved the way for Prime Minister John Diefenbaker's "northern vision." This research program, designed to expedite the development of, in Diefenbaker's words, "those vast hidden resources the last few years have revealed," was announced in 1958 (quoted in Byers 2009). Kenneth Hare (1964) of McGill's Arctic Institute echoed Diefenbaker's call for new research in the North, particularly for the Labrador-Ungava Peninsula, directed toward its great mineral, timber, and hydro-electric potential. Although logistical challenges delayed James Bay hydro-electric exploitation for another decade, timber and mineral extraction began in southern parts of the region. Renewed interest in northern resource extraction generated ample research funds, and the link between commercial and scientific interests was more fully and openly reforged.

Museum- and university-based subarctic ecological work flourished in the 1950s and 1960s (Beals and Shenstone 1968), alongside continued geological investigations (Kranck 1950; Gussow 1953; Lee 1968). The Arctic Institute of North America, then centred at McGill University, led northern science during this period (Adams 1988, 2009). McGill's subarctic research program was centred at Knob Lake research station, near Schefferville, an iron-mining town east of the James Bay region. Quebec's francophone universities opened the Centre d'études nordiques in Great Whale (now Whapmagoostui/Kuujjuarapik), using the abandoned infrastructure of the MCRL base. Research conducted by the Centre d'études nordiques generally focused on Hudson Bay, though James Bay was never entirely ignored.

With calculating eyes on economic potentials, military and other government sources made funding available that opened new opportunities for biological research in James Bay. An inventory of arctic and subarctic insects was launched by the Canadian Department of Agriculture and Canadian Defence Research in 1947 (Freeman and Twinn 1954; Bocking 2009). Survey teams assigned to research stations at Moose Factory, Rupert House, and Great Whale investigated the ecology, relative abundance, and possible control of northern insects, especially mosquitoes and black flies (Macpherson 1964, 16; Jenkins and Knight 1950, 1952). Zoological observations had long been recorded by GSC fieldworkers but were only rarely published. Sustained interest in waterfowl and other birds had been fostered by the 1916 Migratory Bird Convention. Significant contributions to ornithological knowledge were made by Thomas Manning, a zoologist with both the Canadian Wildlife Service and the Canadian Geodetic Survey (Manning 1952, 1981). He also published important papers on James and Hudson Bay hydrological and tidal variations and zoology that ranged from voles to polar bears (Manning 1951, 1971).

Botanical studies for universities, museums, and botanical gardens proliferated at mid-century. Jacques Rousseau travelled from Lake Mistassini to the George River peninsula in the 1940s and early 1950s. His botanical contributions were substantial, as Toby Morantz (2002) points out, and his geographic, ethnobotanical, and ethnographic insights filled long-neglected gaps in scientific knowledge of this region (see Rousseau 1952, 1968; Malaurie and Rousseau 1964). The new 1:50,000 contour maps and aerial photos permitted the mapping of vegetation across northern Quebec (Hare 1950). Organized by the National Museum of Canada, coastal research expeditions in James and Hudson Bays included botanists William K.W. Baldwin (1953) and Ilmari Hustich (1950), from Finland. Hustich's work on vegetative zones and the maritime treeline complemented understandings, developed by Rousseau and Hare, of ecological zones for the interior of the province (Hustich 1949, 1957; Hare 1950; Rousseau 1952, 1968). Longer-term botanical work was also undertaken by Ernest Lepage and Arthème Dutilly, Oblate fathers who travelled both coasts and nearly every large river of Eeyou Istchee between 1943 and 1963 (Dutilly and Lepage 1948; Dutilly, Lepage, and Duman 1958; Payette 1976). William Baldwin (1959) organized a three-thousand-mile "excursion" through forests of the southern James Bay area for an international group concerned with the study of northern botany. The timber industry set its sights farther north, prioritizing botanical analyses of northern vegetation.

Conservation

The conservation movement took shape in Canada at mid-century amid an emerging public discourse regarding wild lands and wildlife. "Ecosystems" and "habitats" were relatively new concepts, entwined not only with ideas about environmental management, but with protection of animals as well. Canadian ecologists began to take on lead roles as advocates for the environment (Bocking 2007). A decline in western Ungava caribou populations prompted the Quebec Ministry of Fish and Game and the federal Departments of Northern Affairs and National Resources to send in a team of ecologists (Banfield and Tener 1958; Bergerud 1967). Their studies attributed the caribou decline to forest fires, a warming climate, and intensified hunting by Crees and Inuit. As Sandlos (2007) and others have shown, Indigenous people were not only absent from these discussions; they were often blamed for the declines. Public interest in the environment grew with the focus on the status of large northern animals such as caribou and polar bears. Concern for the well-being of Indigenous people was not on public agendas.

The story of polar bear protection illustrates the rise of conservation as an international issue. The evolution of new wildlife management policies can be traced in the International Union for Conservation of Nature's

(IUCN) circumpolar agreement on polar bears (Fikkan, Osherenko, and Arikainen 1993). A five-nation Polar Bear Specialist Group was formed in 1968 to "collect [and share] data on the polar bear's natural history as a basis for future management" (IUCN 1970, 2). Funded by the Canadian Wildlife Service and provincial agencies, Canadian biologists kept "constant watch on the research needs and the status of Polar Bears in Canada" (Jonkel 1970, 116). Research begun near Churchill, Manitoba, and on North and South Twin Islands in James Bay was later extended to sites further north. Data were collected for the Twin Islands bear population with some intensity over the first five years, then more sporadically (Stirling and Parkinson 2006; Calvert et al. 1998). Studies by the Polar Bear Specialist Group eventually resulted in a landmark agreement, protecting not only the bears, but also their habitats (Mulrennan 1998). Anne Fikkan, Gail Osherenko, and Alexander Arikainen (1993) point out that though the Canadian polar bear populations were not known to be in grave danger at the time, active engagement in this initiative helped to propel Canada toward the forefront of northern zoology and environmental protection.

During this period, Crees were still marginalized in environmental research. For example, the publications that resulted from the Twin Islands polar bear research do not mention the contributions of Cree research assistants. Fred Blackned (pers. comm., July 2010), a former chief of Wemindji who worked as an assistant for those studies, remembered the experience as satisfying and exciting despite the dangers the team encountered. He was responsible for estimating the weight of each bear prior to tranquilization and sampling procedures; a miscalculation would have had serious consequences for everyone present. According to Blackned, the researchers were friendly companions, but he expressed disappointment that no results of their work were ever returned to him, the other assistants, or the community. The lack of reference to assistants, so noticeable after the mid-1940s, contrasts strongly with the frequent mentions of local guides and informants in early publications, such as those by Low and Preble. The need for constant support from Indigenous guides and research assistants had declined with new research methods and air travel. In addition, my review of studies carried out at the time revealed that acknowledgment of local assistance became less routine. It is possible that an authoritative science relying on reproducible laboratory or field conditions was not very concerned with the long-term observations of local experts who had no "proper" scientific training. Debates on whether Indigenous knowledge should be taken seriously began to surface in the 1970s, when public consciousness of Indigenous rights and land claims rose sharply. Nevertheless, in the preceding decade the scientific community largely ignored Indigenous knowledge and local environmental concerns.

Social Science and Administration

During the 1950s and 1960s, the social sciences took an increasingly important place in northern studies. In an article on modernization in the western Arctic, Farish and Lackenbauer (2009, 519) link this drive to acquire more knowledge about Indigenous people across the North to growing government involvement. Federal aid measures for eastern James Bay Crees were insignificant until the early 1950s. When they were initiated, the strings attached led to significant social change. Social assistance, family allowance, old age pension, and health care required Crees to settle in villages and be educated according to Canadian norms (Vincent and Bowers 1988). As Norman Chance (1968b, 21–22) explains,

> The Canadian government, perceiving the Cree as an impoverished ethnic minority, has promoted a policy of economic integration with the rest of Canada, required all Indians to attend school, either locally or in Government supported boarding schools, offered job-training courses, improved housing and health facilities, and assisted in the development of local fishing cooperatives, sawmills, and other small industries on and off reserves.

Between 1950 and 1970, Crees acquired houses, nursing stations, schools, and incomes. Underlying the policies that brought these changes was the unquestioned assumption that development and assimilation were both desirable and inevitable.

Anthropologists became interested in these changes to traditional ways of life during the 1950s and early 1960s (Kerr 1950; Honigmann 1961, 1962; Knight 1967). In 1964, Norman Chance, quoted above, assembled a multi-disciplinary team, the McGill-Cree Project, funded by McGill

University, the federal Department of Forestry and Rural Development, and the Centre d'études nordiques at Université Laval (Chance 1968a). The project focused on impacts of changing land use and economic development in the southern communities of Eeyou Istchee. Newly established logging and mining operations were disrupting traditional ways of life in that area. Chance's (1968a, 1) aim was to guide development processes in ways that would be "most conducive to economic growth and social well-being." The supposed benefits of modern prosperity were not yet apparent for the Crees who had moved to shanty towns around the mining and lumber towns, seeking seasonal employment.

Project members examined issues such as residential school education and its impacts (Sindell 1968; Wintrob and Sindell 1968; Wintrob 1968; Sindell and Wintrob 1969), increasing environmental stress (Chance 1968b), and changing community life (Pothier 1968). Adrian Tanner (1968), also an early project member, has pointed out that though interventions imposed in the name of development had many traumatic consequences for Crees, development and progress were widely believed to benefit everyone involved. He has criticized Chance's project for its supposition that no alternatives existed, that the transformation of Cree society was inevitable, and that it had already occurred (pers. comm., November 2009). Harvey Feit (pers. comm., October 2009) also imagines a different kind of project, one that could have supported what Crees desired in these new circumstances. Progress and improvement through the power of modern development, however, remained largely unquestioned for most people.

At the end of the decade, Richard Salisbury took over the McGill-Cree Project from Norman Chance. He and a team of graduate students would seek out new ways to work with and write about the James Bay Crees. Richard Preston (1976, 2002), a McMaster University professor, also worked in new ways to examine Cree values, worldviews, local economic strategies, and the development of culturally appropriate school curricula in the community of Waskaganish. These scholars' studies of hunting practices, environmental relationships, land tenure, and the Crees' strategic embrace of state-level institutions would refocus James Bay research and sustain studies with and for the Crees concerning key issues in the future.

DEVELOPING HYDRO-ELECTRIC POWER (1970s)

With the exception of fur traders, hardy adventurers, and the scientists and anthropologists discussed above, southern Quebec and Canada generally ignored the James Bay region until Quebec's aspirations to greater economic and political autonomy began to intersect with its growing expertise in electrical power production. Plans for the James Bay hydro-electric complex were based on an expedient assumption that eastern James Bay, remote and unoccupied, offered the ideal location for large-scale development (Desbiens 2004a, 2004b). Indigenous peoples there were never informed that plans to radically alter their territory were being formulated during the 1960s:

> Mistassini hunters were coming across all sorts of evidence of new activity in their territory at the time. One heard rumours of mysterious strangers out there in the forest. Most likely they were hydrologists from Hydro-Québec doing preliminary studies, but people didn't know who they were. No announcements had been made in the community, nor had anyone sought local consent for research. (Adrian Tanner, pers. comm., November 2009)

In 1971, construction began on what was then the world's largest hydro-electric project. Hydro-Québec had been studying the region's hydrography for a decade, but no information sessions or consultations on the potential impacts of the project were initiated prior to Premier Robert Bourassa's announcement in 1971 that construction had begun. The presence of and implications for Crees and Inuit living on the land were completely ignored. Crees, however, reacted swiftly by launching a court case, aimed at halting the project. Against all odds, they won an injunction against construction, granted by the Montreal Superior Court (Richardson 1977; Peters 1999). The injunction was overturned almost immediately in the Quebec Court of Appeal. This move was followed by an arduous and lengthy process of negotiation with Hydro-Québec, Quebec, and Canada. Negotiations over the next three years led to the signing of the James Bay and Northern Québec Agreement (JBNQA) (Hornig 1999) in 1975 by Crees, Inuit, and both governments. Known as

Canada's first modern land claim agreement, it was a model for later examples, but its implementation was often a contested process.

Quebec created a governing structure for the region in 1971 by forming the James Bay Development Corporation (Société de développement de la Baie James, or SDBJ) to oversee northern development, including mining, forestry, and tourism. A subdivision of SDBJ, the James Bay Energy Corporation (Société d'énergie de la Baie James), was given responsibility for the hydro-electric project. Once again, Indigenous populations who would be directly affected, notes Alan Penn (pers. comm., February 2008), scientific adviser to the Cree Regional Authority, were not involved or taken into account. As James Scott (1998) shows, large-scale manipulations of the environment, in combination with sweeping decisions made from afar, out of tune with local conditions and ignoring local knowledge, often fail or produce unexpected results. Ignoring Cree communities proved a costly error in judgment.

Hydro-Québec and Research

Hydro-Québec's mega project captured the attention of both social and natural scientists. As Eeyou Istchee became simultaneously an object of political contestation and intense scientific investigation, research in the province changed dramatically. In court, biologists and other experts were asked to give evidence about how damming and diversions of large rivers would alter the region's ecosystems. They responded that because so little was known of subarctic Quebec, no evidence could be supplied (McCutcheon 1991). Similar arguments were advanced by environmental groups and other protesters in their calls for environmental assessment processes. According to Harvey Feit (pers. comm., October 2009), who supported and worked with the Crees while a graduate student of Richard Salisbury and then assistant professor at McGill in the 1970s, "It was the protest by Crees and the people who joined with them that created the need for environmental research." Hydro-Québec had begun the project with no intention of examining its potential impacts on the environment, either physical or social. To diffuse the protests and demonstrate the "greatest possible respect" for the environment it was disrupting, the James Bay Development Corporation began to invest heavily in the acquisition of up-to-date knowledge (Hayeur 2001, 8). It was necessary once again to measure, map, and collect new data.

The work of Vincent Gérardin (1980), a forestry engineer who mapped the region's soils, vegetation, geology, and geomorphology for the Canadian Forestry Service during the 1970s, is typical of the studies commissioned at the time. He explains the logistics as follows:

> It was certainly among the best financed ecology projects ever launched. We had funding from the Forestry Service and the Societé de Development de la Baie James. Two base camps were built for us in advance. We had everything we needed to be comfortable and work quickly. In three years we covered an immense territory, nearly one third of the province. First, we used aerial photos to pre-map the different ecosystems. Then we selected representative areas to sample. Six teams of four people went out on foot daily to sample soil, vegetation, and geomorphology. The data was collated during evenings in the camp office, and entered into a data base. (Gérardin, pers. comm., November 2009)

Although Gérardin describes this as a data-gathering exercise rather than an analytical one, the ensuing report "provided tools that are still being used. In fact, you could say that most of the current strategies on protected areas in the north were built on the work we did back then" (ibid.). Diverse studies, some carried out jointly by government and academic scientists, others by consultants, began to augment the slim scientific knowledge base on James Bay ecosystems.[12] Gérardin's group tried to include Crees in the data-gathering process, but according to him, the effort was not well enough thought out to be successful. Most scientists and consultants worked without Cree assistance.

A new "hot spot" for both government- and university-based environmental research, James Bay attracted scientists from a range of disciplines. University-based scholars began working on environmental issues, ranging from geology to biology to marine circulation and sedimentation, as soon as the construction plans were announced.[13] Generous funding became available for students who were interested in James Bay science, although it often came with strings attached. According to Harvey Feit (pers. comm., October 2009), there was enough money "to shape the way that research in Québec universities was done, that's where the money was and it wasn't available for other things." Alan Penn (pers. comm., September 2011) offers a similar assessment of this early phase of Hydro-Québec's research: "Hydro-electric development and more particularly the political role played in Québec by Hydro-Québec itself exerted a major inhibitory effect on research, and discouraged many prominent researchers who did not wish to fall into the trap of being beholden to Hydro-Québec." Though some researchers stayed away, many others were hired directly by Hydro-Québec as consultants who produced an immense quantity of "grey literature" that contributed detailed inventories but little in the way of analysis about the region. Harvey Feit (pers. comm., October 2009) points out that at this stage, "funding was systematically channelled towards surveys of plant and animal distributions and habitat, and directed away from impact assessments." Reports commissioned by Hydro-Québec and the SDBJ tended to answer questions that supported Hydro-Québec's claims to environmental concern. They bolstered the impression that the company was committed to monitoring the impacts, social or biophysical, of hydro-electric development.

The information commissioned by Hydro-Québec was not widely available. This was not atypical for times when industry and government dealt in private, ignoring concerns raised by Indigenous people, environmentalists, and other Canadians (Canadian Arctic Resources Committee, Environmental Committee of Sanikiluaq, and Rawson Academy of Aquatic Science 1991). "Summaries of knowledge acquired" were made available to the public, and selected reports could be accessed in university libraries (Gantcheff, McCormack, and Couture 1982a; Roy, Lemire, and Messier 1985; Comité de la Baie James sur le mercure 1988; Hayeur 2001). Scientific articles and publications written by or for Hydro-Québec were archived in the company's Montreal library, which has only recently opened its doors to the public (by appointment).

Crees and Research

Industry and government held no monopoly on research. Crees themselves undertook the studies needed to defend their lands and rights. During the court case seeking an injunction against the project, many hunters were called upon to answer questions about how the dams would alter their lives (Richardson 1977). Scientists were called by both sides to add details about Cree hunting practices, to present harvest statistics, and to supply information about fauna and ecosystems. Crees enlisted academic researchers as consultants to provide information that would strengthen their position and contribute to the preservation of their way of life.

Many of those researchers, including Harvey Feit, Ignatius La Rusic, Colin Scott, and Toby Morantz, who worked as graduate students with Richard Salisbury at McGill University, are still involved in Cree issues today.[14] Salisbury, director of the McGill-Cree Project, called for attention to impacts on the Crees as the central issue for assessing the project (Salisbury 1986; Young 1999; Feit and Scott 2004). Harvesting and land-use studies (Elberg et al. 1972; Weinstein 1976; La Rusic 1979), ecological assessments (Spence and Spence 1972; Berkes 1980), and research on hunting, fishing, and trapping practices (Feit 1978; La Rusic 1979; Colin Scott 1979) were quickly initiated to add to work undertaken earlier by Feit (1969) and Tanner (1979).

Crees hired their own full-time staff, including anthropologist Brian Craik and geographer Alan Penn, both of whom still work with the Grand Council of the Crees and the Cree Regional Authority, to co-ordinate the work of these consultants. Thus, in the 1970s, Eeyou Istchee became a hot spot not only for studies directed by the interests of Hydro-Québec, but also for research that was increasingly Cree-led. According to Feit (pers. comm., October 2009), "the centre of research began to shift away from government and industry. The work was less well funded but now it was done by Cree and for Cree organizations and under Cree leadership. It was more significant." Crees had redirected the research agenda to address their own concerns. "In doing so, they became responsible for more of the social research, and for a period, more of the wildlife research, being carried out," Feit (ibid.) says, "than any other group."

The assumptions informing Cree-managed studies diverged radically from those associated with Hydro-Québec's research. The developer asserted that any impacts to the land would be inconsequential in the long run (Vincent and Bowers 1988), but changes mattered intensely to people who depended on the land and waters for their food. Despite fundamental differences in understandings of the land, both researchers and developers would have to work in tandem with Cree institutions if future projects were to move forward.

JAMES BAY RESEARCH AFTER THE SIGNING OF THE JBNQA

During the years that followed the advent of the JBNQA, Eeyou Istchee was transformed from an empty space on the research map to one of the most studied regions in northern Canada. Much of that research was managed by Crees for the benefit of their communities. It is impossible to review all of that work in both natural and social sciences here, but the remainder of this chapter will summarize some of the problems that generated research questions after the JBNQA was signed, discuss the proposed Great Whale project of the 1990s, and explore various new research agreements and partnerships that were cemented with increased Cree autonomy during the early twenty-first century.

Research in the Era of JBNQA Implementation

The political landscape remained rocky in Eeyou Istchee after the signing of the JBNQA. Promises made in the agreement were not delivered according to expectations, implementation of many aspects lagged, mechanisms for its practical functioning were lacking, and anticipated job opportunities did not materialize (Feit 2004; Mulrennan and Scott 2005).[15] Hydro-Québec's research program contributed to provincial environmental policy in ways that would benefit the corporation and "greenwash" its image (McCutcheon 1991; Warner 1999). Cree concerns about the impacts of construction, flooding, diversions, and subsequent environmental and social change seemed to be sidelined in this program.

As physical environments, social relations, and traditional lifeways were being transformed by large-scale development projects, the Cree communities acquired responsibility, under the JBNQA, for their own administration, education, health, and income supplement services (Salisbury 1986). Setting up and putting effective services in place required well-informed planning processes backed by research. These studies were initiated and managed by Crees with the assistance of research consultants they had hired from universities and private companies. In education, for example, Crees had to make comprehensive decisions about use of the Quebec curriculum and English, French, or Cree as the language of instruction, teacher training, and constitution of the school board (Murdoch 1975; Preston 1979; Tanner 1981). Research on immediate health crises, as well as studies of traditional medicine, was implemented through the newly created Cree health system (Cree Board of Health and Social Services of James Bay 1980; Grand Council of the Crees 1981; Marshall 1984; Marshall and Cree Regional Authority 1989). The innovative Cree Hunters and Trappers Income Security Program was developed by the Cree Regional Authority and closely monitored in consultation with Cree hunters (La Rusic 1979; Feit 1991; Scott and Feit 1992).

Academics who had been supportive throughout the negotiation process and were now well known in the communities were invited to work with Crees to seek out and evaluate the information needed to sustain decision making. Harvey Feit, Adrian Tanner, and Colin Scott, whose independent academic interests coincided with Cree concerns and who had been working on these issues since the 1970s, contributed to deeper understandings of linked environmental knowledge, cosmology, and hunting practices (Feit 1986, 1988, 1991, 1995; Tanner 1979, 1988; Scott 1989, 1996). The working relationships these researchers developed with the Crees have endured over the decades and became the foundations of effective community-based research partnerships in the present.

Crises resulting from unforeseen impacts of the flooding, particularly methyl-mercury contamination of fish in the reservoirs, kept James Bay in the public eye during the 1980s.[16] The diets of most Crees depended heavily on local fish, so the communities urgently needed solutions (Penn 1978; Roy, Lemire, and Messier 1985; Cloutier 1987). Some of this work was initiated by the Crees; much of it was funded by Hydro-Québec. The mercury issue, potentially damaging to the image of responsible citizenship that Hydro-Québec had worked so hard to project, was carefully managed, but it has yet to receive adequate attention from industry, government, or university researchers (Alan Penn, pers. comm., March 2011; Harvey Feit, pers. comm., March 2011).

The loss of land for Crees, whose values are based in a sense of community and reciprocal social relations with the land, represented a loss of history, culture, identity, and community (Niezen 1993, 1998). Hunting sites, camps, travel routes, and burial grounds drowned as the reservoirs were dammed and filled, and rivers were rerouted. Salvage archaeology projects at selected sites slated for flooding were funded by Hydro-Québec and carried out with Cree collaboration. Local people were asked to suggest sites and provide other information (Adrian Tanner, pers. comm., November 2009). Archaeology projects could not displace the pain of loss, but they coincided with a desire among Crees to record knowledge, place names, and elders' stories for future generations (Denton 2001). Work on the history of the fur trade and Cree participation in that trade, undertaken by Toby Morantz and others, provided a new sense of Cree history (Laliberté 1982; Francis and Morantz 1983; Hanks 1983; Morantz 1983; Kenyon 1986; Colin Scott 1996).

Great Whale and Impact Assessment Processes

Hydro-Québec announced in 1989 that it would begin a second phase of dam construction near the most northern Cree community of Whapmagoostui on the Great Whale River. The project was strongly and effectively opposed by Crees, who enlisted the support of environmental and social justice networks and brought it to international attention (Coon Come 1991, 2004). Under the JBNQA, Great Whale required the broad environmental assessments that were missing in earlier phases of the project.

Hydro-Québec's proposed assessments did not respect Cree concerns over cumulative impacts and did not include community consultations (Grand Council of the Crees n.d.). Without legal proceedings initiated by the Crees and the environmental and human rights issues they raised, the evaluation of social impacts might have been marginalized (Mulrennan 1998; Tanner 1999). As in early confrontations with Hydro-Québec, protection of hunting, and thus of traditional ways of life, was again at centre stage. The Crees were not opposed to all forms of industrial development, but they rejected the Great Whale project as "irrational and disrespectful from a social, economic, moral, and environmental perspective" (Awashish 2005, 175). Their campaign was successful. In 1994, this phase was put on hold. The possibility remains that it could be revived, but the necessity of obtaining Cree consent to any development agenda has been clearly established.

The environmental impact assessment (EIA) process, so much in the news around the Great Whale controversy, continues to have impacts of its own on James Bay research.[17] A highly structured process, EIA is framed by the concerns of government and industry and their reliance "on the authority of science to legitimize its judgments" (Nasr and Scott 2010, 132). Accordingly, Hydro-Québec hires consultants to inventory the changes that follow construction projects and to propose measures for mitigation of any problems (Warner 1999). EIA treats impacts as technical problems. The process splits these problems into small segments for analysis, so that specific changes, perhaps related to roads, wildlife, or salinity levels in James Bay, are each treated separately, rather than as related indicators within a broad framework of overall impacts. It is an approach that tends to leave out the complex, detailed, and nuanced knowledge of those most affected. What results is "an imbalance that endorses political, administrative-bureaucratic, and scientific processes. While better than nothing, these processes are not satisfactory. Research can clarify what people need to know, but those who will be most affected should be making the decisions" (Harvey Feit, pers. comm., 2009). Wren Nasr and Colin Scott (2010) concur that environmental assessment has yet to achieve a meaningful dialogue between Indigenous and scientific knowledge. Crees commissioned a pair of wide-ranging studies to document, in Cree people's own words, the experienced and expected impacts of hydro development (Scott and Ettenger 1994; Naka-shima and Roué 1994). In Cree hands, the assessment of impacts could begin to counter the lack of useful knowledge about social impacts.

The politics of hydro-electric development were still central to James Bay social science research in the 1990s (for a few representative examples, see Coon Come 1991; McCutcheon 1991; Peters 1992; Niezen 1993; Penn 1995; Rosenberg et al. 1997; Hornig 1999). So, too, were community-based studies examining Cree worldviews and knowledge of the land (Berkes 1995; George, Berkes, and Preston 1996; Colin Scott 1996; Ohmagari and Berkes 1997; McDonald, Arragutainaq, and Novalinga 1997). The importance of traditional and local knowledge and consultations with local hunters, trappers, and their community-based organizations is highlighted in these studies. Community decisions on environmental issues now routinely in-volve such consultations. A further step has been to link the resources, knowledge, and skills of Cree and non-Cree experts in sustained partner-ship, as described throughout this volume, to expand opportunities for reaffirming local values and cultural heritage.

Partnership in Research

Across Canada, land claim settlements have meant increased autonomy for Indigenous people and changes in the governance structures of northern territories. Regional governments at many levels are closely involved in research. This varies from directing and carrying out the work to regulating it. Scientists are now often required to engage communities in their research and have had to develop protocols that respect community values and practices. Local people "want to know what researchers are doing, why they are doing it, and what benefits their work might bring to the North. They also want to know that the research will not harm wildlife, the environment, or cultural resources" (Eamer 2005, 29). In the Arctic, formal methods for tracking research and granting permits have been established. In the James Bay Cree territories, guidelines rather than formal permitting procedures are in place. Research proposals must be approved by communities and/or by the Cree Regional Authority. Guidelines for the conduct of research within the framework of community-based partnerships, such as those that sustain this volume, are being jointly developed and revised by community members and researchers (for a brief discussion, see the Introduction to this volume).

Constrained by limited time and funding, as well as lack of training and confidence, researchers sometimes avoid contact with the Cree communities, even when field sites are on Cree lands. These scientists miss out on the benefits of local expertise derived from long-term familiarity with the land, as well as the social benefits of interaction in the communities. Cree hunter Henry Stewart (pers. comm., July 2010) pointed out to me that the presence of these researchers on the land, however, does not go unnoticed. Whereas some disciplines, particularly in the social sciences, have more ready support and training for this type of engagement, researchers in the natural sciences are discovering the value that these collaborations can add to their work. They have sought out Cree expertise and have incorporated local knowledge at various stages (Fraser et al. 2006; Jacqmain et al. 2007, 2008; Chapter 7, this volume). Chapter 2 of this book demonstrates the complexity of collaborative processes and how they involve far more than the simple sharing of information.

Rodney Mark of Wemindji, who has overseen the development of several collaborative community partnerships, some academic, some corporate-based, explains that "research partners add value, and that's what is important: adding value. Hydro[-Québec's] research ... only served the developers" (pers. comm., October 2009; see Chapter 3, this volume). At its best, Mark concludes, "collaboration should result in an ongoing learning process, which you find has resulted in a common vision of how to work together and make decisions." His vision for the community includes recognition of the potential benefits in adapting external tools and methodologies to meet local needs.

As Chief Mark points out above, research funding in the past has too often been linked to development projects and too narrowly focused on specific impacts and changes (Stewart, Dunbar, and Bernier 1993). Integrated designs for research that include understanding cumulative effects of changes to the environment have been overlooked. Cree requests for work of this nature have been ignored by both government and industry. Recent community-based collaborative trans-disciplinary projects, such as the Wemindji Protected Areas Project, are developed to respond to these needs, enhance local research capacity, make new technologies more accessible, and co-ordinate information and artifacts from past studies such as place names, stories, and photographs (Mulrennan, Mark, and Scott 2012).

As experience with research partnerships grows, increased Cree participation and direction at all levels, from hypothesis formation and design of research questions to analysis, promise outcomes with much greater relevance to local situations. At the regional and community levels today, most research has become a negotiated process of collaboration with academic researchers or consultants. At the regional level, an independent Cree research and development institute is now being created to direct, design, and monitor the research carried out in Eeyou Istchee.

CONCLUDING REMARKS

The drive to accumulate up-to-date knowledge of the environment continues today with the province's most recently conceived resource extraction and development scheme. In 2009, Quebec premier Jean Charest promoted a new "vision" for the future, a Plan Nord for James Bay and northern

Quebec. Government publications proclaimed that "this huge expanse of land offers outstanding potential" for "new economic spaces" in energy, mining, forestry, recreation, tourism, and wildlife development (Quebec 2009, 3). Little has changed in the language of economic possibility and resource development since Robert Bourassa's "project of the century," John Diefenbaker's "northern vision," and the Geological Survey's early explorations in the "land of hidden treasure" (Curran 1907; Bourassa 1985; Byers 2009). Little has changed as well in the desire for greater understanding of the northern ecology and environments that are to be developed, exploited, and altered. Quebec continues to insist on the need for an integrated, wide-ranging, and current knowledge base (Gouvernement du Québec 2011).

Despite past experience, Plan Nord, billed as the "project of a generation," was released with almost no prior consultation with First Nations and Inuit. As the provincial government realized it must address this omission and develop collaborative relations with Indigenous leaders, Crees released a "A Cree Vision of the Plan Nord" in which they made their position very clear: any development would require their consent (Grand Council of the Crees 2011). The Quebec government claimed that development would occur in "a socially responsible and sustainable form" and promised to protect 50 percent of northern lands from any industrial development (Gouvernement du Québec 2011). Partnerships forged among northern communities, provincial government, and other stakeholders would jointly define the opportunities and new spaces for collaboration (Quebec 2009). The new regional governance structure includes Crees in a stronger role in regional municipal governance over large portions of their lands (Grand Council of the Crees 2012). Will knowledge, under these circumstances, come more extensively to reflect relations of co-production?

Between the lines of Plan Nord lies the history outlined in this essay. Modern nations, as James Scott (1998) explains, have sought to impose a degree of "legibility" on nature, with the intention of improving conditions for their citizens. The new Dominion of Canada charged geologist-explorers with making legible what was, for them, a vast, uncharted "wilderness." Guided by Crees who knew the land well, they embarked on lengthy expeditions to James Bay to map and measure its breadth, to discover its hidden treasures, and to ensure a bright future for the new country. At the same time, maps and inventories served to assert a Canadian presence in the region (Bocking 2009; Farish and Lackenbauer 2009). Indigenous people, of minor concern to the state during this period, were marginalized in research contexts. Northern research was altered by the changing scientific discourses and practices that accompanied the technological advances of the Second World War. Indigenous people, however, were still relegated to the sidelines as passive observers whose knowledge of their environment was discounted. Improved access to the North allowed scientists a greater presence on the land and made more frequent and shorter fieldwork seasons the norm. Their acquisition of knowledge continued to reflect and support an assertion of southern authority, whether in the management of wildlife or the implementation of social programs.

Further control of the North has been assumed by state and industrial powers through large-scale manipulations of nature, such as the massive reconfiguration of large rivers and waterways to generate electricity. Major engineering projects are designed to solve problems, but they are just as likely to create new ones, especially in environmental and social domains (Josephson 2004). They inevitably produce a whole range of unforeseen results. Announcement of the James Bay hydro-electric projects transformed northern research, shifting responsibility from government to corporate shoulders. Hydro-Québec's data collection projects extended corporate influence to academia. At the same time, the Crees' self-government bodies themselves became active sponsors of research and authors of research agendas. A major impact of northern development has been the empowerment of Indigenous people in research and their involvement at various levels in nearly every aspect of social and natural science studies.

Most recently, Plan Nord (Gouvernement du Québec 2011) called for a broad knowledge base to integrate new data and old. Hydro-Québec expressed similar needs in the 1970s. Existing data are to be incorporated from knowledge bases held by each partner – government, corporate, academic, regional, and local, including Cree entities. New research will investigate the impacts of ongoing and new development. It is intended "to understand and protect the most vulnerable ecosystems and species ... [and to] focus on measures to adapt to climate change and industrial development to ensure that biodiversity is preserved" (Quebec 2009, 15).

Compilation of the knowledge base and the acquisition of new data will be carried out by government agencies and a research consortium, with members from Quebec universities and the private sector (Hydro-Québec, mining, and forestry). Environmental impact assessments are to precede every project, and there is a proposal to include "a no-build" option within this EIA regime. How Indigenous knowledge will be integrated has not been well clarified in this enterprise. Nor has the willingness of all parties to share their data. This is cause for concern.

James Bay Crees and other Indigenous peoples, under-represented for so long in environmental decision making, are taking charge of research and policy development to address their own needs in light of local perspectives and knowledges. Research domains continue to be politically contested and negotiated. In this new era of development and related research relationships, there are indications that collaboration extended beyond the usual parameters can generate productive, long-term conversations and increased local capacity. Ignoring the lessons of past research and development practices cannot be an option. This chapter opened with a quote from Rodney Mark about the changing relationship between Crees and scientists from the South. He is optimistic and ready to move ahead on this new footing, but he is also alert to the many challenges and the need for vigilance. In the past, Chief Mark (pers. comm., October 2009) has said, "research was not for people. It was about how to get the job done and get on with development. It was about impact assessment, social impacts, and socio-economic research. It only served the developers, not the community. But now, research in the community must be always also for the community."

NOTES

1 See Julie Cruikshank's (2005) discussion of the essential support offered by Indigenous people to explorers/researchers in the Pacific Northwest.
2 Stephen Bocking (2009); Michael Bravo and Sverker Sörlin (2002); Trevor Levere (1993); and Graeme Wynn (2007) elaborate these themes in diverse northern settings.
3 Bruce Braun published under the name of Bruce Willems-Braun in 1997.
4 Bibliographies assembled by other researchers have been helpful. Jacques Rousseau (1954) compiled a large and truly comprehensive annotated list for the Mistassini area that contains many early documents from missionaries and fur traders, as well

as later scientific material. Gaëtan Hayeur's *Summary of Knowledge Acquired in Northern Environments from 1970 to 2000* (2001) comments briefly on research related to the James Bay Hydro-Québec project. A 1993 report on potential marine protected areas contains what is probably the most complete bibliography, with nearly eight hundred references cross-indexed by key words (Stewart, Dunbar, and Bernier 1993). The creation of an up-to-date bank of references is a project that deserves much further and sustained attention.

5 See Peter Mulvihill, Douglas Baker, and William Morrison (2001), who dissect some of the problems inherent in constructing frameworks for understanding northern Canadian environmental history.

6 Low's finely detailed maps of thousands of kilometres of the rivers and coasts of James Bay, drawn between 1881 and 1898, proved reliable tools for those who followed him. Adventurous paddlers who retrace his routes still use them and appreciate their accuracy (Finkelstein and Stone 2004).

7 For example, Charles and Alexander Leith (1910), from the University of Illinois, searched the east coast of James and Hudson Bays to locate iron formations. Edward Preble (1902) mounted "biological investigations" for the American Museum of Natural History. Working for the same institution but in search of Cree artifacts, Alanson Skinner (1912) was the first anthropologist to visit James Bay.

8 Flaherty is credited with the "discovery" of the Belcher Islands. His 1918 article in the *Geographical Review* describes a map drawn by a local Inuit HBC navigator, George Weetaltuk, that made this success possible. According to Milton Freeman (1983), the Inuit map was far more accurate than the one later created by Flaherty for publication. The Belcher Islands had been Weetaltuk's home hunting territory before he moved his family to Charlton and Cape Hope Islands.

9 For contemporary discussion of the issue, see Rudolph Anderson (1934); Duncan MacLulich (1937); C.F. Jackson (1938); Charles Elton and Mary Nicholson (1942); and Helen Chitty (1948).

10 For an exploration of these military installations and their legacies, see B. Sistili et al. (2006) and Leonard Tsuji et al. (2006).

11 According to Kenneth Hare (1964, 461–62), "The National Topographic Series maps available in 1948 were simple eight-mile-to-the-inch (1: 506,880) sheets showing streams, lakes, and coastlines – essentially the hydrography. There was no depiction of relief, and only a crude estimate (by spot heights) of altitudes of prominent hill-tops ... Substantial areas, especially in the northwest, were almost devoid of detail; on several sheets only sketches of the main water-courses, transcribed from travelers' notebooks, were available ... In 1948–1951 the extensive program of vertical photography was carried out by the federal government, partly by a wing of the R.C.A.F., partly by private companies under contract ... Since 1951 the Department of Mines and Technical Surveys has completed the eight-mile-to-the-inch hydrography series."

12 See Jean-Pierre Ducruc et al. (1976); Jean-Claude Dionne (1976, 1980); Richard Zarnovican (1978); and Georges Gantcheff, R.J. McCormack, and Armand Couture (1982a, 1982b).

13 See R. Alison (1976); Roger Bider (1976); Claude Hillaire-Marcel (1976); Léon Hardy (1977); I.P. Martini et al. (1980); I.P. Martini (1986); Yves Lambert and J.J. Dodson (1982); Grant Ingram et al. (1986); and Grant Ingram and V.H. Chu (1987).

14 In conversation, Harvey Feit (pers. comm., October 2009) emphasized how much this experience changed his own work: "It is important to say here how open Cree were to this process, how welcoming to researchers and to having researchers who had skills they didn't have come in and work closely with them. I think they were highly selective ... and they sought relationships that would endure over many years. And in these ways, they not only shaped the work but they shaped the people who worked with them. My own life ... has been profoundly shaped by the style of research and collaboration that Cree ... always demonstrated, encouraged, and taught."

15 A substantial body of work examines the JBNQA and its implementation. See Colin Scott (2001); Alan Penn (1995); Adrian Tanner (1999); Brian Craik (2004); Harvey Feit (2004); Monica Mulrennan and Colin Scott (2005); and Caroline Desbiens (2007).

16 The problems caused by the methyl-mercury in Hydro-Québec's reservoirs have yet to be resolved. As Harvey Feit (pers. comm., October 2009) explains, "What Hydro-Québec, Quebec and Canada haven't done is work through all the research to really understand the scientific processes and minimize the effects on the Cree of their actions and recommendations, rather than just to protect themselves. The fact that nearly forty years later we still don't have detailed scientific base information is testimony to the failure and to the control of research in the region."

17 See Miriam Atkinson and Monica Mulrennan (2009) for a discussion of Eastmain-1-A and Rupert Diversion Project, which underwent the required environmental assessment and review process. It was supported by Cree leadership at the regional level, but there was strong criticism of the project and its review at the community level.

WORKS CITED

Adams, Peter. 1988. "Bases for Field Research in Arctic and Subarctic Canada." *Arctic* 41 (1): 64–70.

–. 2009. "InfoNorth: The McGill Axel Heiberg Expeditions: Reconnaissance Year, 1959." *Arctic* 62 (3): 363–69.

Alison, R.M. 1976. "Occurrences of Duck Hybrids at James Bay." *The Auk* 93 (3): 643–44.

Anderson, Rudolph Martin. 1934. *Mammals of the Eastern Arctic and Hudson Bay-Arctic Flora*. Ottawa: J.O. Patenaude.

Atkinson, Miriam, and Monica E. Mulrennan. 2009. "Local Protest and Resistance to the Rupert Diversion Project, Northern Quebec." *Arctic* 62 (4): 468–80.

Awashish, Philip. 2005. "From Board to Nation Governance: The Evolution of Eeyou Tapay-tah-jeh-souwin (Eeyou Governance) in Eeyou Istchee." In *Reconfiguring Aboriginal-State Relations*, ed. M. Murphy, 165–83. Montreal and Kingston: McGill-Queen's University Press.

Baldwin, William K.W. 1953. "Plants from Two Small Island Habitats in James Bay." In *Annual Report of the National Museum of Canada for the Fiscal Year 1951–52*,

Bulletin No. 128, 154–67. Ottawa: Minister of Resources and Development, Department of Resources and Development Canada.

–. 1959. *Botanical Excursion to the Boreal Forest Region in Northern Quebec and Ontario [Guide Book]*. Ottawa: Queen's Printer.

Banfield, Alexander W.F., and J.S. Tener. 1958. "A Preliminary Study of the Ungava Caribou." *Journal of Mammalogy* 39 (4): 560–73.

Beals, Carlyle Smith, and D.A. Shenstone, eds. 1968. *Science, History and Hudson Bay*. Ottawa: Queen's Printer.

Bell, Robert. 1881. "On the Commercial Importance of Hudson's Bay, with Remarks on Recent Surveys and Investigations." *Proceedings of the Royal Geographical Society and Monthly Record of Geography* 3 (10): 577–86.

–. 1887. "Reef Structures on the Attawapiskat River." In *Geological and Natural History Survey of Canada, Annual Report* 2, 27G–28G. Montreal: William Foster Brown & Co.

Bergerud, Arthur T. 1967. "Management of Labrador Caribou." *Journal of Wildlife Management* 3 (4): 621–42.

Berkes, Fikret. 1980. *Productivity of Fisheries in the James Bay Area*. Montreal: Cree Regional Authority.

–. 1986. "Common Property Resources and Hunting Territories." *Anthropologica* 28 (1–2): 145–62.

–. 1995. "Indigenous Knowledge and Resource Management Systems: A Native Canadian Case Study from James Bay." In *Property Rights in a Social and Ecological Context: Case Studies and Design Applications*, ed. S. Hanna and M. Munasinghe, 99–109. Washington, DC: Beijer International Institute of Ecological Economics and the World Bank.

Bider, J. Roger. 1976. "The Distribution and Abundance of Terrestrial Vertebrates of the James and Hudson Bay Regions of Quebec." *Cahiers de géographie du Québec* 20 (50): 393–407.

Bocking, Stephen. 2007. "Science and Spaces in the Northern Environment." *Environmental History* 12 (4): 867–94.

–. 2009. "A Disciplined Geography: Aviation, Science, and the Cold War in Northern Canada, 1945–1960." *Technology and Culture* 50 (2): 265–90.

Bourassa, Robert. 1985. *L'énergie du nord: la force du Québec*. Montreal: Québec/Amérique.

Braun, Bruce. 2000. "Producing Vertical Territory: Geology and Governmentality in Late Victorian Canada." *Cultural Geographies* 7 (1): 7–46.

Bravo, Michael, and Sverker Sörlin. 2002. "Narrative and Practice – An Introduction." In *Narrating the Arctic: A Cultural History of Nordic Scientific Practices*, ed. Sverker Sörlin and Michael Bravo, 3–30. Canton, MA: Science History.

Byers, Michael. 2009. "John Diefenbaker's Northern Vision: 'A New Vision' Speech by John G. Diefenbaker at the Civic Auditorium, Winnipeg, 12 February 1958." Who Owns the Arctic? Arctic Sovereignty and International Relations. http://byers.typepad.com/arctic/2009/03/john-diefenbakers-Northern-vision.html.

Calvert, Wendy, Mitchell Taylor, Ian Stirling, Stephen Atkinson, Malcom A. Ramsay, Nicholas J. Lunn, Martyn Obbard, Cam Elliott, Gilles Lamontagne, and James A. Schaefer. 1998. "Research on Polar Bears in Canada 1993–1996." In *Polar Bears: Proceedings of the Twelfth Working Meeting of the IUCN/SSC Polar Bear Specialist Group, 3–7 February 1997, Oslo, Norway,* ed. A.E. Derocher, G.W. Garner, Nicholas J. Lunn, and Ø. Wiig, 69–91. Gland, Switzerland: IUCN.

Canadian Arctic Resources Committee, Environmental Committee of Sanikiluaq, and Rawson Academy of Aquatic Science. 1991. "Sustainable Development in the Hudson Bay/James Bay Bioregion." *Northern Perspectives* 19 (3). https://web.archive.org/web/20160714171517/http://www.carc.org/pubs/v19no3/2.htm.

Carlson, Hans. 2004. "A Watershed of Words: Litigating and Negotiating Nature in Eastern James Bay, 1971–75." *Canadian Historical Review* 85 (1): 63–84.

–. 2008. *Home Is the Hunter: The James Bay Cree and Their Land.* Vancouver: UBC Press.

Caron, Fabien. 1965. "Albert Peter Low et l'exploration du Québec-Labrador." *Cahiers de géographie du Québec* 9 (18): 169–82.

Castellano, Marlene Brant. 2004. "Ethics of Aboriginal Research." *Journal of Aboriginal Health* 1 (1): 98–114.

Chance, Norman A. 1968a. "The Cree Developmental Change Project: An Introduction." In *Conflict in Culture: Problems of Developmental Change among the Cree,* ed. Norman A. Chance, 1–10. Ottawa: Canadian Research Centre for Anthropology, Saint-Paul University.

–. 1968b. "Implications of Environmental Stress for Strategies of Developmental Change among the Cree." In *Conflict in Culture: Problems of Developmental Change among the Cree,* ed. Norman A. Chance, 11–32. Ottawa: Canadian Research Centre for Anthropology, Saint-Paul University.

Chitty, Dennis, and Charles Elton. 1937. "Canadian Arctic Wild Life Enquiry, 1935–36." *Journal of Animal Ecology* 6 (2): 368–85.

Chitty, Helen. 1948. "The Snowshoe Rabbit Enquiry, 1943–46." *Journal of Animal Ecology* 17 (1): 39–44.

Cloutier, Luce. 1987. "Quand le mercure s'élève trop haut à la Baie James ..." *Acta Borealia: A Nordic Journal of Circumpolar Societies* 4 (1): 5–23.

Comité de la Baie James sur le mercure. 1988. *Rapport d'activités.* Montreal: Comité de la Baie James sur le mercure.

Coon Come, Matthew. 1991. "Environmental Development, Indigenous Peoples and Governmental Responsibility." *International Legal Practitioner* 16: 108–11.

–. 2004. "Survival in the Context of Mega-resource Development: Experiences of the James Bay Crees and the First Nations of Canada." In *In the Way of Development: Indigenous Peoples, Life Projects and Globalization,* ed. M. Blaser, Harvey A. Feit, and Glenn McRae, 153–64. London: Zed Books.

Cooper, John M. 1929. "Canadian Indians Live by Hunting." *Science News-Letter* 16 (448): 286–87.

–. 1933. "The Cree Witiko Psychosis." *Primitive Man* 6 (1): 20–24.

–. 1938. *Snares, Deadfalls, and Other Traps of the Northern Algonquians and Northern Athapaskans.* Anthropological Series No. 5. Washington, DC: Catholic University of America Press.

Cormie, John A. 1917. "The Hudson Bay Route." *Geographical Review* 4 (1): 26–40.

Craik, Brian. 2004. "The Importance of Working Together: Exclusions, Conflicts and Participation in James Bay, Quebec." In *In the Way of Development: Indigenous Peoples, Life Projects and Globalization,* ed. M. Blaser, Harvey A. Feit, and Glenn McRae, 166–86. London: Zed Books.

Cree Board of Health and Social Services of James Bay. 1980. *The Cree Board of Health and Social Services of James Bay Report.* Waswanipi, QC: Cree Board of Health and Social Services of James Bay.

Cruikshank, Julie. 2005. *Do Glaciers Listen? Local Knowledge, Colonial Encounters, and Social Imagination.* Vancouver: UBC Press.

Curran, William Tees. 1907. *Glimpses of Northern Canada: The Land of Hidden Treasure.* Montreal: Cambridge.

Denton, David. 2001. "A Visit in Time: Ancient Places, Archaeology and Stories from the Elders of Wemindji." Nemaska, QC: Cree Regional Authority.

Desbiens, Caroline. 2004a. "Nation to Nation: Defining New Structures of Development in Northern Quebec." *Economic Geography* 80 (4): 351–66.

–. 2004b. "Producing North and South: A Political Geography of Hydro Development in Quebec." *Canadian Geographer* 48 (2): 101–19.

–. 2007. "'Water All around, You Cannot Even Drink': The Scaling of Water in James Bay/Eeyou Istchee." *Area* 39 (3): 259–67.

Dionne, Jean-Claude. 1976. "Miniature Mud Volcanoes and Other Injection Features in Tidal Flats, James Bay, Quebec." *Canadian Journal of Earth Sciences* 13 (3): 422–28.

–. 1980. "An Outline of the Eastern James Bay Coastal Environments." In *The Coastline of Canada,* ed. S.B. McCann, 311–38. Ottawa: Energy, Mines and Resources Canada.

Ducruc, Jean-Pierre, Richard Zarnovican, Vincent Gérardin, and Michel Jurdant. 1976. "Les régions écologiques du territoire de la Baie James: caractéristiques dominantes de leur couvert végétal." *Cahiers de géographie du Québec* 20 (50): 365–91.

Dutilly, Arthème, and Ernest Lepage. 1948. *Coup d'oeil sur la flore subarctique du Québec de la Baie James au Lac Mistassini.* Washington, DC: Catholic University of America.

Dutilly, Arthème, Ernest Lepage, and Maximilian Duman. 1958. *Contribution à la flore des îles et du versant oriental de la baie James.* Washington, DC: Catholic University of America Press.

Eamer, Claire. 2005. *Research Licensing in Northern Canada: An Overview.* Edmonton, AB: Canadian IPY Secretariat.

Elberg, Nathan, Jacqueline Hyman, K. Hyman, and Richard F. Salisbury. 1972. *Not by Bread Alone: The Use of Subsistence Resources among James Bay Cree.* Montreal:

Programme in the Anthropology of Development, Department of Sociology and Anthropology, McGill University.

Elton, Charles, and Mary Nicholson. 1942. "The Ten-Year Cycle in Numbers of the Lynx in Canada." *Journal of Animal Ecology* 11 (2): 215–44.

Farish, Matthew, and P. Whitney Lackenbauer. 2009. "High Modernism in the Arctic: Planning Frobisher Bay and Inuvik." *Journal of Historical Geography* 35 (3): 517–44.

Feit, Harvey A. 1969. "Mistassini Hunters of the Boreal Forest: Ecosystem Dynamics and Multiple Subsistence Patterns." Master's thesis, McGill University.

–. 1978. "Waswanipi Realities and Adaptations: Resource Management and Cognitive Structure." PhD thesis, McGill University.

–. 1986. "Hunting and the Quest for Power: The James Bay Cree and Whitemen in the Twentieth Century." In *Native Peoples: The Canadian Experience*, ed. R. Bruce Morrison and Roderick Wilson, 171–207. Toronto: McClelland-Stewart.

–. 1988. "Forms of Knowing and Managing Northern Wildlife." In *Traditional Knowledge and Renewable Resource Management in Northern Regions*, ed. M. Freeman and L. Carbyn, 72–91. Edmonton, AB: Boreal Institute for Northern Studies.

–. 1991. "Gifts of the Land: Hunting Territories, Guaranteed Incomes and the Construction of Social Relations in James Bay Cree Society." *Senri Ethnological Studies* (Osaka, Japan, National Museum of Ethnology) 30: 223–68.

–. 1995. "Hunting and the Quest for Power: The James Bay Cree and Whitemen in the 20th Century." In *Native Peoples: The Canadian Experience*, ed. R.B. Morrison and C.R. Wilson, 181–223. Toronto: McClelland and Stewart.

–. 2004. "James Bay Crees' Life Projects and Politics: Histories of Place, Animal Partners and Enduring Relationships." In *In the Way of Development: Indigenous Peoples, Life Projects and Globalization*, ed. M. Blaser, Harvey A. Feit, and Glenn McRae, 92–110. London: Zed Books.

–. 2007. "Myths of the Ecological Whitemen: Histories, Science and Rights in North American-Native American Relations." In *Native Americans and the Environment: Perspectives on the Ecological Indian*, ed. M.E. Harkin and D.R. Lewis, 52–92. Lincoln: University of Nebraska Press.

Feit, Harvey A., and Colin H. Scott. 2004. "Applying Knowledge: Anthropological Praxis and Public Policy." In *Ethnography and Development: The Work of Richard F. Salisbury*, ed. M. Silverman, 233–51. Montreal: McGill University Libraries.

Fikkan, Anne, Gail Osherenko, and Alexander Arikainen. 1993. "Polar Bears: The Importance of Simplicity." In *Polar Politics: Creating International Environmental Regimes*, ed. O.R. Young and G. Osherenko, 96–151. Ithaca: Cornell University Press.

Finkelstein, Max, and James Stone. 2004. *Paddling the Boreal Forest: Rediscovering A.P. Low.* Toronto: Natural Heritage Books.

Flaherty, Robert J. 1918a. "The Belcher Islands of Hudson Bay: Their Discovery and Exploration." *Geographical Review* 5 (6): 433–58.

–. 1918b. "Two Traverses across Ungava Peninsula, Labrador." *Geographical Review* 6 (2): 116–32.

–. 1924. *My Eskimo Friends, 'Nanook of the North.'* Garden City, NY: Doubleday Page.

Flannery, Regina. 1935. "The Position of Woman among the Eastern Cree." *Primitive Man* 8 (4): 81–86.

–. 1939. "The Shaking-Tent Rite among the Montagnais of James Bay." *Primitive Man* 12 (1): 11–16.

–. 1995. *Ellen Smallboy: Glimpses of a Cree Woman's Life.* Montreal and Kingston: McGill-Queen's University Press.

Fleming, Sandford, Charles Horetzky, John Macoun, Marcus Smith, James H. Rowan, Walter Moberly, Henry Spencer Palmer, Alexander Mackenzie, Alfred R.C. Selwyn, and George Vancouver. 1874. *Canadian Pacific Railway Exploratory Survey Report: Report of Progress on the Explorations and Surveys up to January, 1874.* Ottawa: Maclean, Roger.

Francis, Daniel, and Toby Elaine Morantz. 1983. *Partners in Furs: A History of the Fur Trade in Eastern James Bay, 1600–1870.* Montreal and Kingston: McGill-Queen's University Press.

Fraser, Dylan J., Thomas Coon, Michael R. Prince, Rene Dion, and Louis Bernatchez. 2006. "Integrating Traditional and Evolutionary Knowledge in Biodiversity Conservation: A Population Level Case Study." *Ecology and Society* 11 (2): Article 4. https://www.ecologyandsociety.org/vol11/iss2/art4/.

Freeman, Milton M.R. 1983. "George Weetaltuk (ca. 1862–1956)." *Arctic* 36 (2): 214–15.

Freeman, T.N., and C.R. Twinn. 1954. "Present Trends and Future Needs of Entomological Research in Northern Canada." *Arctic* 7 (3): 275–83.

Gantcheff, Georges, R.J. McCormack, and Armand Couture. 1982a. *Environmental Studies, James Bay Territory: 1972–1979 Summary Report.* Montreal: Société de développement de la Baie James.

–. 1982b. *Studies of the Marine Environment and Its Fauna.* Montreal: Société de développement de la Baie James.

George, Peter, Fikret Berkes, and Richard J. Preston. 1996. "Envisioning Cultural, Ecological and Economic Sustainability: The Cree Communities of the Hudson and James Bay Lowland, Ontario." *Canadian Journal of Economics/Revue canadienne d'Économique* 29 (l): 356–60.

Gérardin, Vincent. 1980. *L'inventaire du capital-nature du territoire de la Baie-James: les régions écologiques et la végétation des sols minéraux.* Ottawa: Environnement Canada, Lands Directorate, Société de développement de la Baie James.

Gouvernement du Québec (Ministère des Ressources naturelles et de la Faune). 2011. "Building Northern Quebec Together: The Project of a Generation 2011." http://www.environnement.gouv.qc.ca/communiques_en/2011/c20110812-plannord.htm.

Grand Council of the Crees. 1981. *Health Crisis amongst the James Bay Cree of Northern Quebec, Canada.* Val d'Or: Grand Council of the Crees of Quebec.

–. 2011. "A Cree Vision of the Plan Nord." https://www.cngov.ca/wp-content/uploads/2018/03/cree-vision-of-plan-nord.pdf.

–. 2012. "Agreement on Governance in the Eeyou Istchee James Bay Territory between the Crees of Eeyou Istchee and the gouvernement du Québec." https://www.cngov.ca/governance-structure/legislation/agreements/.

–. N.d. "Cree Legal Struggles against the Great Whale Project." Grand Council of the Crees, Cree Regional Authority. http://www.gcc.ca/archive/article.php?id=3 *(site now discontinued)*.

Gussow, W.C. 1953. "Silurian Reefs of James Bay Lowland, Ontario: Geological Notes." *AAPG Bulletin* 37 (10): 2422–24.

Hanks, Christopher. 1983. "An Ethnoarchaeological Approach to the Seasonality of Historic Cree Sites in Central Québec." *Arctic* 36 (4): 350–55.

Hardy, Léon. 1977. "La déglaciation et les épisodes lacustre et marin sur le versant québécois des basses terres de la baie de James." *Géographie physique et Quaternaire* 31: 261–73.

Hare, F. Kenneth. 1950. "Climate and Zonal Divisions of the Boreal Forest Formation in Eastern Canada." *Geographical Review* 40 (4): 615–35.

–. 1952. "Review." *Geographical Review* 42 (1): 165–66.

–. 1964. "New Light from Labrador-Ungava." *Annals of the Association of American Geographers* 54 (4): 459–76.

Hayeur, Gaëtan. 2001. *Summary of Knowledge Acquired in Northern Environments from 1970 to 2000*. Montreal: Hydro-Québec. http://www.hydroquebec.com/data/developpement-durable/pdf/Summary_Knowledge_Northern_Environment_1970_2000.pdf.

Hillaire-Marcel, Claude. 1976. "La déglaciation et le relèvement isostatique sur la côte est de la Baie d'Hudson." *Cahiers de géographie de Québec* 20 (50): 185–220.

Honigmann, John Joseph. 1961. *Foodways in a Muskeg Community: An Anthropological Report on the Attawapiskat Indians*. Ottawa: Northern Co-ordination and Research Centre.

–. 1962. *Social Networks in Great Whale River: Notes on an Eskimo, Montagnais-Naskapi, and Euro-Canadian Community*. Ottawa: Department of Northern Affairs and National Resources.

Hornig, James F., ed. 1999. *Social and Environmental Impacts of the James Bay Hydroelectric Project*. Montreal and Kingston: McGill-Queen's University Press.

Hustich, Ilmari. 1949. "On the Forest Geography of the Labrador Peninsula. A Preliminary Synthesis." *Acta Geographica* 10 (2): 1–63.

–. 1950. "Notes on the Forests of the East Coast of Hudson Bay and James Bay." *Acta Geographica* 11 (1): 2–83.

–. 1957. "On the Phytogeography of the Subarctic Hudson Bay Lowland." *Acta Geographica* 16 (1): 1–48.

Ingram, R. Grant, and V.H. Chu. 1987. "Flow around Islands in Rupert Bay: An Investigation of the Bottom Friction Effect." *Journal of Geophysical Research* 92 (13): 14521–33.

Ingram, R. Grant, B.F. d'Anglejan, S. Lepage, and D. Messier. 1986. "Changes in Current Regime and Turbidity in Response to a Freshwater Pulse in the Eastmain Estuary." *Estuaries and Coasts* 9 (4): 320–25.

IUCN. 1970. *Proceedings of the 2nd Working Meeting of the Polar Bear Specialists.* IUCN Publications New Series Supplementary Paper No. 29. Morges, Switzerland: International Union for Conservation of Nature and Natural Resources.

Jackson, C.F. 1938. "Notes on the Mammals of Southern Labrador." *Journal of Mammalogy* 19 (4): 429–34.

Jacqmain, Hugo, Louis Belanger, Susanne Hilton, and Luc Bouthillier. 2007. "Bridging Native and Scientific Observations of Snowshoe Hare Habitat Restoration after Clearcutting to Set Wildlife Habitat Management Guidelines on Waswanipi Cree Land." *Canadian Journal of Forest Research/Revue canadienne de recherche forestière* 37 (3): 530–39.

Jacqmain, Hugo, Christian Dussault, Réhaume Courtois, and Louis Belanger. 2008. "Moose Habitat Relationships: Integrating Local Cree Native Knowledge and Scientific Findings in Northern Quebec." *Canadian Journal of Forest Research/ Revue canadienne de recherché forestière* 38 (12): 3120–32.

Jenkins, Dale W., and Kenneth Knight. 1950. "Ecological Survey of the Mosquitoes of Great Whale River, Quebec (Diptera, Culicidae)." *Proceedings of the Entomological Society of Washington* 52: 209–23.

–. 1952. "Ecological Survey of the Mosquitoes of Southern James Bay." *American Midland Naturalist* 47 (2): 456–68.

Jonkel, Charles J. 1970. "Some Comments on Polar Bear Management." *Biological Conservation* 2 (2): 115–19.

Josephson, P.R. 2004. *Resources under Regimes: Technology, Environment, and the State.* Cambridge, MA: Harvard University Press.

Kenyon, Walter Andrew. 1986. *The History of James Bay, 1610–1686: A Study in Historical Archaeology.* Toronto: Royal Ontario Museum.

Kerr, A.J. 1950. *Subsistence and Social Organization in a Fur Trade Community.* Ottawa: National Committee for Community Health Studies.

Kindle, E.M. 1925. "The James Bay Coastal Plain: Notes on a Journey." *Geographical Review* 15 (2): 226–36.

Knight, Rolf. 1967. *Ecological Factors in Changing Economy and Social Organization among the Rupert House Cree.* Anthropology Papers, National Museum of Canada No. 15. Ottawa: Queen's Printer.

Kranck, E.H. 1950. "On the Geology of the East Coast of Hudson Bay and James Bay." *Acta Geographica (Helsinki)* 11 (2): 1–71.

Lackenbauer, P. Whitney, and Matthew Farish. 2007. "The Cold War on Canadian Soil: Militarizing a Northern Environment." *Environmental History* 12 (4): 920–50.

Laliberté, Marcel. 1982. *Les schèmes d'établissement cris de la Baie James (contribution à l'étude des sites historiques et préhistoriques).* Quebec City: Direction générale du patrimoine.

Lambert, Yves, and J.J. Dodson. 1982. "Structure et rôle des facteurs physiques dans le maintien des communautés estuariennes de poissons de la baie James." *Le Naturaliste canadien* 109: 8–15.

La Rusic, Ignatius E. 1979. *The Income Security Program for Cree Hunters and Trappers: A Study of the Design, Operation, and Initial Impacts of the Guaranteed Annual Income Programme Established under the James Bay and Northern Quebec Agreement.* Brief Communications Series No. 43, Programme in the Anthropology of Development, McGill University. http://publications.gc.ca/collections/collection _2017/aanc-inac/R5-160-1978-eng.pdf.

Lee, Hulbert A. 1968. "Marine Geology." In *Science, History and Hudson Bay,* ed. C.S. Beals and D.A. Shenstone, 503–42. Ottawa: Queen's Printer.

Leith, Charles K., and Alexander T. Leith. 1910. "An Algonkian Basin in Hudson Bay: A Comparison with the Lake Superior Basin." *Economic Geology* 5 (3): 227–46.

Levere, Trevor. 1993. *Science and the Canadian Arctic: A Century of Exploration, 1818– 1918.* New York: Cambridge University Press.

Lindsay, Debra, 1993. *The Modern Beginnings of Subarctic Ornithology: Northern Correspondence with the Smithsonian Institution, 1856–68.* Winnipeg: Manitoba Record Society.

Low, Albert P. 1888. *Report on Explorations in James' Bay and Country East of Hudson Bay Drained by the Big, Great Whale and Clearwater Rivers.* Montreal: W. Foster Brown.

–. 1896. *Report on Explorations in the Labrador Peninsula along the East Main, Koksoak, Hamilton, Manicuagan and Portions of Other Rivers in 1892–93–94–95.* Ottawa: Geological Survey of Canada.

MacLulich, Duncan. 1937. *Fluctuations in the Number of the Varying Hare (Lepus americanus).* Toronto: University of Toronto Press.

Macpherson, Andrew H. 1964. "Canadian Arctic Biology: Terrestrial." *BioScience* 14 (5): 14–17.

Malaurie, Jean, and Jacques Rousseau. 1964. *Le Nouveau-Québec: Contribution à l'étude de l'occupation humaine.* Paris: Mouton.

Manning, Thomas H. 1951. "Remarks on the Tides and Driftwood Strand Lines along the East Coast of James Bay." *Arctic* 4 (2): 122–30.

–. 1952. *Birds of the West James Bay and Southern Hudson Bay Coasts.* Ottawa: Department of Resources and Development.

–. 1971. *Geographical Variation in the Polar Bear Ursus Maritimus Phipps.* Ottawa: Information Canada.

–. 1981. *Birds of the Twin Islands, James Bay, NWT, Canada.* Ottawa: National Museum of Natural Sciences, National Museums of Canada.

Marshall, Susan. 1984. "The Articulation of the Biomedical and the Cree Medical Systems." Master's thesis, McGill University.

Marshall, Susan, and Cree Regional Authority. 1989. *Healing Ourselves, Helping Ourselves: The Medicinal Use of Plants and Animals by the People of Waskaganish.* Val d'Or: Cree Regional Authority.

Martini, Irenio Peter, ed. 1986. *Canadian Inland Seas.* Amsterdam: Elsevier.

Martini, Irenio Peter, R.I.G. Morrison, W.A. Glooschenko, and R. Protz. 1980. "Coastal Studies in James Bay, Ontario." *Geoscience Canada* 7 (1): 11–21.

McCutcheon, Sean. 1991. *Electric Rivers: The Story of the James Bay Project.* Montreal: Black Rose Books.

McDonald, Miriam, Lucassie Arragutainaq, and Zack Novalinga. 1997. *Voices from the Bay.* Ottawa: Canadian Arctic Resources Committee and Environmental Committee of the Municipality of Sanikiluaq.

Melvill, C.D., Napoleon A. Comeau, and Arthur Reginald Marsden Lower. 1915. *Reports on Fisheries Investigations in Hudson and James Bays and Tributary Waters in 1914.* CIHM/ICMH Microfiche Series No. 81865. Ottawa: J. de L. Taché.

Morantz, Toby. 1983. *An Ethnohistoric Study of Eastern James Bay Cree Social Organization, 1700–1850.* Ottawa: National Museums of Canada.

–. 2002. *The White Man's Gonna Getcha: The Colonial Challenge to the Crees in Quebec.* Montreal and Kingston: McGill-Queen's University Press.

Mulrennan, Monica E. 1998. *A Casebook of Environmental Issues in Canada.* New York: John Wiley and Sons.

Mulrennan, Monica E., Rodney Mark, and Colin Scott. 2012. "Revamping Community-Based Conservation through Participatory Research." *Canadian Geographer* 56 (2): 243–59.

Mulrennan, Monica E., and Colin H. Scott. 2005. "Co-management: An Attainable Partnership? Two Cases from James Bay, Northern Quebec and Torres Strait, Northern Queensland." *Anthropologica* 47 (2): 197–213.

Mulvihill, Peter R., Douglas C. Baker, and William R. Morrison. 2001. "A Conceptual Framework for Environmental History in Canada's North." *Environmental History* 6 (4): 611–26.

Murdoch, John. 1975. *The Old Way.* Wemindji, QC: Cree Way Project.

Nakashima, Doug, and Marie Roué. 1994. *Great Whale Environmental Assessment Community Consultation: Final Report for Chisasibi and Whapmagoostui.* Nemaska: Grand Council of the Crees of Québec/Cree Regional Authority.

Nasr, Wren, and Colin H. Scott. 2010. "The Politics of Indigenous Knowledge in Environmental Assessment: James Bay Crees and Hydro-Electric Projects." In *Cultural Autonomy: Frictions and Connections,* ed. Petra Rethmann, Imre Szeman, and Will Coleman, 132–55. Vancouver: UBC Press.

Niezen, Ronald. 1993. "Power and Dignity: The Social Consequences of Hydro-Electric Development for the James Bay Cree." *Canadian Review of Sociology and Anthropology* 30 (4): 510–29.

–. 1998. *Defending the Land: Sovereignty and Forest Life in James Bay Cree Society.* Boston: Allyn and Bacon.

Ohmagari, Kayo, and Fikret Berkes. 1997. "Transmission of Indigenous Knowledge and Bush Skills among the Western James Bay Cree Women of Subarctic Canada." *Human Ecology* 25 (2): 197–222.

Payette, Serge. 1976. "Hommage à Ernest Lepage, botaniste et explorateur du Nouveau-Québec." *Cahiers de géographie de Québec* 20 (50): 179–81.

Penn, Alan. 1978. *The Distribution of Mercury, Selenium and Certain Heavy Metals in Major Fish Species from Northern Quebec.* Ottawa: Fisheries and Environment Canada.

–. 1995. "The James Bay and Northern Quebec Agreement: Natural Resources, Public Lands, and the Implementation of a Native Land Claim Settlement." Report prepared for the Royal Commission on Aboriginal Peoples.

Peters, Evelyn J. 1992. "Protecting the Land under Modern Land Claims Agreements: The Effectiveness of the Environmental Regime Negotiated by the James Bay Cree in the James Bay and Northern Quebec Agreement." *Applied Geography* 12 (2): 133–45.

–. 1999. "Native People and the Environmental Regime in the James Bay and Northern Quebec Agreement." *Arctic* 52 (4): 395–410.

Pothier, Roger. 1968. "Community Complexity and Indian Isolation." In *Conflict in Culture: Problems of Developmental Change among the Cree,* ed. Norman A. Chance, 33–46. Ottawa: Canadian Research Centre for Anthropology, Saint Paul University.

Preble, Edward Alexander. 1902. *A Biological Investigation of the Hudson Bay Region.* Washington, DC: Government Printing Office.

Preston, Richard J. 1975. "A Survey of Ethnographic Approaches to the Eastern Cree-Montagnais-Naskapi." *Canadian Review of Sociology/Revue Canadienne de Sociologie* 12 (3): 267–77.

–. 1976. *Cree Narrative: Expressing the Personal Meanings of Events.* Mercury Series, Canadian Ethnology Service Paper No. 30. Ottawa: Museum of Man.

–. 1979. "The Cree Way Project: An Experiment in Grass-Roots Curriculum Development." *Papers of the 10th Algonquian Conference/Actes du Congrès des Algonquinistes* 10: 92–101.

–. 2002. *Cree Narrative: Expressing the Personal Meanings of Events.* Carleton Library Series No. 197. Montreal and Kingston: McGill-Queen's University Press.

Quebec. 2009. "Plan Nord – For a Socially Responsible and Sustainable Form of Economic Development." Quebec City. Working document. http://www.plannord. gouv.qc.ca/english/documentation.asp *(site now discontinued).*

Richardson, Boyce. 1977. *Strangers Devour the Land: The Cree Hunters of the James Bay Area versus Premier Bourassa and the James Bay Development Corporation.* Toronto: Macmillan of Canada.

Rosenberg, D., Fikret Berkes, R. Bodaly, R. Hecky, and C. Kelly. 1997. "Large-Scale Impacts of Hydroelectric Development." *Environmental Reviews* 5 (1): 27–54.

Rousseau, Jacques. 1948. "Bataille de sextants autour du lac Mistassini." Ed. Université de Montréal. *L'Action Universitaire* 14 (2): 99–116.

–. 1952. "Les zones biologiques de la péninsule Québec – Labrador et l'hemiarctique." *Canadian Journal of Botany/Revue Canadienne de Botanique* 30 (4): 436–74.

–. 1954. *Essai bibliographique sur la région du lac Mistassini.* Montreal: s.n.

–. 1968. "The Vegetation of the Quebec-Labrador Peninsula between 55 and 60 N." *Naturaliste Canadienne (Québec)* 95: 469–563.

Roy, Dominique, Roger Lemire, and Danielle Messier. 1985. *Évolution du mercure dans la chair des poissons.* Montreal: Société d'énergie de la Baie James Direction ingénierie et environnement.

Salisbury, Richard Frank. 1986. *A Homeland for the Cree: Regional Development in James Bay, 1971–1981.* Montreal and Kingston: McGill-Queen's University Press.

Sandlos, John. 2007. *Hunters at the Margin: Native People and Wildlife Conservation in the Northwest Territories*. Vancouver: UBC Press.

Scott, Colin H. 1979. "Modes of Production and Guaranteed Annual Income in James Bay Cree Society." Master's thesis, McGill University.

–. 1989. "Knowledge Construction among the Cree Hunters: Metaphors and Literal Understanding." *Journal de la Société des Américanistes* 75 (1): 193–208.

–. 1996. "Science for the West, Myth for the Rest? The Case of James Bay Cree Knowledge Construction." In *Naked Science: Anthropological Inquiry into Boundaries, Power, and Knowledge*, ed. L. Nader, 69–86. London: Routledge.

–. 2001. "On Autonomy and Development." In *Aboriginal Autonomy and Development in Northern Quebec and Labrador*, ed. C. Scott, 3–20. Vancouver: UBC Press.

Scott, Colin H., and Kreg Ettenger. 1994. *Great Whale Environmental Assessment Community Consultation: Final Report for Wemindji and Eastmain*. Nemaska: Grand Council of the Crees of Québec/Cree Regional Authority.

Scott, Colin H., and Harvey A. Feit. 1992. *Income Security for Cree Hunters: Ecological, Social and Economic Effects*. Montreal: McGill Programme in the Anthropology of Development, McGill University.

Scott, James C. 1998. *Seeing Like a State: How Certain Schemes to Improve the Human Condition Have Failed*. New Haven, CT: Yale University Press.

Sindell, Peter S. 1968. "Some Discontinuities in the Enculturation of Mistassini Cree Children." In *Conflict in Culture: Problems of Developmental Change among the Cree*, ed. Norman A. Chance, 83–12. Ottawa: Canadian Research Centre for Anthropology, Saint-Paul University.

Sindell, Peter S., and Ronald M. Wintrob. 1969. "Cross-Cultural Education in the North and Its Implications for Personal Identity: The Canadian Case." http://www.eric.ed.gov/ERICWebPortal/contentdelivery/servlet/ERICServlet?accno=ED040776.

Sistili, B., M. Metatawabin, G. Iannucci, and Leonard J.S. Tsuji. 2006. "An Aboriginal Perspective on the Remediation of Mid-Canada Radar Line Sites in the Subarctic: A Partnership Evaluation." *Arctic* 59 (2): 142–54.

Skinner, Alanson. 1912. *Notes on the Eastern Cree and Northern Saulteaux*. New York: The Trustees of the American Museum of Natural History.

Speck, Frank G. 1915. "Some Naskapi Myths from Little Whale River." *Journal of American Folklore* 28 (107): 70–77.

–. 1923. "Mistassini Hunting Territories in the Labrador Peninsula." *American Anthropologist* 25 (4): 452–71.

–. 1935. *Naskapi, the Savage Hunters of the Labrador Peninsula*. Norman: University of Oklahoma Press.

Spence, John, and G. Spence. 1972. *Ecological Considerations of the James Bay Project*. Stockholm: Report for the UN Conference on the Human Environment.

Stewart, D.B., M.J. Dunbar, and L.M.J. Bernier. 1993. *Marine Natural Areas of Canadian Significance in the James Bay Marine Region*. Winnipeg, MB: Canadian Parks Service.

Stirling, Ian, and Claire L. Parkinson. 2006. "Possible Effects of Climate Warming on Selected Populations of Polar Bears (*Ursus maritimus*) in the Canadian Arctic." *Arctic* 59 (3): 261–75.

Stocek, Christine, and Rodney Mark. 2009. "Indigenous Research and Decolonizing Methodologies: Possibilities and Opportunities." In *Indigenous Knowledges, Development and Education*, ed. Jonathan Langdon, 73–96. Rotterdam: Sense.

Tanner, Adrian. 1968. "Occupation and Life Style in Two Minority Communities." In *Conflict in Culture: Problems of Developmental Change among the Cree*, ed. Norman A. Chance, 47–68. Ottawa: Canadian Research Centre for Anthropology, Saint-Paul University.

–. 1979. *Bringing Home Animals: Religious Ideology and Mode of Production of the Mistassini Cree Hunters*. St. John's: Institute of Social and Economic Research, Memorial University of Newfoundland.

–. 1981. *Establishing a Native Language Education Policy: A Study Based on the Views of Cree Parents in the James Bay Region of Quebec*. Val d'Or: Grand Council of the Crees of Quebec.

–. 1988. "The Significance of Hunting Territories Today." In *Native People, Native Lands: Canadian Indian, Inuit and Metis*, ed. B.A. Cox, 60–74. Ottawa: Carleton University Press.

–. 1999. "Culture, Social Change, and Cree Opposition to James Bay Hydroelectric Development." In *Social and Environmental Impacts of the James Bay Hydroelectric Project*, ed. James F. Hornig, 121–40. Montreal and Kingston: McGill-Queen's University Press.

Tsuji, Leonard J.S., Bruce C. Wainman, Ian D. Martin, Jean-Philippe Weber, Celine Sutherland, and Evert Nieboer. 2006. "Abandoned Mid-Canada Radar Line Sites in the Western James Bay Region of Northern Ontario, Canada: A Source of Organochlorines for First Nations People?" *Science of the Total Environment* 370 (2): 452–66.

Vincent, Sylvie, and Garry Bowers. 1988. *Baie James et nord québécois, dix ans après/ James Bay and Northern Quebec: Ten Years After*. Montreal: Recherches Amerindiennes au Québec.

Vodden, Christy. 1992. *No Stone Unturned: The First 150 Years of the Geological Survey of Canada*. https://web.archive.org/web/20081207042708/http://gsc.nrcan.gc.ca/hist/150_e.php.

Warner, Stanley. 1999. "The Cree People of James Bay: Assessing the Social Impact of Hydroelectric Dams and Reservoirs." In *Social and Environmental Impacts of the James Bay Hydroelectric Project*, ed. James F. Hornig, 93–120. Montreal and Kingston: McGill-Queen's University Press.

Weinstein, Martin S. 1976. *What the Land Provides: An Examination of the Fort George Subsistence Economy and the Possible Consequences on It of the James Bay Hydroelectric Project*. Montreal: Grand Council of the Crees.

Willems-Braun, Bruce. 1997. "Buried Epistemologies: The Politics of Nature in (Post) Colonial British Columbia." *Annals of the Association of American Geographers* 87 (1): 3–31.

Wintrob, Ronald M. 1968. "Acculturation, Identification, and Psychopathology among Cree Indian Youth." In *Conflict in Culture: Problems of Developmental Change among the Cree,* ed. Norman A. Chance, 93–104. Ottawa: Canadian Research Centre for Anthropology, Saint Paul University.

Wintrob, Ronald M., and Peter S. Sindell. 1968. "Education and Identity Conflict among Cree Indians." Paper presented at the Annual Meeting of the American Psychiatric Association, Miami Beach, FL, May 1969.

Wynn, Graeme. 2007. *Canada and Arctic North America: An Environmental History.* Santa Barbara: ABC-CLIO.

Young, Oran R. 1999. "Introduction to the Issues." In *Social and Environmental Impacts of the James Bay Hydroelectric Project,* ed. James F. Hornig, 3–18. Montreal and Kingston: McGill-Queen's University Press.

Zarnovican, Richard. 1978. *Les tourbières de la Baie James: la végétation et les sols.* Ottawa: Service canadien des forêts, Direction régionale des terres.

Zaslow, Morris. 1975. *Reading the Rocks: The Story of the Geological Survey of Canada, 1842–1972.* Toronto: Macmillan of Canada in association with the Department of Energy, Mines and Resources, and Information Canada.

PART 2

What to Protect

5

Shoreline Displacement and Human Adaptation in Eastern James Bay
A Six-Thousand-Year Perspective

FLORIN PENDEA, ANDRE COSTOPOULOS,
GAIL CHMURA, COLIN D. WREN, JENNIFER BRACEWELL,
SAMUEL VANEECKHOUT, JARI OKKONEN, EVA HULSE,
and DUSTIN KEELER

On a map, eastern James Bay appears to be nothing more than a typical subarctic coastal plain, sparsely interrupted by a few rugged stretches of rocky hills. This image hints little at its amazing landscape history and the extraordinary socio-environmental system that it comprises. Bringing together peoples and landscapes, this system was forged through continuous transformation in which shoreline displacement, climate change, fire, and human activity played important roles. In this chapter, we present the main events characterizing the landscape evolution of eastern James Bay (Eeyou Istchee). Using paleo-ecological and archaeological evidence found deep in peat bogs and scattered across the mossy soils, and in conversation with our Cree partners, we reconstruct former landscapes and lay the groundwork for a better understanding of the human societies that entered this area as early as six thousand years ago.

PALEOGEOGRAPHY

The present configuration of eastern James Bay and its landscape is the result of a series of events that began about ten thousand years ago. At that time, the earth's climate started to warm, which caused the rapid disappearance of an immense ice sheet – the Laurentide Ice Sheet – that had covered most of northern North America for tens of thousands of years.[1] Deglaciation did not happen all at once. The ice advanced and melted again a few times until it finally disappeared around six thousand years ago. At first, the Laurentide Ice Sheet broke into two masses, along a line that follows today's Harricana River. Its eastern half, now named the Nouveau Québec glacier, occupied the eastern sector of eastern James Bay and retreated progressively northeast (Hardy 1982). The western half, called the Hudson glacier, occupied the western edge of eastern James Bay, and unlike the Nouveau Québec glacier, it re-advanced at least twice before it finally retreated to the northwest.

As the two glaciers melted, their waters formed a very large glacial lake (Lake Ojibway) between about 10,000 and 8,500 years ago. The re-advance of the ice changed the extent of the lake at least twice during its brief existence. Lake Ojibway was formed because the glacier's meltwater was held back by a large northern ice dam. About 8,500 years ago, this dam broke, and the released fresh water mixed with the salt water of the Tyrrell Sea (now Hudson Bay) to the north. After Lake Ojibway suddenly emptied, the waters of the Tyrrell Sea flooded all ice-free lowland areas of eastern James Bay up to the ridge of the Sakami Moraine (Hardy 1982).

As soon as the water reached the Sakami ridge, it began to move back. The heavy glaciers were extremely thick (up to three kilometres), and they had pressed down on the earth's crust. When the ice melted, the earth was released from its weight and began to rise like a metal spring. The shoreline started its long trip away from the Sakami ridge and toward its present location. The release and rise of the earth's crust, which is called isostatic rebound, continues today.

As our Cree partners put it, the land grows. They experience this growth, even today, during a single lifetime. It is not unusual to be surveying or collecting samples through a boggy area and to hear stories about how an elder used to fish there with his grandfather. In the past, the growth was even faster than it is now and would have affected people's lives more strongly (Pendea et al. 2010).

The speed at which the land grows in eastern James Bay is mainly controlled by two forces. The isostatic rebound raises the land and moves the shore westward. The second force is the change in the level of the sea itself, called the eustatic sea level rise. Since the end of the Ice Age, worldwide sea levels have risen. When the ice was first melting, the glaciers suddenly released large quantities of water, and sea levels rose rapidly. The speed of the shore's movement away from the Sakami ridge at any given moment depends on the balance between the rising land (isostatic rebound) and the changing sea level (eustatic sea level rise). During the last eight thousand years (Table 5.1), sea levels have risen approximately five metres. Most of this increase took place between eight and six thousand years ago (Peltier 2002). During the same period, the land in eastern James Bay rose more than three hundred metres (Hillaire-Marcel 1980). Thus, most of the shoreline displacement in the area since the end of the Ice Age is due to isostatic rebound (Tarnocai 1982; Klinger and Short 1996; Glaser, Hansen, et al. 2004; Glaser, Siegel, et al. 2004).[2]

But how do we know when the shore was at various points on the land? For example, how do we know that it was at the Sakami ridge 8,500 years ago? Using radiocarbon (Carbon-14), we can date organic materials found in the soil where there is evidence of a shoreline. The majority of studies conducted on eastern James and Hudson Bays date marine shells from places that are now far from the coast.[3] Despite the widespread abundance of the shells, they are not the best material to use in age calculations of old coastlines. Shells and driftwood – another material commonly used to date shorelines – can be moved around by erosion, and their relation to a particular sea level is often uncertain. In other words, we are not certain that the spot where we found a shell actually was a shore when it came to rest there.

Another problem is that shells tend to give radiocarbon ages that are much older than their real ages. During their lifetime, animals that produce the shells take in particles of very old carbonates from the sea water. When radiocarbon analyses are done, they will indicate an older age than that of the organism. How to correct for this is still subject to much scientific debate, and there are many questions about the validity of some shoreline displacement reconstructions.

To produce a more reliable land uplift curve for Wemindji territory, we tried to use more suitable datable material that had a closer and more obvious relationship with the presence of the shore at a particular place in the past. Our sample collection sites ranged from locations near the present coastline to the James Bay Highway, which lies near Paakumshumwaau (Old Factory Lake), about a hundred kilometres inland (Figure 5.1). We concentrated on a series of ancient bays or isolation basins (Shennan et al. 2000) in the Paakumshumwaashitkw and Maatuskaau (Poplar River) watersheds. These isolation basins form when a body of water that was

TABLE 5.1
Chronology of Tyrrell Sea shoreline evolution during deglaciation in eastern Hudson/James Bay

Study	Rate of emergence				Max Tyrrell Sea level				
	Period (ka cal BP)	Past (m/century)	Present (m/century)	Dating method	Radiocarbon age (BP)	Corrected age (BP)	Age (cal BP)	Error (95% confidence)	Dating method
Hardy (1982)*	8.5–7.5	9	0.2–0.5	¹⁴C-14 shells	7,880 ± 160	7,650	8,574	8,999–8,148	¹⁴C shells
Lajeunesse & Allard (2003)*	8.3–6	7		¹⁴C-14 shells	7,730 ± 120	7,500	8,289	8,539–8,039	¹⁴C shells
Hillaire-Marcel (1980)*	7.4–?	6	1.1	¹⁴C-14 shells					
Lajeunesse & Allard (2003)*	6–5	4	1.6	¹⁴C-14 driftwood, paleosol, and shells					
Hardy (1982)*	4–0	0.9		¹⁴C-14 shells					
Andrews (1970)			1.3	Estimated from uplift curves					
Bégin, Bérubé, & Grégoire (1993)			1.3	Dendrochronology					
Dyke, Moore, & Robertson (2003)*					7,940 ± 240	7,710	8,549	9,149–7,949	¹⁴C shells

* The original Carbon-14 data were corrected for marine reservoir correction by Arthur S. Dyke, Andrew Moore, and Louis Robertson (2003) and calibrated with OxCal3 software (Bronk Ramsey 2001).

initially connected to the sea is progressively isolated as the land rises and the shore moves away. Sediments continue to accumulate in these environments and can provide a long-term record of ecological change, including the actual transition between sea and land (see Appendix A).

HOLOCENE CLIMATE CHANGE IN EASTERN JAMES BAY – A SPECIAL CASE

Together with isostatic rebound, climate change dramatically reshaped the landscape of James Bay, transforming the thick ice fields of the early Holocene (10,000–9,000 years before the present [BP]) to the boreal taiga and the subarctic forest-tundra we know today. To better understand the patterns of climate variability and to get a clearer view of the climate processes at work, we need to extend the region of analysis to the whole eastern seaboard of James and Hudson Bays, and to look at climate patterns throughout northeastern Canada and the North Atlantic. It is clear that at some times in the past, the global climate in northeastern Canada generally was warmer than it is today; at other times, it was colder.[4] Between 7,000 and 4,000 years ago, the climate was warmer than it is today. For the past four thousand years, it has tended to be cooler. Within those long periods, there were also shorter warming and cooling phases.

Less is known about the climatic past of the James and Hudson Bay region than about the climate of many other areas. For example, there are fewer paleoenvironmental and paleoclimatic records for eastern James and Hudson Bays specifically (Figure 5.1 and Appendix A) than for southern or eastern Labrador. But the marine and terrestrial data still offer a fairly clear picture of climate variability during the Holocene in this region because of high spatial resolution and the variety of sources.[5]

The existing marine record of eastern James and Hudson Bays is based primarily on a marine core taken offshore from the Great Whale estuary (A on Figure 5.1). It provides information on the last 8,500 years BP.[6] The main finding from this core is that the warm period that occurred in other regions between 7,000 and 4,000 years ago was not as pronounced in eastern James and Hudson Bays. At the same time, it shows that there was no intense cooling later on. It seems that eastern James and Hudson Bays had a more stable climate in the past eight thousand years than some of the

FIGURE 5.1

Overview map showing position of paleo-shorelines, cores, and archaeological sites

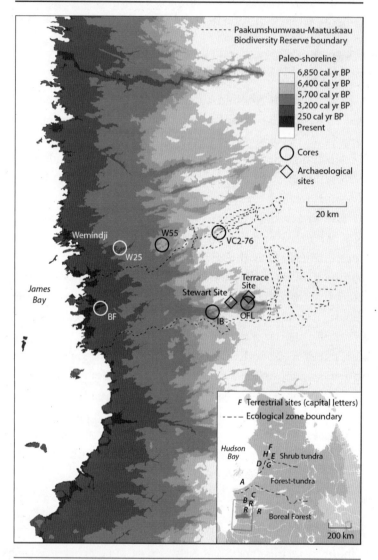

Note: Younger shorelines are represented with progressively darker shades. The cores used to reconstruct shoreline history are marked with capital letters and/or numbers. Paakumshumwaau (Old Factory Lake) is marked as OFL. The inset map shows ecological zones and pollen data presented in Appendix A. Capital letters in the inset represent the terrestrial sites used in the paleoclimate reconstructions discussed in Appendix B.

Source: Elevation and provincial boundary data is from Natural Resources Canada (1998, 2007).

neighbouring regions and relative to the rest of the world in general. The terrestrial record largely agrees with this (see Appendix B).

In summary, both terrestrial and marine records show that, unlike other regions, eastern James and Hudson Bays had no major warm period, except in the northern inland zone, where the forest meets the tundra. With respect to the late Holocene cooling, the marine record, permafrost dynamics, and some pollen evidence suggest a short-lived cooling trend centred around 1,000 years BP, but this is not confirmed by macrofossil records at treeline locations or by variations in beetle assemblages from boreal peatlands.

THE HUMAN RECORD

How did humans both shape and adapt to this rapidly changing environment? In general, archaeologists agree that for the Canadian Shield, which includes eastern James Bay and most of the boreal forest zone, people in the past were fewer and more mobile, in terms of where they lived, than they are now (Wright 1968, 1972b; Martijn and Rogers 1969). Basing their arguments on the available information, most archaeologists believe that the first inhabitants of the Shield, and of eastern James Bay (Eeyou Istchee), were hunter-gatherers who lived in small groups of twenty or twenty-five at most, were scattered over large territories, and didn't stay long in any one place. They camped in different spots at different times of the year to hunt and gather the resources that were in season. According to these models, the human population of the boreal forest began to increase, to settle down in larger groups, and to live in one place for longer periods only since the time of the fur trade.

However, recent work has questioned those old assumptions about the simplicity of the past in the eastern Canadian boreal forest (Denton 1998; McCaffrey 2006; Bracewell 2015). In some other boreal regions, population size, density, and mobility increased or decreased at various times in the past. We have evidence from Finland (Hulse and Costopoulos 2007; Vaneeckhout 2008; Costopoulos et al. 2012; Vaneeckhout, Okkonen, and Costopoulos 2012), Sweden (Broadbent 1979), Norway (Hood 1995), Russia (Weber and Bettinger 2010), and Labrador (Fitzhugh 1978) that at some times in the past, boreal forest hunters lived in large groups and formed villages in which they remained for several generations. Later, they stopped building villages and resumed living in small, mobile groups.

The history of these other boreal forest regions is one of change between larger populations that moved around less and smaller populations that moved around more often. When we started working at Paakumshumwaau (Old Factory Lake), our review of the archaeological literature revealed that the story it told about the peoples of the boreal forest in eastern North America was quite different, indicating thousands of years without change in mobility or population density. Naturally, because of our previous experiences in northern Europe and northern Asia, we asked: Why should it be different?

Our Archaeological Survey

We conducted our survey work at Paakumshumwaau during the summers between 2005 and 2009. The size of our team varied from year to year, and individual members came and went, but it always consisted of archaeologists and our Cree partners. Only some of the Cree team members had previous archaeological experience, but their knowledge of the landscape and of hunting and gathering was invaluable. After each day of survey or excavation, we spoke about our findings in camp and created hypotheses about what we were uncovering. This led to discussions about where and how to continue our work to best collaboratively improve our understanding of human history in the region.

The Terrace Site

Our largest archaeological site was on a terrace five to eight metres above Paakumshumwaau, on a peninsula that forms a choke point between the two large segments of the lake. Because of the orientation of the lake and the dominant wind patterns, the peninsula is a logical stopping point for travellers today. The wind often makes it difficult to paddle on one or the other part of the lake. Its two portions are rarely choppy at the same time, so people often stop at the peninsula to wait for the wind to change so they can travel onward. Nowadays, they often camp on the beach that lies at the foot of the Terrace Site, and they use the site itself for hunting and berry picking. We found it because our Cree partners brought us to see the campsites on the beach. Being curious about the terrace above, we took a look.

FIGURE 5.2
Archaeological remains at the Terrace Site on Paakumshumwaau

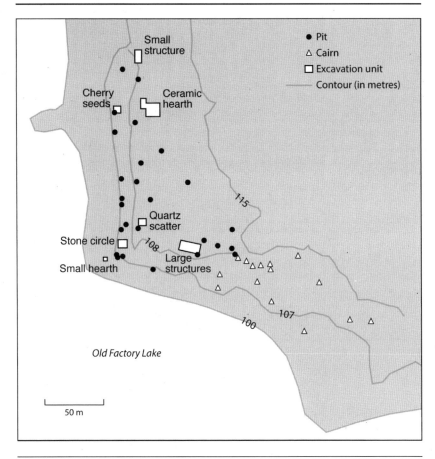

The Terrace Site has several surprising features that suggest its history is more complex than the standard model for the Canadian Shield. First, the site is large. The archaeological remains cover the entire terrace (Figure 5.2), which consists of more than twenty thousand square metres (about five acres). Second, the site is organized, with various kinds of remains situated in differing areas. In the southeastern portion, we found stone cairns that are probably burials. The southwestern zone features a 7.5-metre-long stone structure that could have been related to housing (Figure 5.2). It contains evidence of subsistence in the form of (surprisingly)

unburnt animal bones and other indications of domestic life, such as tool making, use, and maintenance. The edge of the terrace has a row of pits, which are about two square metres and one metre deep. More pits lie in the middle of the terrace. We found a small structure and some scatters of pottery in the northern part of the site. A third indication that this site was more than a temporary camp is that some of its structures represent a significant investment of energy. They are made of stones or boulders, which must have been carried up from the lakeshore or brought from a slope lying a hundred metres to the northeast, where they can still be found. If energy investment is an indication of how much time people spend in a place, this site was occupied by hunters and gatherers who were less mobile than expected for the standard model for the Canadian Shield, at least during some parts of its history.

Dating the Occupations

When did humans first use this terrace? It is important to date the occupations so that we can understand if this site represents people living by the coast or if the James Bay shoreline had already moved far to the west and what that tells us about the ecological stability of Old Factory Lake. However, we encountered several problems in attempting to date the occupations of the terrace. First, there was little stratification. All the material we found was on the surface, except for objects that were purposely buried in storage pits by the inhabitants. In such pits, the oldest material can easily be mixed with the youngest. Simply looking at how deeply an object is buried will not reveal whether it is old or young.

Second, we found very little organic material that could be used for Carbon-14 dating. Like all the other finds, the organic material that could be dated was on the surface. It went through at least one major forest fire in 1989 and probably went through many other major fires over the centuries since it was deposited on the terrace. The new carbon from the forest fires made any dates from surface finds very suspect. The only buried organic objects were a number of pin cherry seeds *(Prunus pennsylvanica)* in an otherwise empty storage pit that we excavated. They were deposited in the pit between 370 and 290 years ago (330±40 14C years BP, BETA-257584),[7] when Old Factory Lake was about seventy-five kilometres from the James Bay coast.

Some large pottery fragments found at the northern end of the terrace can be dated on stylistic grounds to about 450 to 250 years ago, roughly the same period as the cherry seeds. Their decorative pattern was common in the St. Lawrence River Valley and the eastern Great Lakes at that time and was usually associated with Huron groups. Our chemical testing of the clay showed that it was very similar to that used for pottery around Lake Huron and the north of Lake Superior (Clark, Neff, and Glascock 1992), not at all similar to the clays from the Old Factory Lake region (Wren 2010).

The charcoal signal in a core recovered about one kilometre away from the site raised the possibility that the lake has a much longer and continuous occupation history than we expected from the standard Canadian Shield model. Charcoal particles from both natural fires and those made by humans (fireplaces or forest clearings) become airborne and then fall near the fire (Figure 5.3). The large bog on the southern shore of Paakumshumwaau has accumulated 2.5 metres of peat over the last 5,600 years, and this charcoal has become integrated in its peat layers. The obvious problem in interpreting this charcoal signal is how to separate the charcoal that was produced by natural forest fires from that of human-induced fires. The answer lies in the differences between natural forest fires and small-scale human-produced forest burning or fireplaces. Natural fires tend to burn vast tracts of land, thus creating the same charcoal signal over hundreds of kilometres. Conversely, fires set by humans to clear land for habitation or those in the fireplaces generate a charcoal signal that is typically weaker and more localized (within a few kilometres).

For this reason, we looked at the charcoal signal from the Old Factory bog, which is 2.5 kilometres away from the Terrace Site, and then we looked at bogs farther away in areas that were apparently uninhabited. The Old Factory bog signal differed strikingly from that of another site (W25) situated fifty kilometres northwest of Old Factory Lake. The W25 signal contained spikes of charcoal particles in some peat layers, separated by layers that contained no charcoal. This is what one would expect from a natural fire regime, which is characterized by generalized regional fire cycles alternating with periods when the forest regenerates and fires do not occur. At Old Factory Lake, however, the fire signal did not conform to this pattern. Although it too contained several spikes of charcoal, these

FIGURE 5.3
Schematic drawing of the charcoal signal of *a*) natural cyclic fires (such as in core W25) and *b*) background charcoal signal from human campfires, plus natural cyclic fires (such as in the Old Factory Lake core).

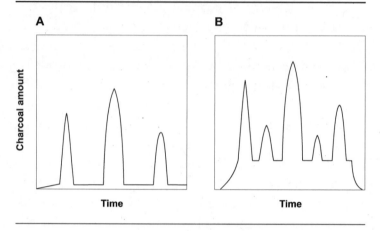

FIGURE 5.4
Schematic drawing of the Paakumshumwaau charcoal signal, with a possible interpretation of the archaeological remains as representing continuous occupation

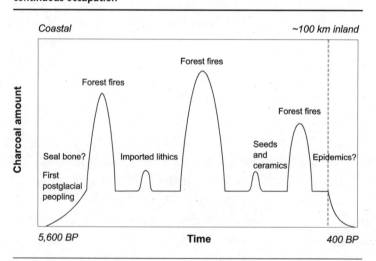

were linked together by lower levels of charcoal in an almost continuous signal from the time the site was a seashore, 5,600 years ago, until about the 1600s AD. It is highly unlikely that natural forest fires occurred continuously at Old Factory Lake during the last 5,600 years, so we interpreted this continuous signal as being from human activity. Of course, the faint signal produced by human burning was superimposed on the stronger but discontinuous natural signal, and the result was a continuous spiked trend of charcoal deposition (Figure 5.4).

Curiously, the continuous part of the charcoal signal in the Old Factory Lake core ended abruptly about four hundred years ago (Figure 5.4). There is a potentially very important coincidence between its disappearance and the smallpox epidemics that began to devastate Iroquois and Huron villages in the eastern Great Lakes and St. Lawrence Valley regions in 1634. During the next ten years, as much as an estimated 60 percent of the Huron and Mohawk population died of the disease (Warrick 2003; Snow 1996). Because they are decorated in typical seventeenth-century Huron patterns, the ceramics found at the Terrace Site strongly suggest contact with the Great Lakes and St. Lawrence Valley regions. This, combined with the fact that the charcoal signal ends abruptly around the same time period, suggests that the 1634 smallpox epidemic might have been felt even in eastern James Bay. This could have led to a dramatic drop in population that can be connected to the equally dramatic drop in the charcoal signal: fewer people using the area resulted in fewer campfires and land clearing fires after the 1630s. There is conflicting evidence in the historical records about whether the great epidemics of the mid-1600s affected eastern James Bay (Morantz 2006, 162), and the earliest Hudson's Bay Company records for the region date from the 1670s, about forty years later (Morantz 1984). But the sudden disappearance of a 5,200-year-long charcoal signal suggests that the area might indeed have been affected by the epidemics. Another possible explanation is that people simply moved to the coast of James Bay to take advantage of the emerging fur trade or to seek shelter from the raids of Iroquois and Huron groups who were fighting for control of the inland fur trading routes.

From 6,850 to 5,400 years ago, when the charcoal signal began, we have one other circumstantial indication that the terrace was a beach at sea level and may have been occupied while Paakumshumwaau was just an inlet off of James Bay. We recovered a small seal phalanx, or flipper bone, from one of the stone structures, but it has not provided enough material for radiocarbon dating. Considering its size, the bone could belong to either a species of freshwater river and lake seal (making it more recent) or a small marine species (making it older). While we don't have freshwater seals in the region today, some freshwater species still exist in northern European lakes and rivers, as in Lake Ladoga in northwestern Russia, for example. It is also possible that a marine seal killed on the coast was carried inland but, since they don't have much meat, flippers tend not to be carried far from a kill site. So perhaps the site was occupied both more recently when the ceramics arrived and much earlier when Paakumshumwaau was an inlet off James Bay

Spatial Distribution of the Remains

Near the southern tip of the terrace, overlooking the lake, we found a large stone structure. It consists of small broken stones in two clusters, each about 3 metres long, which combine to make a structure of about 7.5 metres long. A continuous layer of bone fragments sat just below the layer of stones. The identifiable bones were mostly caribou, a few came from smaller mammals, and one was the seal phalanx. Around and inside the stone clusters were the remains of stone tool making, using both local quartz and imported stone. In the summer of 2008, we found a source of quartz near the eastern end of the lake that had been exploited. The imported stone was in smaller flakes than the quartz flakes and probably represented debitage from the sharpening and maintenance of tools such as knives and projectile points (arrow heads, for example). Some of it came from the Nastapoka area on Hudson Bay. We found only one complete tool of imported stone in that sector of the terrace, but there were forty-two flakes that ranged in weight from a hundredth of a gram to several hundred grams.

The only pottery came from the northern sector of the terrace. We found a few pieces associated with a small cluster of broken stones and a few flakes. Many big pieces of pottery were scattered to the south of this structure, near a few unbroken stones that may have been part of a differently shaped domestic building. We found no stone tools or flakes there.

The site also included at least seven stone cairns, possibly as many as a dozen, which were on the eastern slope of the point, well separated from the rest of the remains. All were oval, about 2 metres long, 1.5 metres wide, and about 60 centimetres high. At least one was surrounded by a circle of large stones. Most were oriented north-northeast to south-southwest. One, the lowest on the slope, was much larger than the others. It was about 3.5 metres long and over a metre high. The rocks used in the cairns were much larger and more rounded than those in the domestic structures.

Explaining the Distribution Pattern

Clearly, the northern and southern ends of the terrace differ in terms of archaeological remains (pottery and a small stone structure versus stone tools and a large stone structure). The same is true for the eastern and western sides (stone cairns versus storage pits). The question is: Why? So far, we have considered two possibilities. Perhaps the differences represent differences in use. The people who lived in the southern part, where we found evidence of subsistence (bone and stone tools), could have been engaged in specialized activities that differed from those related to pottery, in the northern part. Although the stone structures at either end of the terrace differ in size, their similarity in construction suggests that they may have been erected at about the same time. The other possibility is that the differences in remains could represent differences in time. Perhaps the southern and northern portions of the terrace were not used simultaneously.

Since the charcoal signal suggested that Old Factory Lake had been occupied for a very long period, we tested the possibility that the differences represented time. The first clue was that stone tool use is rare in the area with the ceramics, and there are no ceramics in the area with most of the stone tools. When iron and steel tools became available through trade, Indigenous populations in the boreal forest rapidly stopped making stone tools and stopped importing the necessary materials. About the same time, copper kettles started to replace local pottery. The pottery-rich sectors of the terrace could be empty of stone tools simply because their occupants were using metal tools obtained through trade with Europeans.

The southern sector of the terrace looks like an older occupation than the northern sector. It is dominated by evidence of stone tool making, which indicates that traded metal was not available when it was inhabited. It has no pottery, though the work of David Denton (1989) on the Eastmain River shows that pottery was available in the James Bay area as early as about 1,500 years ago. But we do not know how widespread it was or how often people used it. Farther south, Indigenous peoples in the St. Lawrence Valley and the eastern Great Lakes were making and using pottery as early as three thousand years ago (Wright 1972a). In any case, the absence of pottery on the southern tip of the terrace and the presence of high-quality imported stone material there suggest that it was inhabited earlier than the northern end, with its pottery and less frequent stone tools.

The possible burial cairns and the large stone structure in the southern part of the terrace are common features in the archaeological record for the Labrador coast. There, similar structures are between five and four thousand years old and are associated with the Maritime Archaic adaptation (Fitzhugh 1985). For the Maritime Archaic people, marine and coastal resources provided a plentiful and predictable food base, and bigger, more permanent structures than the more mobile groups (Fitzhugh 1978). Intriguingly, Maritime Archaic people used mica, possibly as decoration, and we found some pieces of mica in the southern sector of the terrace but none in the northern sector. A maritime adaptation has never been formally identified on the James Bay side of Quebec, but it wouldn't be particularly surprising that peoples of both northern coasts of Quebec shared similar adaptations at the same time and under roughly similar environmental conditions.

Perhaps a maritime adaptation has never been detected in the James Bay area either because much of the archaeological surveying was connected with the hydro-electric diversion and damming projects or because it focused on recent historical sites on the current coast. The area in which the Terrace Site is located, between the present-day coast and the reservoirs east of the Sakami Moraine, remains largely unexplored by archaeologists.

Recent excavations at a site called Sander's Pond (EhGo-1), at the southern end of James Bay near the Cree community of Waskaganish, found evidence of people living near what was probably the James Bay shore about four thousand years ago. The team found at least two circular structures, each about two metres across. Again, the animal bone was very fragmented, but a mix of large and medium-sized mammals was identified, as was a fish vertebra. Although the team could not use the fragmented bone to confirm that Sander's Pond was a coastal site where marine foods were being eaten, the stone tools showed some similarity to those produced during Newfoundland's Maritime Archaic adaptation, including both ground stone methods and a rock known as Ramah chert, which comes from northern Labrador (Denton 2014; Izaguirre and Denton 2015). The connection to Newfoundland and Labrador and the presence of people living near the James Bay coast four thousand years ago strongly suggest that people used marine resources throughout James Bay's past.

CONCLUSION

During the last six thousand years, the James Bay region has experienced continuous landscape alteration via shoreline retreat and vegetation change. Our reconstruction of eastern James Bay's sequence of shoreline displacement, climate change, and fire shows that the area had a highly dynamic environment, to which people constantly had to adapt. However, in some places, such as at Old Factory Lake, local conditions were more predictable. This may have allowed people to develop larger and more stable communities than elsewhere. The extensive and diverse archaeological remains at Old Factory Lake suggest that the occupation history of the boreal forest of eastern James Bay, and possibly the Canadian Shield in general, is more complex than has usually been acknowledged. At the very least, we have a site of unusual density, size, and organization for highly mobile, very low-density hunter-gatherers.

The question of when Old Factory Lake was first inhabited remains unresolved. The 5,600-year-long charcoal signal raises an interesting possibility of much earlier occupation than previously thought. And certainly, people occupied the Terrace Site intensively as early as 450 years ago. But since we have no dates for certain sectors of the terrace, particularly the largest structure and the cairns in the southern sector, we cannot yet confirm that the charcoal signal was from human activity at the site or elsewhere on Old Factory Lake. The local community has not yet decided whether the cairns will be further studied, so we have no additional information on those either.

A very early occupation would not be particularly surprising, given what we know of the past in similar regions in northern Europe and northern Asia. Their human populations were very responsive to environmental change during the past ten thousand years, readily altering their settlement and subsistence patterns according to their circumstances. Would it be such a surprise if this were true of the people of eastern James Bay?

APPENDIX A: THE RECONSTRUCTION OF THE LONG-TERM SHORELINE DISPLACEMENT IN EASTERN JAMES BAY

Our reconstruction of former coastlines required two major steps. First, we obtained a series of sediment cores that contained the critical transition from marine to terrestrial environments across a land gradient of increasing age. Second, we used palynology to identify each developmental stage from subtidal to tidal (salt marsh) and further to terrestrial ecosystems (forest, fen, bog). We collected organic materials associated with the buried tidal sediments (seeds, leaves, and other plant or insect parts) that are most likely to belong to the tidal environment, which in turn is the most representative for the mean sea level at various times. The elevation measurements, together with the accelerator mass spectrometry (AMS) radiocarbon dates from the tidal materials, allowed us to reconstruct a shoreline displacement model for Wemindji territory during the last 6,500 years (Figure 5.1). In addition, we reconstructed a relative sea level curve for the region (Figure 5A.1), which allowed comparison with other sea level curves reconstructed in the Hudson/James Bay area (Pendea et al. 2010). Our major finding is that the rate of shoreline displacement in the last 6,500 years varied from 3.4 centimetres per year between 6,500 and 5,500 calibrated years BP to 2.3 centimetres per year between 5,500 and 3,400 calibrated years BP and then to around 1.4 centimetres per year between 3,400 and 250 calibrated years BP. In the last three hundred years, the shoreline displacement rate was around 2.3 centimetres per year, which is significantly higher than reported by Hardy (1982) at around 0.5 centimetres per year. Our results show some similarities with other shoreline displacement reconstructions, but also important differences. For instance, one of the most recent relative sea level curves for Hudson Bay (Lajeunesse and Allard 2003) shows that the rates of shoreline displacement were progressively lower starting from eight thousand years ago to the present.

FIGURE 5A.1

Vertical shoreline displacement curve for eastern James Bay in the past 6,500 years

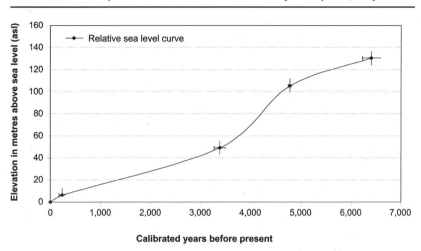

Note: Error bars take into account the theoretical GPS elevation error and accelerator mass spectrometry radiocarbon calibration intervals.

Our curve (Figure 5A.1) suggests that this trend is applicable in James Bay also, but there are differences in the rates of shoreline displacement and in the timing of emergence of various areas.

APPENDIX B: TERRESTRIAL RECORD IN JAMES AND HUDSON BAYS

The terrestrial records consist mainly in pollen-based climatic inferences (Gajewski and Garralla 1992; Gajewski, Payette, and Ritchie 1993; Richard 1995; Sawada et al. 1999; Kerwin et al. 2004) and permafrost development studies (Allard and Seguin 1987; Lavoie and Payette 1995; Payette and Delwaide 2000) complemented by reconstructions based on plant macrofossils and charcoal (Payette and Gagnon 1985; Lavoie and Payette 1996) and fossil beetles (Lavoie and Arseneault 2001). The study by Claude Lavoie and Dominic Arseneault (2001) is particularly important, as fossil beetles (Coleoptera) are ideal indicators of past temperatures because many species are highly specialized to narrow temperature ranges, have the ability to migrate rapidly from one place to another, and can be identified to a high taxonomic resolution (Elias 1997).

The pollen records throughout the James and Hudson Bay region vary from site to site. The pollen data from several sites (Gajewski and Garralla

1992; Gajewski, Payette, and Ritchie 1993; Richard 1995) ranging from the present-day boreal forest (C on Figure 5.1) to the forest-tundra transition (G on Figure 5.1) and the tundra zone (E, F, and H on Figure 5.1) were used to reconstruct climate variability in the region by means of a transfer function (Sawada et al. 1999). The boreal forest site (Gajewski, Payette, and Ritchie 1993) shows a warming trend throughout the mid- to late Holocene, without any evidence for a mid-Holocene maximum and a subsequent late Holocene cooling. This trend is fairly similar to that for the tundra site (H on Figure 5.1), which shows little temperature variation throughout the last 6,500 years. This is in contrast to the G site (Figure 5.1), situated not far to the south in the forest-tundra zone, where the paleotemperature warmed significantly between 3,000 and 1,000 years BP and cooled in the last millennium (Gajewski, Payette, and Ritchie 1993). The pollen evidence from a present coastal tundra site (D on Figure 5.1; Gajewski and Garralla 1992) shows an increase of spruce pollen between 3,500 and 800 years BP, which might suggest a climate-induced invasion of spruce forests into this area in response to warming. However, comparable amounts of spruce pollen were recorded in modern samples from this area, even though almost no spruce trees grow there today. This evidence led Konrad Gajewski and Silvina Garralla (1992) to conclude that spruce pollen was being transported there far from the south. Other pollen records from the forest zone in the eastern James Bay region (the Rs on Figure 5.1) were not used in statistical paleoclimate reconstructions (Richard 1995), but the general conclusion is that around 6,000 years BP the regional vegetation was more dense and more diversified than it is today. Pierre J.H. Richard (1995) suggests that this vegetation pattern at 6,000 years BP could be related to both soil conditions (rich mineral soils left by the retreating glaciers) and milder temperatures.

The latest paleoclimate reconstruction, using an improved transfer function for pollen-climate relationships, indicates that around 6,000 years BP, there is no evidence for a mid-Holocene thermal maximum in the eastern James and Hudson Bay region and that July temperatures were actually slightly cooler than at present (Kerwin et al. 2004). However, Michael W. Kerwin et al. (2004) document a faint warming (+0.5 °C) along the eastern coast of James Bay during a later interval (5,500–2,000 years BP) but state that this warming was minimal in comparison with the rest of northeastern North America. Moreover, east of the James Bay coast, toward central Labrador, this regional Holocene warming was even less pronounced, if it existed at all. With respect to the late Holocene cooling, Kerwin et al. (2004) state that in the eastern James and Hudson Bay region, the cooling was much delayed (around 1,000 years BP as opposed to 4,000–2,000 years BP in other regions), short (about five hundred years), and minimal (-0.5°C).

This pattern of minimum to no climate change during the mid- to late Holocene is confirmed by plant remains (macrofossils) and charcoal evidence. Claude Lavoie and Serge Payette (1996) used tree macrofossils to determine changes in the geographical position of the northern treeline in Labrador. If the magnitude of climate change is high enough, northern treelines migrate either to the north, when climates warm, or to the south, when growing-season temperatures decline. Studies in Keewatin (north-central Canada) report major movements (50–350 kilometres) of the forest limit since 5,000 years BP (Kay 1979). There were no such changes in the position of the treeline during the last three thousand years in northern Labrador or the eastern James and Hudson Bay region (Lavoie and Payette 1996). Before 3,000 years BP, the Labrador treeline was kept south of its present location by remnants of continental ice caps. As the land became free of ice, it migrated to the north, without any subsequent movements.

Important evidence for a lack of major climate variability during the mid- to late Holocene in eastern James Bay comes from a climate reconstruction based on fossil beetles (B on Figure 5.1; Lavoie and Arseneault 2001). It provides a more direct estimate of climate than pollen records do, as it is a vegetation-independent climatic record that does not have the inherent interpretation problems of pollen records, related to long-distance transport and plant immigration lags. The beetle assemblages recovered from three peat bogs in eastern James Bay show little variation since 5,000 years BP (Lavoie and Arseneault 2001). The reconstructed July temperatures during the last five thousand years were between 14.5°C and 17.5°C, with a variation range of 3°C, smaller than the one recorded at a nearby weather station for the period 1977–96 AD (11.7°C to 16.2°C). The mean July temperature at 5,000 years BP was slightly higher than that of today, but the difference is probably related to the tendency of the transfer function to overestimate reconstructed temperatures for high-latitude locations

(Elias 1997) rather than to warmer temperatures five thousand years ago. A slightly different picture of climate variability was recorded a few hundred kilometres north of B site at the transition between forest and tundra zones of eastern Hudson Bay (Lavoie, Elias, and Payette 1997). The beetle assemblages recorded there show a warming trend between 5,800 and 2,000 years BP, in agreement with pollen-based July temperature reconstructions (Kerwin et al. 2004), suggesting a possible higher sensitivity of forest-tundra ecosystems to climate signals.

Many permafrost studies conducted in the arctic and subarctic zones of Quebec/Labrador show the relatively recent origin of permafrost structures such as palsas and peat plateaus (Allard and Seguin 1987; Lavoie and Payette 1995; Payette and Delwaide 2000), which could sustain the idea of a late Holocene cooling. Serge Payette (2001) compiled forty-five radiocarbon ages from palsas (permanently frozen peat mounds whose diameter is less than a hundred metres) and peat plateaus (permanently frozen peat mounds with a diameter of greater than a hundred metres) in northern Quebec and concluded that permafrost initiation began around 3,200 years BP and happened more frequently after 2,000 years BP. Payette (2001) indicates that permafrost growth was particularly strong during the Little Ice Age at the forest-tundra transition of northern Quebec. However, temperature is not the only controlling factor for permafrost initiation, and thus paleotemperature inferences based on permafrost can be equivocal. Precipitation levels and biological factors could be equally important. Many studies show a link between the development of sphagnum peat layers and the initiation of permafrost (Robinson and Moore 2000; Zimmermann and Lavoie 2001). Dense forest cover can also favour the initiation of permafrost because the canopy can intercept the snow, thus allowing greater heat loss from the soil during winter (Zoltai 1972; Robinson and Moore 2000). Matti Seppälä (1986) demonstrates the importance of a thin snow cover in the formation of ice lenses, which is the first step in the development of palsas and peat plateaus. A succession of dry winters (not necessarily cold) can create suitable conditions for permafrost growth. The latest permafrost-inferred paleoclimatic reconstruction from northern Quebec used borehole temperature depth profiles to infer recent ground temperature variations (Chouinard, Fortier, and Mareschal 2007). It found evidence for a cooling episode during the sixteenth and seventeenth centuries

AD (the Little Ice Age). The authors found no evidence that the Medieval Warm Period occurred in the area under study.

NOTES

1 Studies on the Late Wisconsinan-Holocene paleogeography and landscape evolution of the Hudson/James Bay area provide information on the nature and extent of Laurentide deglaciation events, the evolution of the proglacial lakes, and the marine shore dynamic (see Lee 1960; Vincent and Hardy 1977; Hardy 1977, 1982; Dionne 1980; Shilts 1982; Dyke and Prest 1987; Veillette 1994; Dyke, Moore, and Robertson 2003; Pendea et al. 2010). For eastern James Bay, Léon Hardy (1982) reports four main events associated with the early to mid-Holocene deglaciation: the retreat of the Nouveau Québec glacier, the glacio-lacustrine episode associated with Cochrane glacial re-advance (Ojibway lacustrine stage), the Ojibway Lake drainage, and the Tyrrell marine phase.

2 Features of the early Holocene Tyrrell Sea transgression and the subsequent coastline retreat, caused by isostatic rebound, are summarized by several authors (Lee 1960; Andrews 1970; Hillaire-Marcel 1980; Hardy 1982; Andrews and Peltier 1989). Maximum postglacial marine transgression occurred around ~8,500 cal years BP, followed by a rapid fall in relative sea level due to isostatic rebound in the early part of the deglaciation, which significantly decreased toward the present day. Present rates of shoreline emergence for eastern Hudson/James Bay vary between 1.6 metres per century and 0.2 metres per century (Table 5.1). This variation is mainly due to spatial differences in the Hudson Bay region, with different ice loads and rates of unloading during the early Holocene that result in variable rates of isostatic adjustment throughout the region (Hillaire-Marcel 1980). The radiocarbon-based projections of recent rates of shoreline displacement (Andrews 1970; Andrews and Peltier 1989) are confirmed by ecological studies on seaward migration of coastal ecosystems on newly emerged coastlines (Bégin, Bérubé, and Grégoire 1993; von Mörs and Bégin 1993).

3 Isotope fractionation and marine reservoir corrections are critical to obtaining accurate ages for shell. Dyke, Moore, and Robertson (2003) review all chronological data and apply the latest marine reservoir corrections (Table 5.1). The corrections reveal that most paleogeographical events are younger than previously reported.

4 Evidence from numerous marine and terrestrial sites across the North Atlantic region indicates that the Holocene was characterized by several major warm and cold oscillations, in which temperatures were either warmer or colder than at present (see Webb, Bartlein, and Kutzbach 1987; Koç, Jansen, and Haflidason 1993; Kerwin et al. 1999, 2004; Kaufman et al. 2004). During the mid- to late Holocene (7,000–0 years BP), most evidence indicates that two important climate oscillations occurred, known as the mid-Holocene warm period (Kerwin et al. 2004), between 7,000 and 4,000 years BP, and the late Holocene cooling or Neoglaciation (Denton and Porter 1970), between 4,000 and 0 years BP. During the last two millennia, some paleorecords suggest the existence of a second warming episode, known as the Medieval Warm

Period (MWP), between the ninth and fourteenth centuries AD, and a second cold interval, termed the Little Ice Age (LIA), between the fourteenth and nineteenth centuries AD (e.g., Lamb 1965; Keigwin 1996). However, these climatic oscillations had a restricted, regional character (Hughes and Diaz 1994), and recent reviews (Mann 2007) note that the MWP and LIA should not be seen as broad-scale climate change events. Also, most paleoclimate reconstructions have a millennial-scale resolution and are therefore unable to reconstruct short climate events such as the MWP and LIA.

5 Dinoflagellate cysts, pollen, plant remains, fossil beetles, microscopic charcoal, and tree rings.

6 The marine sediments were analyzed for dinoflagellate cysts assemblages, and the paleoclimate data were derived using a transfer function. Dinoflagellate cyst assemblages reveal small amplitude sea surface temperature (SST) variations around seven degrees Celsius in summer, a value close to the modern average (de Vernal and Hillaire-Marcel 2006). There is no evidence for a mid-Holocene thermal maximum or for any significant temperature trends between 8,000 and 2,000 years BP. This evidence confirms an earlier study (Sawada et al. 1999), but Anne de Vernal and Claude Hillaire-Marcel (2006) do not report any trend for the period 2,000 years BP to the present. Michael Sawada et al. (1999) reconstruct a cooling trend (~1°C) between 2,000 and 1,000 years BP, after which the SST rose toward present levels. This cooling trend could correspond with the late Holocene Neoglaciation in the North Atlantic, but it seems to be of a much shorter duration and its initiation much delayed. Most probably, the late Holocene cooling was not particularly intense around the Labrador Peninsula. This hypothesis is sustained by SST reconstructions from the Atlantic seaboard of Labrador (Levac and de Vernal 1997), where dinoflagellate cyst and pollen assemblages in offshore sediments indicate no clear cooling trend during the late Holocene.

7 By convention, when archaeologists say "years ago," they mean number of years before 1950.

WORKS CITED

Allard, Michel, and Maurice K. Seguin. 1987. "The Holocene Evolution of Permafrost near the Tree-Line, on the Eastern Coast of Hudson Bay (Northern Québec)." *Canadian Journal of Earth Sciences* 24 (11): 2206–22.

Andrews, John T. 1970. "Present and Postglacial Rates of the Uplift for Glaciated Northern and Eastern North-America Derived from Post-Glacial Uplift Curves." *Canadian Journal of Earth Sciences* 7 (2): 703–15.

Andrews, John T., and Richard W. Peltier. 1989. "Quaternary Geodynamics in Canada." In *Geology of Canada: Quaternary Geology of Canada and Greenland,* ed. R.J. Fulton, 541–72. Ottawa: Geological Survey of Canada.

Bégin, Yves, Dominique Bérubé, and Martin Grégoire. 1993. "Downward Migration of Coastal Conifers as a Response to Recent Land Emergence in Eastern Hudson Bay, Québec." *Quaternary Research* 40 (1): 81–88.

Bracewell, Jennifer Maddick. 2015. "The Infertile Crescent Revisited: A Case (Study) for the History of Archaeology." *Bulletin of the History of Archaeology* 25 (2): Article 3. http://doi.org/10.5334/bha.257.

Broadbent, Noel. 1979. *Coastal Resources and Settlement Stability: A Critical Study of a Mesolithic Site Complex in Northern Sweden.* Uppsala: Uppsala University Institute of North European Archaeology.

Bronk Ramsey, Christopher. 2001. "Development of the Radiocarbon Program OxCal." *Radiocarbon* 43 (2A): 355–63.

Chouinard, Christian, Richard Fortier, and Jean-Claude Mareschal. 2007. "Recent Climate Variations in the Subarctic Inferred from Three Borehole Temperature Profiles in Northern Québec, Canada." *Earth and Planetary Science Letters* 263 (3–4): 355–69.

Clark, Canti P., Hector Neff, and Michael D. Glascock. 1992. "Neutron Activation Analysis of Late Woodland Ceramics from the Lake Superior Basin." In *Chemical Characterization of Ceramic Pastes in Archaeology,* ed. Hector Neff, 255–67. Madison: Prehistory Press.

Costopoulos, André, Samuel Vaneeckhout, Jari Okkonen, Eva Hulse, Ieva Paberzyte, and Colin D. Wren. 2012. "Social Complexity in the Mid-Holocene Northeastern Bothnian Gulf." *European Journal of Archaeology* 15 (1): 41–60.

Denton, David. 1989. "Archaeological Resources in the Eastmain Area, James Bay, Quebec." Cree Regional Authority.

–. 1998. "From the Source to the Margins and Back." *Paleo-Québec* 27: 17–32.

–. 2014. "Archéologie et paysages près de Waskaganish sur la côte de la Baie James." Paper presented at the Colloque annuel de l'Association des Archéologues professionnels du Québec, Trois-Rivières, QC, May 1.

Denton, George H., and Stephen C. Porter. 1970. "Neoglaciation." *Scientific American* 222 (6): 100–10.

de Vernal, Anne, and Claude Hillaire-Marcel. 2006. "Provincialism in Trends and High Frequency Changes in the Northwest North Atlantic during the Holocene." *Global and Planetary Change* 54: 263–90.

Dionne, Jean-Claude. 1980. "An Outline of the Eastern James Bay Coastal Environments." In *The Coastline of Canada,* ed. Samuel Brian McCann, 311–38. Ottawa: Geological Survey of Canada.

Dyke, Arthur S., A.J. Moore, and L. Robertson. 2003. *Deglaciation of North America.* Open File 1574, 2 map sheets, 1 CD-ROM. Ottawa: Geological Survey of Canada.

Dyke, Arthur S., and Victor K. Prest. 1987. "Late Wisconsinan and Holocene History of the Laurentide Ice Sheet." *Géographie physique et Quaternaire* 41 (2): 237–63.

Elias, Scott A. 1997. "The Mutual Climatic Range Method of Palaeoclimate Reconstruction Based on Insect Fossils: New Applications and Inter-hemispheric Comparisons." *Quaternary Science Reviews* 16 (10): 1217–25.

Fitzhugh, William. 1978. "Maritime Archaic Cultures of the Central and Northern Labrador Coast." *Arctic Anthropology* 15 (2): 61–95.

–. 1985. "Early Maritime Archaic Settlement Studies and Central Coast Surveys." *Archaeology in Newfoundland and Labrador* 5: 48–84.

Gajewski, Konrad, and Silvina Garralla. 1992. "Holocene Vegetation Histories from Three Sites in the Tundra of Northwestern Québec, Canada." *Arctic and Alpine Research* 24 (4): 329–36.

Gajewski, Konrad, Serge Payette, and James C. Ritchie. 1993. "Holocene Vegetation History at the Boreal Forest – Shrub Tundra Transition in North-Western Québec." *Journal of Ecology* 81 (3): 433–43.

Glaser, Paul H., Barbara C.S. Hansen, Donald I. Siegel, Andrew S. Reeve, and Paul J. Morin. 2004. "Rates, Pathways and Drivers for Peatland Development in Hudson Bay Lowlands, Northern Ontario, Canada." *Journal of Ecology* 92 (6): 1036–53.

Glaser, Paul H., Donald I. Siegel, Andrew S. Reeve, Jan A. Janssens, and David R. Janecky. 2004. "Tectonic Drivers for Vegetation Patterning and Landscape Evolution in the Albany River Region of the Hudson Bay Lowlands." *Journal of Ecology* 92 (6): 1054–70.

Hardy, Léon. 1977. "La déglaciation et les épisodes lacustre et marin sur le versant québécois des basses-terres de la baie de James." *Géographie physique et Quaternaire* 31 (3–4): 261–73.

–.1982. "Le Wisconsinien supérieur à l'est de la Baie James (Québec)." *Le Naturaliste Canadien* 109: 333–51.

Hillaire-Marcel, Claude. 1980. "Multiple Component Post-glacial Emergence, Eastern Hudson Bay, Canada." In *Earth Rheology, Isostasy and Eustasy,* ed. N.A. Morner, 215–30. Toronto: Wiley.

Hood, Bryan C. 1995. "Circumpolar Comparison Revisited: Hunter-Gatherer Complexity in the North Norwegian Stone Age and the Labrador Maritime Archaic." *Arctic Anthropology* 32 (2): 75–105.

Hughes, Malcolm K., and Henry F. Diaz. 1994. "Was There a 'Medieval Warm Period' and if so, Where and When?" *Climatic Change* 26 (2–3): 109–42.

Hulse, Eva, and André Costopoulos. 2007. "Spatial Patterning within a 5000 Year Old Structure in Northern Finland." In *Space – Archaeology's Final Frontier? An Intercontinental Approach,* ed. R. Salisbury and D. Keeler, 109–33. Newcastle: Cambridge Scholars.

Izaguirre, Dario, and David Denton. 2015. "Résultats de l'intervention archéologique sur le site EhGo-l, Waskaganish, Québec: Rapport d'étape, saison 2014." Gouvernement de la Nation crie et la Nation crie de Waskaganish, Waskaganish, QC.

Kaufman, Darrell S., Thomas A. Ager, John Anderson, Patricia M. Anderson, John T. Andrews, Pat J. Bartlein, Linda B. Brubaker, Larry L. Coats, Les C. Cwynar, Mathieu L. Duvall, and Arthur S. Dyke. 2004. "Holocene Thermal Maximum in the Western Arctic (0–180 W)." *Quaternary Science Reviews* 23 (5): 529–60.

Kay, Paul A. 1979. "Multivariate Statistical Estimates of Holocene Vegetation and Climate Change, Forest-Tundra Transition Zone, NWT, Canada." *Quaternary Research* 11 (1): 125–40.

Keigwin, Lloyd D. 1996. "The Little Ice Age and Medieval Warm Period in the Sargasso Sea." *Science* 274 (5292): 1504–8.

Kerwin, Michael W., Jonathan T. Overpeck, Robert S. Webb, and Katherine H. Anderson. 2004. "Pollen-Based Summer Temperature Reconstructions for the Eastern Canadian Boreal Forest, Subarctic, and Arctic." *Quaternary Science Reviews* 23 (18–19): 1901–24.

Kerwin, Michael W., Jonathan T. Overpeck, Robert S. Webb, Anne de Vernal, David H. Rind, and Richard J. Healy. 1999. "The Role of Oceanic Forcing in Mid-Holocene Northern Hemisphere Climatic Change." *Paleoceanography* 14: 200–10.

Klinger, Lee F., and Susan K. Short. 1996. "Succession in the Hudson Bay Lowland, Northern Ontario, Canada." *Arctic and Alpine Research* 28: 172–83.

Koç, Nalân, Eystein Jansen, and Haflidi Haflidason. 1993. "Paleoceanographic Reconstructions of Surface Ocean Conditions in the Greenland, Iceland and Norwegian Seas through the Last 14 Ka Based on Diatoms." *Quaternary Science Reviews* 12 (2): 115–40.

Lajeunesse, Patrick, and Michel Allard. 2003. "Late Quaternary Deglaciation, Glaciomarine Sedimentation and Glacioisostatic Recovery in the Rivière Nastapoka Area, Eastern Hudson Bay, Northern Québec." *Géographie physique et Quaternaire* 57 (1): 65–83.

Lamb, Hubert H. 1965. "The Early Medieval Warm Epoch and Its Sequel." *Palaeogeography, Palaeoclimatology and Palaeoecology* 1: 13–37.

Lavoie, Claude, and Dominic Arseneault. 2001. "Late Holocene Climate of the James Bay Area, Québec, Canada, Reconstructed Using Fossil Beetles." *Arctic, Antarctic, and Alpine Research* 33 (1): 13–18.

Lavoie, Claude, Scott A. Elias, and Serge Payette. 1997. "Holocene Fossil Beetles from a Tree Line Peatland in Subarctic Québec." *Canadian Journal of Zoology* 75 (2): 227–36.

Lavoie, Claude, and Serge Payette. 1995. "Analyse macrofossile d'une palse subarctique (Québec nordique)." *Canadian Journal of Botany* 73: 527–37.

–. 1996. "The Long-Term Stability of the Boreal Forest Limit in Subarctic Québec." *Journal of Ecology* 77: 1226–33.

Lee, Hulbert A. 1960. "Late Glacial and Post-glacial Hudson Bay Sea Episode." *Science* 131 (3413): 1609–11.

Levac, Elisabeth, and Anne de Vernal. 1997. "Postglacial Changes of Terrestrial and Marine Environments along the Labrador Coast: Palynological Evidence from Cores 91-045-005 and 91-045-006, Cartwright Saddle." *Canadian Journal of Earth Sciences* 34 (10): 1358–65.

Mann, Michael E. 2007. "Climate over the Past Two Millennia." *Annual Review of Earth and Planetary Sciences* 35: 111–36.

Martijn, Charles A., and Edward S. Rogers. 1969. *Mistassini-Albanel: Contributions to the Prehistory of Québec*. Collection Nordicana No. 25, Centre d'études nordiques. Quebec City: Université Laval.

McCaffrey, Moira 2006. "Archaic Period Occupation in Subarctic Québec: A Review of the Evidence." In *The Archaic of the Far Northeast,* ed. David Sanger and M.A. Priscilla Renouf, 161–90. Orono: University of Maine Press.

Morantz, Toby. 1984. "Economic and Social Accommodations of the James Bay Inlanders to the Fur Trade." In *The Sub-Arctic Fur Trade: Native Economic and Social Adaptations,* ed. Shepard Krech, III, 55–79. Vancouver: UBC Press.

–. 2006. "In the Land of Lions: The Ethnohistory of Bruce G. Trigger." In *The Archaeology of Bruce Trigger: Theoretical Empiricism,* ed. R.F. Williamson and M.S. Bisson, 142–73. Montreal and Kingston: McGill-Queen's University Press.

Natural Resources Canada. 1998. National Topographic Database: 1:50 000 scale, edition 3.0. Online database. http://ftp.geogratis.gc.ca/pub/nrcan_rncan/vector/ntdb_bndt/.

–. 2007. Canadian Digital Elevation Data: 1:50 000 scale, edition 3.0. Online database. http://ftp.geogratis.gc.ca/pub/nrcan_rncan/archive/elevation/geobase_cded_dnec/.

Payette, Serge. 2001. "Les processus et les formes périglaciaires." In *Écologie des tourbières du Québec-Labrador,* ed. Serge Payette and L. Rochefort, 199–239. Quebec City: Les Presses de l'Université Laval.

Payette, Serge, and Ann Delwaide. 2000. "Recent Permafrost Dynamics in a Subarctic Floodplain Associated with Changing Water Levels, Québec, Canada." *Arctic and Alpine Research* 32 (3): 316–23.

Payette, Serge, and Réjean Gagnon. 1985. "Late Holocene Deforestation and Tree Regeneration in the Forest-Tundra of Québec." *Nature* 313: 570–72.

Peltier, W. Richard 2002. "On Eustatic Sea Level History: Last Glacial Maximum to Holocene." *Quaternary Science Reviews* 21 (1–3): 377–96.

Pendea, I. Florin, André Costopoulos, Colin Nielsen, and Gail L. Chmura. 2010. "A New Shoreline Displacement Model for the Last 7 Ka from Eastern James Bay, Canada." *Quaternary Research* 73 (3): 474–84.

Richard, Pierre J.H. 1995. "Le couvert végétal du Québec–Labrador il y a 6000 ans BP: Essai." *Géographie Physique et Quaternaire* 49: 117–40.

Robinson, Stephen D., and Timothy R. Moore. 2000. "The Influence of Permafrost and Fire upon Carbon Accumulation in High Boreal Peatlands, Northwest Territories, Canada." *Arctic, Antarctic, and Alpine Research* 32 (2): 155–66.

Sawada, Michael, Konrad Gajewski, Anne de Vernal, and Pierre J.H. Richard. 1999. "Comparison of Marine and Terrestrial Holocene Climatic Reconstructions from Northeastern North America." *Holocene* 9 (3): 267–77.

Seppälä, Matti. 1986. "The Origin of Palsas." *Geografiska Annaler* 68A: 141–47.

Shennan, Ian, Kurt Lambeck, Ben P. Horton, Jim B. Innes, Jerry M. Lloyd, Jenny J. McArthur, Tony A. Purcell, and Mairead M. Rutherford. 2000. "Late Devensian and Holocene Records of Relative Sea-Level Changes in Northwest Scotland and Their Implications for Glacio-Hydro-Isostatic Modelling." *Quaternary Science Reviews* 19 (11): 1103–36.

Shilts, William W. 1982. "Quaternary Evolution of the Hudson/James Bay Region." *Le Naturaliste Canadien* 109: 309–32.

Snow, Dean R. 1996. "Mohawk Demography and the Effects of Exogenous Epidemics on American Indian Populations." *Journal of Anthropological Archaeology* 15: 160–82.

Tarnocai, Charles. 1982. "Soil and Terrain Development on the York Factory Peninsula, Hudson Bay Lowland." *Le Naturaliste Canadien* 109: 511–22.

Vaneeckhout, Samuel. 2008. "Sedentism on the Finnish Northwest Coast: Shoreline Reduction and Reduced Mobility." *Fennoscandia Archaeologica* 25: 61–72.

Vaneeckhout, Samuel, Jari Okkonen, and André Costopoulos. 2012. "Paleoshorelines and Prehistory on the Eastern Bothnian Bay Coast (Finland): Local Environmental Variability as a Trigger for Social Change." *Polar Geography* 35 (1): 1–13.

Veillette, Jean J. 1994. "Evolution and Paleohydrology of Glacial Lakes Barlow and Ojibway." *Quaternary Science Reviews* 13 (9–10): 945–71.

Vincent, Jean-Serge, and Léon Hardy. 1977. "L'évolution et l'extension des lacs glaciaires Barlow et Ojibway en territoire québécois." *Géographie physique et Quaternaire* 31: 357–72.

von Mörs, Iris, and Yves Bégin. 1993. "Shoreline Shrub Population Extension in Response to Recent Isostatic Rebound in Eastern Hudson Bay, Québec, Canada." *Arctic and Alpine Research* 25: 15–23.

Warrick, Garry 2003. "European Infectious Disease and Depopulation of the Wendat-Tionontate (Huron-Petun)." *World Archaeology* 35 (2): 258–75.

Webb, Thomson, III, Patrick J. Bartlein, and John E. Kutzbach. 1987. "Climatic Change in Eastern North America during the Past 18,000 Years: Comparisons of Pollen Data with Model Results in North America and Adjacent Oceans during the Last Deglaciation." In *The Geology of North America*, vol. 3, ed. William F. Ruddiman and Herb E. Wright, Jr., 447–62. Boulder: Geological Society of America.

Weber, Andrzej W., and Robert Bettinger. 2010. "Middle Holocene Hunter-Gatherers of Cis-Baikal, Siberia: An Overview for the New Century." *Journal of Anthropological Archaeology* 29 (4): 491–506.

Wren, Colin D. 2010. "Identifying Interactions on Prehistoric Eastern James Bay through Ceramic Sourcing." Paper presented at the 75th conference of the Society for American Archaeology, St. Louis, MO, April 14–18.

Wright, James V. 1968. "The Boreal Forest." In *Science, History and Hudson Bay*, ed. C.S. Beals, 55–68. Ottawa: Queen's Printer.

–. 1972a. *Ontario Prehistory*. Ottawa: National Museum of Canada.

–. 1972b. *The Shield Archaic*. Ottawa: National Museum of Canada.

Zimmermann, Claudia, and Claude Lavoie. 2001. "A Paleoecological Analysis of a Southern Permafrost Peatland, Charlevoix, Québec." *Canadian Journal of Earth Sciences* 38 (6): 909–19.

Zoltai, Stephen C. 1972. "Palsas and Peat Plateaus in Central Manitoba and Saskatchewan." *Canadian Journal of Forest Research* 2: 291–302.

6

Patterns on the Land
Paakumshumwaau through the Lens of Natural History

JAMES W. FYLES, GRANT R. INGRAM,
GREGORY M. MIKKELSON, FLORIN PENDEA,
KATHERINE SCOTT, and KRISTEN WHITBECK

The patterns of Paakumshumwaau are those of a patchwork quilt. We find a mix of textures, patterns, and hues mostly in greens and browns that arise from its uplands and lowlands, rocks and forests, bogs, lakes, and rivers. The western edge dissolves into shades of aquamarine before it dips beneath the deep blue of James Bay. Like many an heirloom quilt, Paakumshumwaau, as we will call it here, displays signs of a long and eventful history, the original pattern showing clearly but in varying tones as patches and colours change over time.

Readers of this book may have noticed that the name Paakumshumwaau is used in various ways, implying the lake or the watershed depending on author and context. In this chapter we adopt the broadest definition. The river and lake are central, but where do they begin and end? Their waters extend to the tiny rivulets and even the soil pores across the landscape, and to the heights of land that surround it. These waters are connected directly to the waters that flow through the trees and other vegetation, which are, in turn, coupled with the waters of the atmosphere that both removes and returns water. Much of what occurs in the watershed is under the influence of lands and waters elsewhere. Rainfall, animals, birds, pollen, wind, fire, people; all connect the land influenced by the river to surrounding areas. The English language does not have a term that effectively captures such a focused, integrated entity defined by such undefinable boundaries. The concepts of watershed and biodiversity reserve identify boundaries, but do not recognize the insufficiency of those boundary definitions. Thus, in this chapter, we borrow Paakumshumwaau as the best name we know of to express the river, the land, and the set of relationships in which they are engaged and embedded. We use the name with apologies to those who would apply a more precise definition.

Our most treasured quilts, like those made in Wemindji, are pieced from worn clothing, old favourites that find new life in the patterns and borders. Each quilt block comes with a story: the faded denim tells of a young hunter's first goose; the bright calico conjures a mother's apron; the rough plaid speaks of strong coffee and campfire smoke. Drawn together around such a quilt, we name each piece, share its meanings, pass them along, and conjure up stories from other times and places.

The Paakumshumwaau quilt is much the same. Each patch, each valley and hilltop, animal, plant, tree, and rock, has a name and a meaning for those who know their stories. Some names and meanings are known from personal experiences, many have been learned from others, and some we interpret from their likeness to other places and times.

The Wemindji Protected Areas Project described in this volume has been about sharing stories of the land and about discovering new stories as we explored how those we brought with us resonated with those the land told. Our purpose in writing this chapter is to share some of what we learned to build an understanding and appreciation of the area's natural heritage. We describe some of the patterns of the Paakumshumwaau quilt and some of the forces through which they have been shaped. Our stories are told from our "Western" scientific perspectives on geology, soils, plants, and ecosystems. Our approach is to combine knowledge of plant communities and ecological function of marine and terrestrial ecosystems to give a broad view of landscape and coastal patterns. We focus on the physical features of the land and its vegetation, largely omitting the animals, not because they are unimportant but because they are discussed in Chapter 7 of this book.

Our chapter differs in some respects from others in this part of the book. It is not the product of primary scientific investigation; collecting the necessary detailed data to support such a discussion was beyond the scope of the Wemindji Protected Areas Project. Nor is it strictly a literature review; little has been published that is specific to the ecology of Paakumshumwaau, and our treatment of the more general literature is not exhaustive. Instead, this chapter is a synthesis of knowledge and ideas that have been focused through the lens of the Wemindji Protected Areas Project.

This chapter is a story – a natural history story – following in the long tradition of First Nations peoples and later nations' field ecologists and geologists, sitting around campfires, on hilltops overlooking vistas, or on the beach after a long paddle, telling about what they have seen and what they think it means. It is a story of what we have seen and how we can bring meaning to those observations, with the hope that such an understanding might help bring about a better future. It is a story for visitors to Paakumshumwaau. We are all visitors.

For readers who have visited Paakumshumwaau many times, we hope that our story will add something to your already rich experience and will go along with the stories you yourselves tell. For readers who intend to visit, we hope that our story helps you to know what to expect and will enrich your visit and your appreciation of the land. For researchers who work in the area, we hope that the story will help you to embed your research in a larger context and encourage you to tell stories of your own. For those who will carry the responsibility of plotting the future of the Paakumshumwaau-Maatuskaau Biodiversity Reserve and its surrounding lands, we hope that the story will raise issues that will help guide your thinking and actions.

Paraphrased, the "principle of incompatibility" (Zadeh 1973) states that as complexity increases, meaningful statements lose precision and precise statements lose meaning. Clearly, if anything, the Paakumshumwaau-Maatuskaau Biodiversity Reserve is complex; meaningful statements about it are inherently imprecise. This can be tough for scientists and managers, who strive for precision. But in this respect, the stories told by ecologists about the land are not so different from the stories told by the Crees: they strive for meaning in the face of complexity and are, by necessity, imprecise. This is something that Paakumshumwaau has helped us learn.

Fundamental to this discussion, as to the research project from which it emerged, is the idea that how we see or perceive the land is as important as what we see. Which of the many possible futures of this protected area will actually play out? If the biodiversity reserve is to meet the needs and expectations of generations to come, its future will emerge from the shared understandings of many, drawing on multiple perspectives. Our exploration of concepts in this chapter is intentionally broad, but it naturally reflects our own biases. We have tried, however, to listen closely and to integrate our views with other ways of seeing and understanding the land. We thank the Cree elders and our colleagues for their insight, wisdom, and patience in helping us to see in new ways (see the Introduction, this volume).

A BOREAL QUILT

The quilt-like pattern of the Paakumshumwaau landscape resembles that of many boreal landscapes around the world. Under conditions of poor soils and constraining climate, with frequent disturbance by fire, boreal landscapes have little capacity or opportunity to converge to a uniform dominant vegetation. Each wrinkle in the geological substrate, each tear in the fabric of vegetation history, shows clearly and fades slowly. Drawn by erratic disturbance and strengthened by plant-soil feedbacks, the boundaries between patches are often clearly defined and easily visible, something that is less apparent farther south. At the seashore, the pattern extends beneath the bay, driven by tides and coastal processes that re-distribute sediments and nutrients, and emerging in the communities of organisms that blanket the ocean floor.

This patchwork pattern is typical of the Canadian Taiga Shield eco-zone (Wiken 1986), of which Paakumshumwaau is a part. The circumboreal taiga is the region where the Arctic meets the forest. It stretches from Labrador to Alaska and extends from the forests of Siberia through to Scandinavia. In Canada, this zone largely rests on the ancient geological formations of the Canadian Shield.

A quilt of quilts, the pattern extends across many scales. Small patches make up bigger patches, which are themselves part of even bigger ones, which are part of grander patterns (Elkie and Rempel 2001). This nested patchiness reaches from the level of molecules to the whole region. The flows of energy, materials, and organisms that constitute the ecosystem at any scale are influenced by processes occurring at larger and smaller scales. Linkages across scales lead to the ecological integration that is funda-mental to Paakumshumwaau.

Despite clear boundaries across sharp ecological gradients, the pieces of the quilt are interdependent, each one influencing and being influenced by its neighbours (Hansson 1992). Water, animals, seeds, nutrients, heat, microbes, soil, spores, fire, insects, and many other materials and organisms move between patches, influencing the nature of the ecosystems within them. Changes in one patch will very probably bring about changes in adjacent patches but can cause more subtle changes even in distant areas. The movements of fire across the landscape, which depend on the ability of the forest to carry or extinguish a blaze, are but one example of this integration that is relevant to Paakumshumwaau.

This concept of integration is vitally important because, though a designation of protected area may impose a regulated boundary line on a map, Paakumshumwaau remains connected with surrounding lands and is still under the influence of distant conditions and events, whether they are ecological, socio-economic, or political in nature (Hansen and DeFries 2007). Likewise, Paakumshumwaau itself will continue to influence neighbouring and distant lands, whether through its influence on disturbance, species migrations, recreational opportunity, protected area policy, or the changed perspectives of the people it touches. The web of life is wide.

THE QUILTERS: FORCES BEHIND THE PAAKUMSHUMWAAU ECOSYSTEM

Quilts are sometimes community projects, reflecting the work of many hands and sometimes extending over generations. Each quilter brings a different touch to choosing the fabric, the pattern, and the stitch, details that are ultimately seen in the evolving piece. So too, Paakumshumwaau reflects many forces working in and around each other that have influenced the patterns on the land over the millennia.

It is an ecosystem that reflects the combined influence of a diversity of factors acting over many scales of time and space. Fundamental patterns in ecosystem variations arise from the underlying bedrock geology, modified by the actions of ice and water, and overlain by deposits of materials transported from elsewhere. These basic patterns were altered as the land emerged from the ocean and terrestrial ecosystems formed (for a detailed discussion, see Chapter 5, this volume). Early in their development, ecosystem patterns strongly reflect the basic biophysical environment, but as they develop they are modified by cyclical, iterative relationships between soils, plants, and animals. All are under the influence of climate and fire, the most prevalent agents of disturbance in boreal regions.

Geology and Landform Development

The geology of Paakumshumwaau is typical of the Canadian Shield, dominated by ancient acidic metamorphic rocks, with more recent igneous intrusions (Wiken 1986). Geology is expressed in landscape patterns through its effects on topography and the capacity of soils to provide available mineral nutrients such as phosphorus, potassium, calcium, and magnesium, and to buffer against changes in soil acidity. Although the relationships between geological and ecological patterns in Paakumshumwaau have not yet been studied extensively, our observations indicate that the hills around Old Factory Lake are derived from granitic rocks that are rich in potassium feldspars (Goad and Černý 1981). These areas correspond with the best developed deciduous forests in the region, consistent with the high nutrient demands of, for instance, aspen (Perala and Alban 1982). Hence, the regional bedrock geology may have played an important role in defining the ecosystem patterns in contemporary landscapes.

Early in its development, Paakumshumwaau lay beneath the fresh waters of glacial Lake Ojibway, which ponded south of the great Laurentide Ice Sheet as it retreated to the north and east. About 8,500 years ago, the ice dam was breached, rapidly releasing the lake water into the salt water of the Tyrrell Sea (Hillaire-Marcel, Occhietti, and Vincent 1981). This drainage reduced the water level by three hundred metres. In the process, the Tyrrell Sea inundated the lowlands to the east of James Bay as far as the Sakami Moraine, a ridge of sand and gravels adjacent to what is now the James Bay Highway. Implications of these events are also discussed by Ionel Florin Pendea et al. (2010) and in Chapter 5 of this volume. Extrapolating from similar circumstances in the Nastapoka area farther north on the Hudson Bay coast (Lajeunesse and Allard 2003), we suggest that the ice would have become grounded on the hills around Old Factory Lake, leaving behind aprons of sands and gravels on the hillsides.

Relieved of its great weight of kilometres of ice formed during the periods of glaciation, dry land began to emerge from the sea, while glacial sediments were moved and sorted by waves and coastal currents (Hillaire-Marcel, Occhietti, and Vincent 1981). Beach and bay formations like those along the coastline today would have been present during the history of emergence across the entire Paakumshumwaau landscape. As the land rose, beaches became sandy and gravelly ridgelines on hillsides. The coastal basins, with their muddy bottoms of silts and clays, became estuaries, shallow inlets, or salt marshes, which eventually became freshwater ponds and lakes as they separated from the ocean. Gradually filling with sediment and peat, shallow lakes became fens and, subsequently, bogs when poor soil drainage and level topography caused the water to stagnate (Pendea et al. 2010).

Climate

Climate is a distinctive feature of any landscape and is certainly so for boreal regions. Long, cold winters and short, often hot and dry summers generate conditions that are appropriate for certain communities of organisms. Many species that occur south of James Bay do not exist in Paakumshumwaau, and many of its species are not found in the subarctic and arctic climates to the north (see also Chapter 7, this volume, on mammalian species). A key question about changing climate is how it affects the distribution of species along these north-south gradients (Chapin et al. 1995).

Gradients in average and extreme temperatures, rainfall, snowfall, and their seasonal patterns underlie many of the broad patterns in the Paakumshumwaau quilt. Cold as it might be for bathers, James Bay has a moderating influence on the climate of coastal areas, raising temperatures in fall and spring, and lowering them in summer. On many days of the year, fog blankets the coast, generated by warm humid air over cold waters (Plamondon-Bouchard 1975). Away from the coast, the land heats and cools faster than the water of the bay, generating greater extremes of temperature that can drive strong winds inland. Convective lifting of sea air by the heat of rocky ridges and thinly vegetated soils inland probably increases electrical storm activity and heightens the frequency of fire away from the shore (Parisien and Sirois 2003).

In the area around Old Factory Lake, the hilly topography generates smaller-scale climate patterns. South-facing slopes receive more direct sun than north-facing slopes, giving them a warmer, more southern climate. North-facing slopes are characteristically colder, retaining snow earlier in the fall and longer into the spring. High-elevation ridges are colder and more windblown than the valley bottoms, as witnessed by the arctic vegetation that grows on many ridgetops. Narrow valleys cutting across ridgelines funnel the cold air downward, creating cold pockets in depressions and enclosed valleys (Lomolino, Riddle, and Brown 2005).

Time since Emergence as a Factor in Ecosystem Development

Over time, things happen. Two otherwise identical beach ridges, differing by a few hundred or a thousand years since the day they emerged from the sea, will function as very different ecosystems (Vitousek 2004). The older ridge has had more time for change than the younger one. Weathering of minerals, addition and decomposition of organic matter, shifting of phosphorus from more available to less available forms, accumulation of nitrogen, development of soil horizons, and leaching of nutrients are all slow processes that modify ecosystem properties over time (Chadwick et al. 1999).

The land of Paakumshumwaau represents a time sequence of ecological development, with the oldest ecosystems inland and the youngest on the coast. Future landscapes currently lie beneath the bays and estuaries. Today's islands will be tomorrow's hilltops. Only recently have studies been undertaken of the eastern James Bay time sequence from the perspective of terrestrial ecosystem development since emergence (see Pendea et al. 2010; Chapter 5, this volume). Clearly, however, time since emergence is a gradient that underlies all the patterns in the Paakumshumwaau quilt.

Soil Organic Matter Accumulation

The layer of organic matter blanketing Paakumshumwaau also plays a key role in defining the patterns of ecosystems. This layer stores and releases water and nutrients, protects plant roots, seeds, and soil organisms from fire, stabilizes soil against erosion, insulates it against heating and cooling, and acts as a source of fuel to feed intense fires (Bonan and Shugart 1989). It arises from dead plant tissues that decompose at rates determined by the plant species, tissue of origin, and microclimatic conditions in the forest floor (Pare et al. 2006). The slowly decomposing residues produced by moss and needle-leafed evergreens promote accumulation, which can lead to peat formation in wetter areas and ignitable fuel in dryer areas (Simard et al. 2009). The more rapid decomposition of deciduous foliage stimulates biological activity and forest floor turnover, resulting in the lower amounts of more humified deposits that are typical of stands of aspen and birch (Mitton and Grant 1996). Release of nutrients during decomposition

supports plant growth. Vegetation that produces high-nutrient, easily de-
composed residues generates the active forest floors preferred by the species
that grow there. In contrast, species that produce low-nutrient residues
foster slow decomposition and nutrient release, and tend to be tolerant of
nutrient-poor conditions (Zhang et al. 2008). Hence, the "walking on
sponge" feeling we get in a black spruce forest indicates slow growth and
nutrient poverty, relative to the thin black layer in birch stands.

Fire

More than any other factor, fire has changed the fabric of Paakumshumwaau. From coastal forests, where fire virtually never occurs, to inland
stands, where it is a regular visitor, the signs of this destructive and rejuven-
ating force are everywhere on the land. Fires, like people, share fundamental similarities, yet each one maintains a sense of "self," with characteristics
and qualities that distinguish it from the others. The fire regime, or collective
attributes that contribute to fire behaviour and patterns on the landscape,
is credited as the main driver of ecosystem dynamics in the boreal forest
(Bonan and Shugart 1989; Weber and Flannigan 1997). Fire organizes the
forest's physical and biological attributes through its role in structuring
plant communities via the continual "resetting" of successional dynamics,
or the sequence of arrival of new plant species. Therefore, a snapshot view
of Paakumshumwaau at any time, be it present, past, or future, is not so
much a result of fire in a physical sense, but rather of the fire regime (Weber
and Flannigan 1997). Playing a heavy hand in the cycle between birth and
death of biota, fire creates and maintains a forest mosaic of plant communities in differing stages of renewal and development.

A key trend across the Paakumshumwaau ecosystem is the gradient of
fire regime from the coast inland. Coastal islands and promontories almost
never burn, partly because they are often bathed in fog and thus are moist,
partly because they are beyond the reach of fires ignited elsewhere, and
partly because James Bay itself cannot feed a fire that is driven by the pre-
vailing wind (Beaulieu and Allard 2003; Parisien and Sirois 2003). Farther
inland is drier and hotter, and is perhaps exposed to more lightning. At
the moment, fire dominance in inland areas is very evident on either side
of the James Bay Highway, still marked by the massive fire of 1989, but the
fire regime was ever thus.

Fire is both a creative and a destructive force. In many respects, it rejuvenates the boreal ecosystem, but it also immolates established vegetation and removes valuable nutrients. It can also lead to erosion and changes in the quality of lake and river water (Swanson 1981). Many facets of the paradoxical relationship between humans, ecosystems, and fire are explored elsewhere (Scott and Fyles submitted). Given that the forests of Paakumshumwaau can be highly flammable and that humans often set fires, management of fire-related activity will be important as the biodiversity reserve and its surrounding lands develop. The quandary of fire will have to be faced. When should fires be allowed to burn, and when should they be extinguished? Can and should fire be managed to promote its positive effects while minimizing the negative? Or should we simply recognize that fire has always been a part of the land and accept its presence, absence, and impacts (Weber and Flannigan 1997)?

Connectivity and Isolation

Many features of the Paakumshumwaau quilt arise because of the ways in which patches are joined together. This stitching allows movement of organisms, materials, and energy from one patch to another. Close proximity of similar habitats promotes flow, as do linear features such as streams and roads. River valleys, in particular, are preferred migration routes for a wide range of organisms due to their gentle eroded gradients, rich, moist soils supporting lush vegetation, and currents of water and wind (Naiman, Decamps, and Pollock 1993). An important ecological feature of the eastern James Bay region is that rivers generally run east-west and thus do not act as corridors from south to north. This contrasts with the north-south orientation of the James Bay Highway, which enables movement from the south.

But this landscape, though well stitched, is not fully connected. Rivers, lakes, sea, cliffs, and high ridgetops, although preferred or facilitating habitat for some, block the movement of many animals and plants. Even patches of differing vegetation may isolate one patch from other similar habitats. This isolation or lack of connectivity imparts its own influences on the ecosystems, both locally and across the landscape (Turner and Romme 1994).

Isolation is most easily envisioned on the offshore islands, whose terrestrial ecosystems are separated from others by stretches of water that are not easily crossed. But this effect also applies to other isolated ecosystems: arctic tundra on the summits of hills and ridges, bogs isolated in expanses of upland, and unburned stands in the midst of new burns. In all of these cases, the nature of the isolated ecosystem is determined, in part, simply by its isolation and its distance from similar habitats with which it can exchange (MacArthur 1972).

Isolation is a key feature that regulates the involvement of humans in Paakumshumwaau ecosystems. Land travel can be difficult, resulting in the seclusion of many areas simply due to their inaccessibility. Historically, waterways were the main access routes, and areas far from navigable water received few human footprints, except in winter. A main impact of recent technology has been through increasing access via aircraft, snowmobile, and roads. Roads in particular provide linear connections that allow humans and other organisms to bridge physical and distance barriers (Trombulak and Frissell 2000).

Biotic Interactions

Organisms in an ecological community interact with each other in myriad ways. The nature of a community often arises as much from these interactions as from the effects of external or abiotic factors (Haloin and Strauss 2008). More details on biotic interactions are given in Chapter 7 of this volume; only a few examples are offered here for illustration.

Interactions among plants are processes that contribute to determining the boundaries between patches in the Paakumshumwaau quilt. Plants compete for light, water, and nutrients. Plants that swiftly occupy sites following a fire can pre-empt supplies of these resources and exclude other species. Sprouting from roots, a common adaptation of boreal broadleafed plants such as aspen, may produce patches dominated by a single species (Payette 1983). The chemicals produced in the roots and leaves of *Kalmia* and Labrador tea may interfere with the growth of other species (Yamasaki et al. 1998).

Communities of herbaceous, herbivorous, and carnivorous species that constitute food webs are in a constant state of interaction that controls their abundance. Plant productivity imposes bottom-up control on the web, whereas top carnivores impose top-down control (Elmhagen and

Rushton 2007). Changes in one population are reflected in others, often with a time lag that produces cyclical variation, the boreal hare-lynx cycle being a classic example (Peckarsky et al. 2008). Introductions of new species result in population readjustment across the food web. We can only speculate at the food-web-wide effects of the near extinction of beaver during the early 1900s or the arrival of moose in the 1950s.

Each species in a web of biotic interaction is also under the influence of abiotic factors, and each will respond differently to changes in weather conditions or disturbance by fire. Since each species is also reacting to the adjustments of other species, the result is a complex multiple causality pattern of response. This is inherent to ecosystems and makes it difficult to predict the response of populations and communities to changing environments.

Human Influences

Humans have been part of the Paakumshumwaau landscape for a very long time. Its shorelines have probably been a rich source of food and material resources since they emerged from the bay. As shown in Chapter 5 of this volume, significant human habitations at Old Factory Lake predate 3,500 years BP and testify to the ability of the area to support human needs over many centuries.

The role of humans as engineers of the environment (see Chapter 9 for several examples) has grown with the adoption of new technologies over time. Chronicling the role of humans in producing the Paakumshumwaau quilt is well beyond the scope of this chapter and is undertaken in various ways in other chapters. It is sufficient to say, however, that human activity will probably be the largest force shaping the land and its inhabitants in coming decades. The land will respond, as it always has. But the human footprint can easily extend beyond its capacity for response.

FABRIC THREADS: KEYSTONES OF PAAKUMSHUMWAAU

Ecosystems consist of a diversity of species and non-living components. Each component is tied into a web of relationships that allows the ecosystem to function, to continue, and to respond to change. Although all are part of the fabric, some are critical to the existence of the rest. These are the keystones, species, and components that more than any others determine the patterns of the landscape. Although the technical use of the term "keystone" by ecologists has been subject to controversy and refinement over the years (Davic 2003), we employ it in a general way and extend it beyond its usual confines in community ecology. Several primary keystones of Paakumshumwaau are described below.

Eelgrass *(Zostera marina)*

Eelgrass is a critical species in the ecology of the James Bay coast. In shallow, muddy bays, it forms extensive meadows that provide habitat for fish and invertebrates and food for passing waterfowl, particularly ducks and geese. Along northern coasts around the world, eelgrass plays a similar role as the base of a diverse ecosystem.

Eelgrass and the ecosystem it supports are products of complex interactions between James Bay and the rivers that feed its estuaries (Lalumière et al. 1994). Oceans supply the range of salinity required for good growth and also redistribute sediments, depositing silt in sheltered bays. Rivers deliver fine sediments from the land to provide optimal eelgrass habitat. However, they also deliver fresh water that can dilute the salinity to the point where eelgrass cannot survive. Under very sheltered conditions, where summer water temperatures exceed 20°C and salinity is high, eelgrass can be overgrown by algae and infested with a microbial wasting disease (Muehlstein 1992). Hence, despite forming the solid foundation of a significant ecosystem, eelgrass itself rests in a delicate ecological balance.

Eelgrass meadows along the eastern coast of James Bay have declined significantly in extent. This has been variously attributed to climate change, isostatic rebound, and the wasting disease. However, Frederick Short (2007) argues that the only cause that withstands critical evaluation is diminished salinity due to the increased freshwater inflow to James Bay, produced by hydro-electric developments on the La Grande River. Whether these observations and analyses apply to the coastal portion of Paakumshumwaau requires further exploration, given that the plume of the La Grande River is a considerable distance to the north and has a northerly extension, and the increased flows have coincided with decreased flows in the Opinaca and Eastmain Rivers south of Paakumshumwaau.

White Spruce *(Picea glauca)*

White spruce is unusual among boreal species in that it reproduces only by seed and is not well adapted to fire. For white spruce to be present in a regenerating burned forest, living trees must grow nearby to supply the necessary seeds (Galipeau, Kneeshaw, and Bergeron 1997). Consequently, white spruce is dependent on the patchiness of forest disturbance and can be eliminated by frequent, widespread fire. It is not surprising that the species is abundant only along the coast, where fire frequency is low. Interestingly, white spruce very commonly mixes with aspen across Canada but does not appear to be present in the well-developed aspen forests around Old Factory Lake.

Black Spruce *(Picea mariana)*

Black spruce dominates forests and forested wetlands over much of the Paakumshumwaau region. This evergreen grows in a wide range of conditions, from dry uplands to the deep peat of forested muskeg. It is well adapted to fire. Heat from a burning canopy stimulates the tree to release seed from cones that cluster in a characteristic bulging "head" at its top, ensuring a new crop of seedlings on the burnt seedbed. In the absence of fire, branch tips pressed to the ground by snow can root and form new trees (Farrar 1995). The clumps of spruce that are characteristic of lichen woodlands are generated in this manner.

Black spruce copes with the difficult boreal living conditions by being conservative in all respects. Retaining its needles for a dozen or more years allows it to accumulate and hold the nutrients that it gradually obtains from cold and nutrient-poor soils. Even when presented with good conditions, such as warm temperatures or added nutrients, black spruce has a muted response. These conservative features seem to allow it to grow well with sphagnum and the nutrient-poor peaty soils it produces.

Jack Pine *(Pinus banksiana)*

A classic example of a fire-adapted species, jack pine is common inland, where it often forms pure stands on sand and rocky soils. Like those of black spruce, its seeds are retained for many years inside "serotinous" cones, which require the heat of fire to open (Farrar 1995). This ensures that seed will be released onto freshly burned seedbeds, where competition from other plants will be low. Without fire, the tree is far less likely to reproduce. Short-lived compared to white and black spruce, jack pine can be eliminated from the forest if fire frequency is low.

Mosses

Mosses are easily overlooked next to the more showy, towering, or charismatic forest plants, but they play key, often defining, roles in the Paakumshumwaau forest. Many moss species grow in the region, most of which are little known. For this discussion, we focus on two groups – sphagnum *(Sphagnum)* and feather-moss (*Hylocomium, Pleurizium* and other genera) the ecology of which is reasonably well known.

Sphagnum is the genus of peat-forming moss species. It is a dominant plant in bogs, along with ericaceous shrubs, and is the characteristic groundcover species of many black spruce forests. Productive under moist conditions, these plants produce dead residues with very low decomposition rates, which lead to the accumulation of organic matter and the production of peat (Bona et al. 2013). During decomposition, sphagnum generates acidity, which reduces nutrient availability to plants. The peat also acts as an insulating blanket that can keep the soil cold well into the growing season. Thus, the thick accumulation of cold, nutrient-poor peat presents poor growing conditions for many plant species.

Sphagnum peat has a high water-holding capacity and can remain moist throughout the growing season, even on sites where drainage is not particularly poor. As they decompose, subsurface peat layers are compacted by snow loads and organic accumulation above. Compacted layers reduce the flow of water and can generate a water table closer to the surface than that originally present on the site. Thus, sphagnum seems to create the moist conditions that make it productive.

Feather-mosses (*Pleurizium schreberii, Hylocomium splendens,* and others) commonly grow in boreal forests. In Paakumshumwaau, they are an important component of both the coastal white spruce forests and the black spruce forests inland. Feather-moss foliage is often associated with algae that can capture nitrogen from the air and convert it to a form useful to plants (DeLuca et al. 2002). This process, over time, can raise the fertility of the soil and lead to increased plant growth in these generally nitrogen-poor forests.

Lichens

Seen from an aircraft, lichen woodlands are an outstanding feature of the northern landscape, with tiny clumps of dark spruce in a sea of white lichen. Up close, the lichens sit like small white clouds on the ground. When soaked with morning dew or after a rain, they are like soft sponges. But when they are dry, they become crisp and can be crushed to powder underfoot. Footprints made on a sunny afternoon can still be seen years later. Often thought of as pioneers that establish soon after fire, lichens can virtually monopolize a site for decades due to their sheer density or because they produce chemicals that inhibit the growth of other plants (Morneau and Payette 1989). In areas where deep snow does not accumulate, lichens are a critical winter food for caribou. Because lichen carpets can so easily be crushed and destroyed by human activity, they may require specific management attention as summer use of the biodiversity reserve increases.

The Blueberry Family *(Ericaceae)*

Species of the blueberry family grow as low shrubs in boreal regions. Various combinations of species from this family dominate shrub canopies in all the spruce and pine forests of Paakumshumwaau. They are also, along with low willows, the dominant groundcover of heathlands on coastal headlands and islands.

These are clonal plants that regenerate actively from root stocks following fire. Unless fire completely consumes the organic layer in which most of the roots lie, new green shoots will appear almost immediately after fire. *Kalmia* inhibits the growth of associated tree seedlings and reduces microbial activity and nitrogen availability in soil, possibly due to the production of allelochemicals (Yamasaki, Fyles, and Titus 2002). Though less is known about inhibitory chemical effects in other species, work outside North America suggests that species in this family share the capacity to modify their environments. In circumstances of reduced tree-seed availability following fire, as when an immature forest burns, *Kalmia* and *Ledum* in particular have the capacity to rapidly regenerate and fully occupy a site, leading to the formation of heath.

Aspen *(Populus tremuloides)*

Few descriptions of vegetation in the James Bay region mention aspen, except in passing, because of its sparse and discontinuous distribution. A unique feature of Paakumshumwaau is that aspen is abundant and dominant on the hills around Old Factory Lake. There it forms significant stands along with paper birch and a diverse understory of deciduous shrubs and herbs.

Aspen has a unique set of traits that result in its distinctive ecological behaviour across North America (Farrar 1995), even in the Paakumshumwaau region, which is near the northern limit of its distribution. Its tiny, tufted seeds are produced prolifically and are widely dispersed by the wind, but they die within three weeks if they do not find a moist germination site. Once established, however, aspen spreads by clonal growth and resprouts vigorously from roots if a stand is burned or cut. We speculate that the large stands around Old Factory Lake are the continuation of forests established thousands of years ago.

Aspen and its associated deciduous plant species produce foliage litter that is nutrient rich and low in lignin, compared to the conifers. Decomposition is relatively rapid, and aspen stands are characteristically associated with a well-decomposed organic forest floor. The resultant ecosystem is thus rich in nutrients, an unusual circumstance in the northern boreal region. Aspen is dominant in many forests in the clay belt south of James Bay but largely disappears not far north of Matagami. However, it is an effective pioneer of human disturbances, and it follows the roadsides and old gravel pits of the James Bay Highway all the way to Radisson.

Riparian Zones

The zones of vegetation bordering rivers, streams, and lakes play key ecological roles that make them disproportionately significant, relative to the area they occupy. Bodies of water are important connectors in the landscape, used preferentially by plants and animals to move from place to place. Riparian vegetation provides cover for animals seeking water and, because of nutrients deposited in sediment, often has an abundance of palatable and nutrient-rich food plants. In Paakumshumwaau, riparian vegetation is usually dominated by willow *(Salix)* and alder *(Alnus)* species, which are important browse for beaver and moose.

It is not surprising that humans also use riparian areas preferentially. Rivers remain important travel routes in the region, and much human activity has historically concentrated close to the water. Such activity can alter riparian areas, sometimes to the detriment of the animals that use them and sometimes to their benefit. If it increases significantly in the future, management policy for the biodiversity will be required to focus on riparian use.

The Forest Floor

The organic material that accumulates on the soil plays a keystone role in boreal forests because it is the seedbed from which forests sprout following fire. Many tree species, characteristically pine and spruce, germinate and grow much better on mineral soil than on organic matter. Many other boreal plants resprout from roots, in which case the presence of a moist organic layer may be crucial to their post-fire success. The amount and character of the organic matter remaining after fire has an important influence on which species become established. Whether the organic layer burns or not depends on its degree of decomposition and its moisture content. Thus, it is influenced by the vegetation that is burned, the site moisture regime, and the weather in the days and weeks before the fire (Greene et al. 2007). It is the many combinations of these factors and their influence on forest composition that create the fine detail of the Paakum-shumwaau quilt.

QUILT PATTERNS: GRADIENTS OF CHANGE ACROSS THE LANDSCAPE

Gradients, patterns that change gradually with distance or time, are everywhere in Paakumshumwaau. They are the rule rather than the exception. Some are visible close at hand, whereas others are apparent only from a distance. Some are so clear that anyone could bump into them, whereas others are subtle and only apparent to a careful observer. But the rule in Paakumshumwaau is that, whether we run or wait, we will see change. A Cree story captures some of the patterns of change from inland to the bay.

> Long ago, but not so long ago, far into the inland of Paakumshumwaau,
> legendary Wolverine fought with Giant Skunk, who had frightened all
> of the other animals. Wolverine was fierce and strong in tooth and claw,
> and the battle was fearsome. But as the conflict flared, Wolverine was

sprayed in the eyes and blinded by Skunk's musk. Wolverine knew that he must wash his eyes in the sea or else be blind forever, so he abandoned the fight and rushed toward the coast. He ran blindly, feeling his way as best he could but bumping into trees at every step. Despite his haste, Wolverine stopped at each tree just long enough to ask its name; "Who are you?" he asked. Each tree called back, as trees always do, "I am *wishkui* (white birch)," "I am *uchisk* (jack pine)," "I am *iiyaahtikw* (black spruce)," "I am *miitus* (poplar)," "I am *waachinaakin* (tamarack)," "I am *iyaashiht* (balsam fir)." Finally, Wolverine heard "I am *minihiikw* (white spruce). I am the only tree that stands next to the bay." And so Wolverine knew that he was headed in the right direction, even though he could not see. When at last he reached the bay, Wolverine plunged into the water to wash his eyes and regain his sight. In doing so he made the water salty for ever after.[1]

From the Arctic to the Boreal: The Leading Edge of the Outer Islands

Transitions to the coast from the outer islands of Weston, Old Factory, Pebble, and Solomon's Temple mark a change from the Arctic to the boreal. The outermost islands are subject to the strong winds of James Bay. These islands are mainly the coarse gravels of what were undersea shoals not many centuries ago, with sand dunes blown up along their edges and old beach ridges marching downslope to the water (Scott et al. 2009). They are too exposed to the northwest winds to support trees, and only low or creeping shrubs, hardy grasses, and other low-growing arctic species survive in their sheltered areas. The arctic tundra element of this vegetation is most evident on the outermost locations.

Closer to the coast, the island forests, often drier, lichen carpeted, and quite open, grow on coarse, thin soil, particularly in the protected lee of rocky hills. Denser spruce-moss forests, similar to those on the coast, appear in less well-drained areas, occasionally with southern species such as mountain ash *(Sorbus)* in protected south-facing nooks. Spruces on exposed sides of the islands and coastal promontories are small, twisted by the force of sea winds. In sheltered areas, white spruce may grow quite large and apparently quite quickly, compared to black spruce (Hustich 1950).

From Shoreline to Forest Edge: The Birthing Edge of Paakumshumwaau

Some of the most intricate patterns of the Paakumshumwaau quilt occur at its western edge, where it rolls out of the waves and into the forest. At low tide, bands of brown *Fucus* and green algae line the rocky beaches. In muddy bays, creeping alkaligrass *(Puccinellia phryganodes)* and marsh cinquefoil *(Argentina egedii)* inhabit the low marsh, which extends from the water's edge into the high marsh of tall sedges and grasses (Dignard et al. 1991). The community of herbs and grasses at the top of the beach is dominated by robust marram grass *(Ammophila breviligulata)*. Very similar marram grass communities grow on beaches around the world. Beyond the reach of all but the highest tides, sweet gale *(Myrica gale)*, willow, and alder give a shrubby welcome and then grade into the white spruce forest on higher ground. The bay-to-forest transition thus presents a finely detailed zonation, played out over as little as twenty metres. But it is worth reflecting on the notion that this narrow band has at one time touched each point in the Paakumshumwaau quilt. Over millennia, as the land rose from the bay, these communities, or something very similar, occupied the always emerging edge (Sayles and Mulrennan 2010; Chapter 9, this volume).

Inland from the Coast: Changing Colours in Paakumshumwaau

Changes in the forest "observed" by Wolverine in the story above are easily seen on today's landscape. The coastal strip of forest is dominated by white spruce (Hustich 1950; Parisien and Sirois 2003). The success of white spruce and its associated species in this forest may arise from the higher humidity and cooler growing-season temperatures at sites close to the bay (Beaulieu and Allard 2003; Gamache and Payette 2004), and the lower occurrence of fires in coastal versus inland areas (Caccianiga and Payette 2006). The fragmentation and dissection of the coast by inlets, bays, and rivers also contribute to a lower frequency and intensity of fire in this zone than farther inland. Large stands of tall balsam poplar *(Populus balsamifera)* also appear within a few kilometres of the Paakumshumwaau River mouth, perhaps benefitting from the nutrient-rich sediments, as well as the humid and fire-protected conditions. Although forest fires rarely reach the coast, a few well-drained sections of coastline in the bays north of the river recently burned to the shore.

Often less than a kilometre inland, the forest shifts in dominance to black spruce, forming another strip parallel to the coast. About twenty kilometres wide, this strip is characteristically uniform closed-canopy black spruce, with an understory of blueberry and Labrador tea on variously organic, poorly drained soils (Gamache and Payette 2004). Within this strip, areas that were previously rocky beaches support open-canopy black spruce–lichen woodlands, sometimes with scattered jack pine and aspen.

Farther inland, the forest is variously dominated by jack pine and black spruce, the composition depending on local topography and soils, which interact with fire frequency. Analysis of wood preserved in peat profiles in the zone suggests that these two species have shifted back and forth in dominance for millennia, both continuously present (Desponts and Payette 1993).

GRADIENTS OF CHANGE ACROSS DECADES AND MILLENNIA

All of Paakumshumwaau, from the Sakami Moraine to the coast, was once the sea floor of James Bay. Once washed by coastal surf, the land slowly rose beyond the reach of salt water. Each point on the land took its own path of development, from salt to fresh, from sea floor to forest. If we could hover above Paakumshumwaau and fast-forward through the last few millennia, we would see the land emerging from James Bay and the characteristic colours arising from the broad patterns of developing vegetation: the blue of the bay, the white beach and outcrop strip, the brilliant green of the coastal strip grading into darker colours of white spruce canopies and patchy black spruce and pine as the coast became interior. But the scene would not unfold as a uniform set of colours. It would change constantly.

We might expect that each location, as it rose from the ocean, would follow a pattern of change over time that generally follows the spatial pattern of change that we see from the coast to inland areas. Locations far from the coast might be expected to have supported white spruce long ago, then transitioned to dense black spruce forest and eventually to jack pine as continuing uplift brought them far enough from the coast for fires to become frequent. In terrestrial ecosystems, however, decomposition and fire destroy traces of past plant communities, so we can only speculate about what actually happened during the centuries or millennia after the sea floor became land (Foster, Knight, and Franklin 1998).

In marine and aquatic ecosystems, however, a trace of the past is recorded by the pollen, grains, and fossil parts of plants and animals laid down in calm bays and lakes during the accumulation of sediment and peat, as discussed in Chapter 5 of this volume. Sediment profiles of Paakumshumwaau show clear patterns of change over millennia, with the shift in biota signalling changing conditions as salty tidal flats separated from the sea, became brackish estuaries, and eventually evolved into rich freshwater fens. A shift to peat accumulation that isolates plants from the nutrient-rich sediment below leads to further transition to nutrient-poor bog.

The transition from tidal flat to fen to bog or forest is not simply a replacement of one vegetation community with another, but marks an important change in ecosystem functions (Huston and Smith 1987). The tidal marshes and fens are rich in nutrients that allow for the development of high-biomass-producing communities of grass and herbs, which become prime habitat for many wildlife species, particularly waterfowl, beaver, and moose. The subsistence economy of Wemindji Crees is thus bound to the fate of these nutrient-rich ecosystems (Sayles and Mulrennan 2010). In this respect, the resource richness of the Paakumshumwaau coast is marked by a fragile balance between the nutrient-rich ecosystems born in the wake of the retreating sea and those lost through ecological succession to nutrient-poor bogs and forests.

In our fast-forward view of Paakumshumwaau, a close look at forested areas would show them flicking from dark green to black and then to bright yellow-green before returning to the darker shades. Various patches would follow their own sequence, timing, and palette of colours. If we looked very closely, we might see that the quick change from dark green to black came with a puff of smoke from forest fires, for that shimmering colour change is the succession of plant communities that emerge following fire.

Ecologists around the world have studied plant succession. It is often described as a process that follows a disturbance, in which pioneer species quickly colonize the disturbed area and create conditions that enable later successional species to establish themselves. In the absence of fire, the later species will eventually out-compete the pioneers and become dominant, until the next fire moves through (Hansen et al. 2001). But in Paakumshum-waau, as in much of the boreal, the pattern is different.

Fire is so common in the boreal forest that almost all of its plant species have adaptations that allow them to re-establish immediately following fire. Thus, most Paakumshumwaau forests arise from the ashes, with most of the plant species from the previous stand. Certainly, some classic pioneers are present: willow, aspen, and the fireweeds (*Chamerion* spp.) produce tiny seeds in puffs of down that are carried on the wind for kilometres to colonize new sites. Some herbs, such as bleeding hearts (*Dicentra* spp.), can store seed in the forest floor to germinate decades later, following a fire. But to walk through most Paakumshumwaau forests a year or two after a fire is to see a mirror of the forest that was there before. Willow, aspen, the blueberry family, and many herbs sprout from roots, growing shoots far faster than they would from geminating seed. Pine and black spruce seedlings arise from seed that rained down from the scorched cones in burned canopies. The changing shades of green appear because fast-growing species, such as willow, dominate sooner and slower-growing taller species dominate later. However, the species composition changes little over the life of the stand and is maintained, potentially across many fire cycles.

Forest dynamics on the islands and coastal strip differ from those farther inland. In these areas, the forest burns only infrequently, and the species often lack adaptations to fire. Colonization following fire is much more dependent on seed arriving from unburned adjacent patches. White spruce is a good example of a species that must seed in from the fire's edge, and there are many others. Of course, the coastal areas are often moist, and the fires that do occur usually fail to travel far, so recolonization from edges and unburned patches is effective (Caccianiga and Payette 2006). However, since human activity may concentrate on coastal areas, special attention to fire in this zone will be important as the biodiversity reserve develops.

Like the changing pattern of richness in development from wetland to forest, the shift from fast-growing herbs and shrubs to coniferous trees results in a changing in habitat and food resources for many animals. A newly blackened forest is a desert for most wildlife, but the new shoots of fast-growing sprouts are so nutrient rich, easily reached, and highly palatable that the early years after a fire can provide excellent feeding habitat for many animals and birds. As the trees become tall and the canopy closes, many palatable species decline. Except for some, such as the crossbills that specialize in cracking cones to eat the seeds, a dense forest of tall trees is a hungry place.

IMPLICATIONS OF THE NATURAL HISTORY FOR THE PAAKUMSHUMWAAU BIODIVERSITY RESERVE

The land is a product of its natural history, reflecting the myriad influences of climate, geology, biota, disturbance, and human presence. The natural history of Paakumshumwaau and the history of its nature contribute insights to inform the concept and management of the reserve as it develops.

The history speaks of a landscape that is dynamic, supporting a suite of factors and species that generate and effectively respond to change. The forces of change are seen everywhere in Paakumshumwaau: landscapes arising after glaciation, forests regenerating after fire, species arriving from elsewhere, human activity growing in scope and intensity. In this dynamic landscape, much of what we see today could be different in a year or a decade. Each turn in the river, each hill slope, each animal or plant has a past and a future.

The history also speaks of a landscape that, though not static, maintains a suite of consistent patterns in space and over time. The details of the quilt may change, but its patterns and colours are drawn from a palette that changes little. Many features of Paakumshumwaau have remained the same for eons. Or perhaps more truthfully, their rate of change has been imperceptibly slow from a human perspective. The shape of the land, of its hills and valleys, is much the same now as when it emerged from the ocean; soils that were sandy when the beaches that spawned them were invaded by forest are still sandy today. Arguably, the ways of the plants and animals, their ecological requirements, and their roles in the eco-system have remained largely the same since they arrived. Many of these features impart a degree of stability to Paakumshumwaau. The scope of constant change is bounded within a range of variation (Holling and Meffe 1996). Forest dynamics follow similar cycles conditioned by the available suite of species to regenerate. Repeating patterns emerge, reflecting the underlying influences of the slow variables. Order and chaos co-exist on this land.

Paakumshumwaau can be tough and resilient. It has responded to the creation of new earth and destruction by fire. Its species are adapted to harsh climate, poor soil, and scarce food. The land continues to change and thrive, though only within the realm of its experience.

But under novel conditions, Paakumshumwaau may be fragile. A warming climate provides opportunities for new species that perhaps hitchhike from the south with human visitors, and the footprints and tire tracks left by those visitors may push the land into a set of conditions it has not known before (Skarpaas and Shea 2007). Just as the shores will continue to rise from the bay, the tide of human presence will continue to rise, creating new opportunities and challenges. Even the designation of the biodiversity reserve raises awareness of the region and drives interest and human presence on the land. Government policies aimed at opening access to resource exploitation in the North will accelerate this change.

If the natural history tells us anything, it is the story of large forces that have great influence over small and local details. Paakumshumwaau is not an island. It is connected geographically, climatically, ecologically, economically, and socially to lands and communities beyond its boundaries. This theme echoes the concept of Eeyou Istchee and the embeddedness of the Cree Nation itself in the land, as described in the Introduction to this volume. On the basis of their work with mammals, Murray Humphries, Jason Samson, and Heather Milligan also reach this conclusion (Chapter 7, this volume). Though we may draw an administrative line on a map to designate a protected area, the land remains connected. The quilt has no edge.

The management of Paakumshumwaau must recognize this connectedness. Part of its future rests within the boundaries of the biodiversity reserve, along with its local management authority and knowledge institutions. Much, however, is dependent on processes and decisions in other places and contexts. Those who concern themselves with the patterns of the Paakumshumwaau quilt and their future unfolding must look beyond its immediate borders. As we have learned from Rodney Mark, the Wemindji elders, our Cree colleagues, and from Paakumshumwaau itself, relationship and respect must govern our footprints on the land.

NOTE

1 This story, told by John Blackned in the late 1960s, was recorded by Richard Preston in his book *Cree Narrative* (2002, 159–62). We retell it here, with appreciation and respect for the tale and its many tellers. According to the elders, every telling provides an occasion for tellers and listeners to acquire new understandings. For another version of the story, see Chapter 10, this volume.

WORKS CITED

Beaulieu, Nancy, and Michel Allard. 2003. "The Impact of Climate Change on an Emerging Coastline Affected by Discontinuous Permafrost: Manitounuk Strait, Northern Québec." *Canadian Journal of Earth Sciences* 40 (10): 1393–404.

Bona, Kelly, James W. Fyles, Cindy Shaw, and Werner A. Kurz. 2013. "Are Mosses Required to Accurately Predict Upland Black Spruce Forest Soil Carbon in National-Scale Forest C Accounting Models?" *Ecosystems* 16: 1071–86.

Bonan, Gordon B., and Herman H. Shugart. 1989. "Environmental Factors and Ecological Processes in Boreal Forests." *Annual Review of Ecology and Systematics* 20: 1–28.

Caccianiga, Marco, and Serge Payette. 2006. "Recent Advance of White Spruce (*Picea glauca*) in the Coastal Tundra of the Eastern Shore of Hudson Bay (Québec, Canada)." *Journal of Biogeography* 33 (12): 2120–35.

Chadwick, Oliver A., Louis A. Derry, Peter M. Vitousek, Barry J. Huebert, and Lars O. Hedin. 1999. "Changing Sources of Nutrients during Four Million Years of Ecosystem Development." *Nature* 397 (6719): 491–97.

Chapin, F. Stuart, Gaius R. Shaver, Anne E. Giblin, Knute J. Nadelhoffer, and James A. Laundre. 1995. "Responses of Arctic Tundra to Experimental and Observed Changes in Climate." *Ecology* 76 (3): 694–711.

Davic, Robert D. 2003. "Linking Keystone Species and Functional Groups: A New Operational Definition of the Keystone Species Concept." *Conservation Ecology* 7 (1): r11. http://www.ecologyandsociety.org/vol7/iss1/resp11/.

DeLuca, Thomas H., Olle Zackrisson, Marie-Charlotte Nilsson, and Anita Sellstedt. 2002. "Quantifying Nitrogen-Fixation in Feather Moss Carpets of Boreal Forests." *Nature* 419 (6910): 917–20.

Desponts, Mireille, and Serge Payette. 1993. "The Holocene Dynamics of Jack Pine at Its Northern Range Limit in Québec." *Journal of Ecology* 81 (4): 719–27.

Dignard, Normand, R. Lalumière, A. Reed, and M. Julien. 1991. *Les habitats côtiers du nord-est de la baie James*. Ottawa: Service canadien de la faune.

Elkie, Philip C., and Robert S. Rempel. 2001. "Detecting Scales of Pattern in Boreal Forest Landscapes." *Forest Ecology and Management* 147 (2): 253–61.

Elmhagen, Bodil, and Stephen P. Rushton. 2007. "Trophic Control of Mesopredators in Terrestrial Ecosystems: Top-Down or Bottom-Up?" *Ecology Letters* 10 (3): 197–206.

Farrar, John Laird. 1995. *Trees in Canada*. Ottawa: Canadian Forest Service.

Foster, David R., Dennis H. Knight, and Jerry F. Franklin. 1998. "Landscape Patterns and Legacies Resulting from Large, Infrequent Forest Disturbances." *Ecosystems* 1 (6): 497–510.

Galipeau, Christine, Daniel Kneeshaw, and Yves Bergeron. 1997. "White Spruce and Balsam Fir Colonization of a Site in the Southeastern Boreal Forest as Observed 68 Years after Fire." *Canadian Journal of Forest Research* 27: 139–47.

Gamache, Isabelle, and Serge Payette. 2004. "Height Growth Response of Tree Line Black Spruce to Recent Climate Warming across the Forest-Tundra of Eastern Canada." *Journal of Ecology* 92 (5): 835–45.

Goad, Bruce E., and Petr Černý. 1981. "Peraluminous Pegmatitic Granites and Their Pegmatite Aureoles in the Winnipeg River District, Southeastern Manitoba." *Canadian Mineralogist* 19 (1): 177–94.

Greene, David F., et al. 2007. "The Reduction of Organic-Layer Depth by Wildfire in the North American Boreal Forest and Its Effect on Tree Recruitment by Seed." *Canadian Journal of Forest Research/Revue canadienne de recherche forestière* 37 (6): 1012–23.

Haloin, Jon R., and Sharon Y. Strauss. 2008. "Interplay between Ecological Communities and Evolution." *Annals of the New York Academy of Sciences* 1133 (1): 87–125.

Hansen, Andrew J., and Ruth DeFries. 2007. "Ecological Mechanisms Linking Protected Areas to Surrounding Lands." *Ecological Applications* 17 (4): 974–88.

Hansen, Andrew J., Ronald P. Neilson, Virgina H. Dale, Curtis H. Flather, Louis R. Iverson, David J. Currie, Sarah Shafer, Rosamonde Cook, and Patrick J. Bartlein. 2001. "Global Change in Forests: Responses of Species, Communities, and Biomes." *Bioscience* 51 (9): 765–79.

Hansson, Lennart. 1992. "Landscape Ecology of Boreal Forests." *Trends in Ecology and Evolution* 7 (9): 299–302.

Hillaire-Marcel, Claude, Serge Occhietti, and Jean-Serge Vincent. 1981. "Sakami Moraine, Québec: A 500-km-Long Moraine without Climatic Control." *Geology* 9 (5): 210–14.

Holling, Buzz, and Gary Meffe. 1996. "Command and Control and the Pathology of Natural Resource Management." *Conservation Biology* 10 (2): 328–37.

Hustich, Ilmari. 1950. "Notes on the Forests on the East Coast of Hudson Bay and James Bay." *Acta Geographica* 11: 1–83.

Huston, Michael, and Thomas Smith. 1987. "Plant Succession: Life History and Competition." *American Naturalist* 130 (2): 168–98.

Lajeunesse, Patrick, and Michel Allard. 2003. "The Nastapoka Drift Belt, Eastern Hudson Bay: Implications of a Stillstand of the Québec–Labrador Ice Margin in the Tyrrell Sea at 8 Ka BP." *Canadian Journal of Earth Sciences* 40 (1): 65–76.

Lalumière, Richard, Danielle Messier, Jean-Jacques Fournier, and C. Peter McRoy. 1994. "Eelgrass Meadows in a Low Arctic Environment, the Northeast Coast of James Bay, Québec." *Aquatic Botany* 47 (3): 303–15.

Lomolino, Mark V., Brett R. Riddle, and James H. Brown. 2005. *Biogeography.* 3rd ed. Sunderland, MA: Sinauer Associates.

MacArthur, Robert H. 1972. *Geographical Ecology.* New York: Harper and Row.

Mitton, Jeffry B., and Michael C. Grant. 1996. "Genetic Variation and the Natural History of Quaking Aspen." *Bioscience* 46 (1): 25–31.

Morneau, Claude, and Serge Payette. 1989. "Postfire Lichen-Spruce Woodland Recovery at the Limit of the Boreal Forest in Northern Québec." *Canadian Journal of Botany* 67 (9): 2770–82.

Muehlstein, Lisa K. 1992. "The Host-Pathogen Interaction in the Wasting Disease of Eelgrass, *Zostera marina.*" *Canadian Journal of Botany* 70 (10): 2081–88.

Naiman, Robert J., Henri Decamps, and Michael Pollock. 1993. "The Role of Riparian Corridors in Maintaining Regional Biodiversity." *Ecological Applications* 3 (2): 209–12.

Pare, David, R. Boutin, Guy R. Larocque, and Frédéric Raulier. 2006. "Effect of Temperature on Soil Organic Matter Decomposition in Three Forest Biomes of Eastern Canada." *Canadian Journal of Soil Science* 86 (2): 247–56.

Parisien, Marc-André, and Luc Sirois. 2003. "Distribution and Dynamics of Tree Species across a Fire Frequency Gradient in the James Bay Region of Québec." *Canadian Journal of Forest Research* 33 (2): 243–56.

Payette, Serge. 1983. "The Forest Tundra and Present Tree-Lines of the Northern Québec-Labrador Peninsula." *Nordicana* 47: 3–23.

Peckarsky, Barbara L., et al. 2008. "Revisiting the Classics: Considering Nonconsumptive Effects in Textbook Examples of Predator-Prey Interactions." *Ecology* 89 (9): 2416–25.

Pendea, Ionel Florin, André Costopoulos, Colin Nielsen, and Gail Lois Chmura. 2010. "A New Shoreline Displacement Model for the Last 7 Ka from Eastern James Bay, Canada." *Quaternary Research* 73 (3): 474–84.

Perala, Donald A., and David H. Alban. 1982. *Rates of Forest Floor Decomposition and Nutrient Turnover in Aspen, Pine, and Spruce Stands on Two Different Soils.* US Forest Service Research Paper NC-277S. St. Paul, MN: U.S. Dept. of Agriculture, Forest Service, North Central Forest Experiment Station.

Plamondon-Bouchard, Monique. 1975. "Charactéristiques et fréquence des nuages bas à Poste-de-la-Baleine en 1969." *Cahiers de géographie du Québec* 19 (47): 311–30.

Preston, Richard J. 2002. *Cree Narrative: Expressing the Personal Meaning of Events.* Montreal and Kingston: McGill-Queen's University Press.

Sayles, Jesse S., and Monica E. Mulrennan. 2010. "Securing a Future: Cree Hunters' Resistance and Flexibility to Environmental Changes, Wemindji, James Bay." *Ecology and Society* 15 (4): 22. http://www.ecologyandsociety.org/vol15/iss4/art22/.

Scott, Katherine, Véronique Bussières, Sylvain Archambault, Wren Nasr, Jim Fyles, Kristen Whitbeck, Henry Stewart, and Colin Scott. 2009. "Augmenting Information for a Proposed Tawich National Marine Conservation Area Feasibility Assessment, James Bay Marine Region: Cultural and Bio-ecological Aspects." Montreal,

Wemindji Protected Area Project. http://wemindjiprotectedareapartnership. weebly.com/uploads/1/3/9/6/13960624/scott_et_al_2009_report_for_parks_ canada.pdf.

Scott, Katherine, and James Fyles. Submitted. "Listening to Trees: Post-fire Ecology in Wemindji Territory." In *Dialoguing Knowledges: Finding Our Way to Respect and Relationship*, ed. Colin Scott, Peter Brown, and Jessica Labrecque. Vancouver: UBC Press.

Short, Frederick T. 2008. "Report to the Cree Nation of Chisasibi on the Status of Eelgrass in James Bay: An Assessment of Hydro-Quebec Data Regarding Eelgrass in James Bay, Experimental Studies on the Effects of Reduced Salinity on Eelgrass, and Establishment of James Bay Environmental Monitoring by the Cree Nation." Durham, NH, Jackson Estuarine Laboratory, University of New Hampshire.

Simard, Martin, Pierre Y. Bernier, Yves Bergeron, David Pare, and Lakhdar Guerine. 2009. "Paludification Dynamics in the Boreal Forest of the James Bay Lowlands: Effect of Time since Fire and Topography." *Canadian Journal of Forest Research/ Revue canadienne de recherche forestière* 39 (3): 546–52.

Skarpaas, Olav, and Katriona Shea. 2007. "Dispersal Patterns, Dispersal Mechanisms, and Invasion Wave Speeds for Invasive Thistles." *American Naturalist* 170 (3): 421–30.

Swanson, Frederick J. 1981. "Fire and Geomorphic Processes." In *Fire Regimes and Ecosystem Properties: Proceedings of the Conference. General Technical Report WO-GTR-26. Dec 11, 1978*, ed. Harold A. Mooney, T.M. Bonnicksen, Norman L. Christensen, Jr., James E. Lotan, William A. Reiners, 401–20. Washington, DC: USDA Forest Service.

Trombulak, Stephen C., and Christopher A. Frissell. 2000. "Review of Ecological Effects of Roads on Terrestrial and Aquatic Communities." *Conservation Biology* 14 (1): 18–30.

Turner, Monica G., and William H. Romme. 1994. "Landscape Dynamics in Crown Fire Ecosystems." *Landscape Ecology* 9 (1): 59–77.

Vitousek, Peter Morrison. 2004. *Nutrient Cycling and Limitation: Hawai'i as a Model System*. Oxford: Princeton University Press.

Weber, Michael G., and Mike D. Flannigan. 1997. "Canadian Boreal Forest Ecosystem Structure and Function in a Changing Climate: Impact on Fire Regimes." *Environmental Reviews* 5 (3–4): 145–66.

Wiken, Edward. 1986. *Terrestrial Ecozones of Canada*. Ecological Land Classification Series No. 19, Lands Directorate, Environment Canada. Ottawa: Environment Canada.

Yamasaki, Stephen H., James W. Fyles, Keith N. Egger, and Brian D. Titus. 1998. "The Effect of *Kalmia angustifolia* on the Growth, Nutrition, and Ectomycorrhizal Symbiont Community of Black Spruce." *Forest Ecology and Management* 105 (1–3): 197–207.

Yamasaki, Stephen H., James W. Fyles, and Brian D. Titus. 2002. "Interactions among *Kalmia angustifolia*, Soil Characteristics, and the Growth and Nutrition of Black

Spruce Seedlings in Two Boreal Newfoundland Plantations of Contrasting Fertility." *Canadian Journal of Forest Research* 32 (12): 2215–24.

Zadeh, Lofti A. 1973. "Outline of a New Approach to the Analysis of Complex Systems and Decision Processes." *IEEE Transactions on Systems, Man and Cybernetics* 3 (1): 28–44.

Zhang, Deqiang, Dafeng Hui, Yiqi Luo, and Guoyi Zhou. 2008. "Rates of Litter Decomposition in Terrestrial Ecosystems: Global Patterns and Controlling Factors." *Journal of Plant Ecology* 1 (2): 85–93.

7

The Mammals of Wemindji
in Time, Space, and Ways of Knowing

MURRAY M. HUMPHRIES, JASON SAMSON,
and HEATHER E. MILLIGAN

The boreal forest is North America's largest forested biome, yet the diversity of its mammals is similar from one region to the next (Krebs, Boutin, and Boonstra 2001; Henry 2002). Furthermore, because most mammals occupy the boreal forest year-round (other than migratory caribou) and respond to long-term succession with altered abundance rather than absolute appearance and disappearance, their diversity has remained relatively constant over historical time. Thus, the broad brush of mammal diversity paints a picture of homogeneity, with the mammals in Wemindji territory not much different from those in other boreal regions of Canada or from those of past decades and centuries.

Local knowledge of boreal mammals, whether from Indigenous or scientific perspectives, is filled with contrasts of abundance and scarcity, cyclical and complex forms of change, and uniqueness and irreplaceability. How can the boreal forests of Quebec and Yukon be the same, when in both regions, radically different plant and animal communities are encountered by moving only a few kilometres inland from the coast, or from one watershed to the next, or from a south-facing to a north-facing hillside? How can constancy over time even be contemplated in a forest characterized by such extreme seasonality, regular population cycles, irregular presence of overwhelming numbers of migratory caribou, and transformational cycles of forest fire and succession?

This chapter does not attempt to reconcile these opposing views of boreal mammals, but instead seeks to address both realities. These opposing perspectives are components of both scientific and traditional knowledge of boreal ecology and can be partially explained by differences in scales of comparison. Although it is true that broadly similar points of comparison are often seen as quite distinct when examined in greater detail, lessons of scale in ecology remind us that variation increases with the spatial and temporal scale of comparison (Levin 1992). Thus, if we consider only scale, we would expect components of the boreal forest to be recognized as broadly similar at a local scale and vastly different at a continental scale. The fact that broad-scale comparisons of boreal ecology tend to highlight similarity, whereas local-scale comparisons stress contrasts, speaks to something more complex and intriguing about the perception of the boreal forest that is shared by traditional knowledge holders and ecological scientists.

In this chapter, we catalogue the thirty-six mammal species known or expected to be present in Wemindji territory and their importance to the Wemindji Crees. Then we review some of the spatial and temporal compartments and integrations that shape the ecology of Wemindji mammals and their harvest by the Crees. Finally, we examine some of the current and potential threats to Wemindji mammals, and the extent to which these are addressed by the Paakumshumwaau-Maatuskaau Biodiversity Reserve. Mammals and the manner in which the Crees depend on them are embodied in the design of the protected area in distinct and diverse ways. As we describe later, these include the ecological importance of watersheds for species such as beaver and moose, the effectiveness of rotational harvest practice within watersheds, and the many important contributions of the inland-to-coastal transition to biodiversity and ecological integrity. There is a paucity of scientific research on mammal ecology in Wemindji territory but an abundance of local knowledge held by members of the Wemindji community. Thus, this chapter's content is derived from general scientific knowledge of mammal and boreal ecology, complemented by the specific ecological research we conducted in Wemindji territory and, most significantly, by the traditional knowledge of Cree hunters and trappers who motivated our research approaches, guided us in the field, and shared some of their knowledge of mammal ecology. We hope that this blending of perspective and experience allows us to review mammal ecology in a way that resonates with the scientific and traditional knowledge from which it is derived. We identify the few situations in which traditional and scientific knowledge contradict, but we avoid attributing concepts to either knowledge base if we believe that a general consensus exists.

MAMMALS OF WEMINDJI TERRITORY

Thirty-six species of wild mammals are known or presumed to occur in the traditional territory of the Wemindji Cree First Nation, including humans, five marine species, and thirty-one terrestrial species (Table 7.1). Most are well known, frequently encountered, strongly integrated in local knowledge and traditional activities, well differentiated from similar species in the region, and locally identified within species by gender, age, behaviour, or habitat use.

TABLE 7.1

Species of wild mammals with known or probable occurrence in Wemindji territory

Species	Cree name	Taxonomy	Local status
Moose	*Muus*	Artiodactyla, Cervidae, *Alces alces*	A
Caribou	*Atihkw*	Artiodactyla, Cervidae, *Rangifer tarandus*	A
Arctic fox	*Waapihchaashiish*	Carnivora, Canidae, *Alopex lagopus*	A
Gray wolf	*Mihiihkin*	Carnivora, Canidae, *Canis lupus*	A
Red fox	*Mihchaashiu*	Carnivora, Canidae, *Vulpes vulpes*	A
Canada lynx	*Pishiu*	Carnivora, Felidae, *Lynx canadensis*	A
Wolverine	*Kuihkuhaachaau*	Carnivora, Mustelidae, *Gulo gulo*	D
Northern river otter	*Nichikw*	Carnivora, Mustelidae, *Lontra canadensis*	A
American marten	*Waapishtaan*	Carnivora, Mustelidae, *Martes americana*	A
Ermine	*Sihkus*	Carnivora, Mustelidae, *Mustela erminea*	A
Least weasel		Carnivora, Mustelidae, *Mustela nivalis*	B
American mink	*Achikaash*	Carnivora, Mustelidae, *Mustela vison*	A
Striped skunk	*Shikaakw*	Carnivora, Mephitidae, *Mephitis mephitis*	A
Walrus	*Wiipichiu*	Carnivora, Odobenidae, *Odobenus rosmarus*	D
Ringed seal	*Aahchikw*	Carnivora, Phocidae, *Phoca hispida*	A
Bearded seal	*Mishtaahchikw*	Carnivora, Phocidae, *Erignathus barbatus*	A

Species	Cree name	Taxonomy	Local status
American black bear	*Chishaayaakw*	Carnivora, Ursidae, *Ursus americanus*	A
Polar bear	*Waapiskw*	Carnivora, Ursidae, *Ursus maritimus*	A
Beluga	*Waapimaakw*	Cetacea, Monodontidae, *Delphinapterus leucas*	A
Hoary bat	*Pihkwaachiish*	Chiroptera, Vespertilionidae, *Lasiurus cinereus*	B
Little brown bat	*Pihkwaachiish*	Chiroptera, Vespertilionidae, *Myotis lucifugus*	B
Cinereus shrew	*Chiinishjuwaayaapiku-shiish*	Insectivora, Soricidae, *Sorex cinereus*	B
Pygmy shrew		Insectivora, Soricidae, *Sorex hoyi*	C
Star-nosed mole	*Naasipaatinischaasiu*	Insectivora, Talpidae, *Condylura cristata*	B
Snowshoe hare	*Waapush*	Lagomorpha, Leporidae, *Lepus americanus*	A
American beaver	*Amiskw*	Rodentia, Castoridae, *Castor canadensis*	A
North American porcupine	*Kaakw*	Rodentia, Erethizontidae, *Erethizon dorsatum*	A
Southern red-backed vole		Rodentia, Muridae, *Myodes gapperi*	B
Meadow vole		Rodentia, Muridae, *Microtus pennsylvanicus*	B
Muskrat	*Wichishkw*	Rodentia, Muridae, *Ondatra zibethicus*	A
Deer mouse	*Aapikushiish*	Rodentia, Muridae, *Peromyscus maniculatus*	B
Eastern heather vole		Rodentia, Muridae, *Phenacomys ungava*	C
Northern bog lemming		Rodentia, Muridae, *Synaptomys borealis*	C
Northern flying squirrel	*Shihchaayuwiniku-chaash*	Rodentia, Sciuridae, *Glaucomys sabrinus*	A
Woodchuck		Rodentia, Sciuridae, *Marmota monax*	A
Red squirrel	*Anikuchaash*	Rodentia, Sciuridae, *Tamiasciurus hudsonicus*	A

Note: The authors were unable to verify Cree names for all of the above species. Local status is classified according to the following four categories. (A) Well-known and frequently encountered species, strongly integrated in local knowledge and traditional activities. Differentiated from similar species in the region and frequently identified by more detailed, within-species categories based on gender, age, behaviour, or habitat use. (B) Known to be present locally but encountered less frequently or less integrated in local knowledge and traditional activities. Less commonly differentiated from similar species or subdivided into within-species categories. (C) Species likely to be present locally but typically not differentiated from similar species in the region. (D) Well-known species, historically present in the region, but now rarely encountered. Despite contemporary scarcity, remains strongly integrated in local knowledge and traditional activities.
Sources: Based on local knowledge and Don E. Wilson and Sue Ruff (1999).

Harvest Data

The hunting and trapping of wild mammals is a fundamental component of Crees' respect for and relationship with their traditional territory (Cree Trappers Association 1989). An ability to respectfully and sustainably kill a wide diversity of mammal species in differing seasons and locations is a highly valued skill (Berkes 2008). Harvest data supply a somewhat impoverished perspective of the importance of wild mammals to Wemindji Crees because they include only the reported numbers of individuals killed and are biased toward furbearer and big game species of economic importance. Nevertheless, the data compiled by the Cree Trappers Association during the period 2005–06 provided a snapshot of the prevalence of mammal species in the reported big game and furbearer harvest from Wemindji territory. The number of moose was twice that of black bears and more than three times that of caribou. Multiplying the number of individuals harvested by the average masses for each species (Silva and Downing 1995) suggests that moose harvest accounted for eight and twelve times more biomass than bear and caribou harvest, respectively. The furbearers included ten species but were numerically dominated by beaver and marten (Figure 7.1). Multiplying the number of individuals harvested, by the average May 2006 sale price, as reported by Fur Harvesters Auction in North Bay, Ontario, marten accounted for almost 60 percent of total pelt value from the territory, followed by beaver at 27 percent, lynx at 7 percent, and all other species collectively at 6 percent. Additional mammals that are regularly harvested in the territory (such as snowshoe hare and porcupines) are excluded from the above reporting because they are neither big game nor sold as pelts.

Snow-Tracking Surveys

Snow-tracking surveys completed by the authors and community collaborators in Wemindji territory during the winter of 2008 provided a general indication of the mammal species most frequently encountered during that season. From January to March, following the general methodology of Jani Pellikka, Hannu Rita, and Harto Lindén (2005), we snowshoed triangles at forty-eight sites in inland areas. Each triangle was three kilometres in length and divided into six five-hundred-metre segments. We recorded every snow-track and animal sign that crossed our path, along

FIGURE 7.1

Furbearer species as represented in the total number of pelts harvested from Wemindji territory

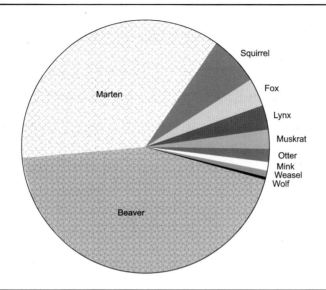

Source: "Trapline Capture Report: Summary for all Communities." July 1, 2005 to June 30, 2006. Cree Trappers Association.

with the general vegetation cover, snow depth, and tracking conditions for each five-hundred-metre segment. Triangles were visited on only one occasion; therefore, the tracks identified were those that had accumulated since the last snowfall. If snowfall exceeded one to two centimetres in depth, we suspended our surveys for at least twenty-four hours to allow tracks to re-accumulate. For reasons of safety, we avoided placing triangles on cliffs and creek crossings, and we also avoided lakes because tracks were often windblown there. These surveys detect only those animals that are active on top of the snow (i.e., the supranivean zone) and that are heavy enough to routinely leave species-identifiable tracks in a wide variety of snow conditions. Thus, the technique frequently omits hibernators (bears, woodchucks, skunks), mammals smaller than red squirrels (mice, voles, shrews, if they are active only under the snow and/or the snow layer is too firm or wind-blown to register their supranivean activity), and semi-aquatic mammals confined under ice cover in winter (beaver, muskrats). Nevertheless, it can detect a wide array of furbearers and ungulates that are of particular value to the local community. On average across all forty-eight triangles, we encountered red squirrels, snowshoe hare, and marten on the majority of the segments; caribou, porcupine, ermine, moose, and red foxes on 25–50 percent of segments; and lynx, wolf, mink, and otter on less than 10 percent of segments (Figures 7.2 and 7.3). The average number of species encountered per segment was 5.4.

We designed our snow-tracking surveys to document the wildlife diversity in areas that Cree tallymen identified as important to protect, especially in the large portion of Wemindji territory that lies east of the Paakumshumwaau-Maatuskaau Biodiversity Reserve. We compared our diversity findings for these areas with those from non-identified areas distributed along major roads. Land-use mapping interviews, conducted in August 2007 by other McGill researchers involved in the protected area project, identified wildlife areas that tallymen would like to see protected from industrial development (mining and hydro) as well as non-Cree hunting. All were outside the biodiversity reserve and thus represented potential future locations for expansion of the reserve or for the creation of different forms of protection. Accordingly, we refer to them as suggested protected areas (SPAs).

With the collaboration, often the accompaniment, and in one case the independent efforts of the local tallymen, we surveyed thirty-one sites in eight SPAs. Their locations were chosen by the tallymen or their representatives, based on their wildlife abundance and diversity, their proximity to winter bush camps where we stayed, and the trails used to access the sites. We surveyed an additional seventeen control sites, placed every twenty kilometres along the James Bay Highway and Trans Taiga Road. To situate the triangles away from the road, we positioned them so that one of their points lay >100 metres from the roadside. We encountered more species in SPA sites (5.6 ± 2.3 species per five-hundred-metre segment) than in control sites (4.9 ± 2.3l; $t334 = 2.63$, $p < 0.01$). We detected moose and porcupine in about 50 percent of SPA triangle segments and in less than 20 percent of the control segments (moose: $t54 = 3.64$, $p < 0.01$; porcupine: $t54 = 4.65$, $p < 0.001$; Figure 7.3). There were also non-significant tendencies ($p > 0.11$ in all cases) for marten to be more common and snowshoe hare and caribou to be less common in SPA triangles than in control triangles.

FIGURE 7.2

Results of snow-tracking surveys, Wemindji territory, winter 2008: Overall encounter frequency of mammal species

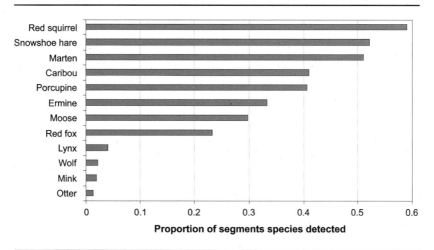

FIGURE 7.3

Results of snow-tracking surveys, Wemindji territory, winter 2008: Differences in encounter frequency of key mammal species between suggested protected areas and control sites

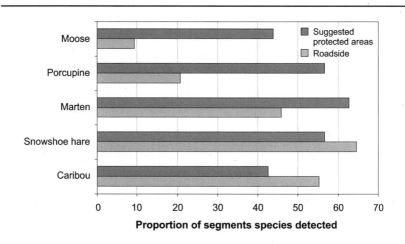

All species detected across our forty-eight triangles occurred in at least one SPA site, but we never encountered mink, otter, and flying squirrels in control sites.

THE ROLE OF SPACE AND TIME IN ANIMAL ECOLOGY

Ecological systems can be envisioned as consisting of compartments whose strongly interacting components are integrated to varying degrees by interactions that span the compartments (McCann et al. 2006). These compartments occur in both space and time. For example, aquatic and terrestrial food webs can be perceived as spatial compartments and interfaces within boreal ecosystems (Naiman and Décamps 1997). Freshwater macrophytes, phytoplankton, zooplankton, and fish make up part of the aquatic compartment. Shrubs, hare, and lynx make up part of the terrestrial compartment. Beaver, moose, and mink integrate across these compartments. Winter and summer can be seen as temporal compartments in boreal ecosystems (Pruitt 1978). Flowering plants, deciduous leaves, migratory birds, and hibernating mammals make up part of the summer compartment. Migratory caribou, the wolves that follow them, and birds that are present only in winter make up part of the winter compartment. Residents that are present and active year-round, such as moose, hare, marten, and lynx, integrate across these seasonal compartments. In general, we must take care never to imply that compartments are contained within barriers, such that they represent closed systems, or to suggest that any species is entirely confined to a given compartment. On the other hand, we must not focus on integrations and integrators to the extent that we fail to recognize the predominance and significance of within-compartment interactions.

Spatial Compartments and Integrations Affecting Wemindji Mammals

Wemindji's mammal assemblage is reflective of a territory situated at a four-way intersection of sea, land, forest, and tundra. Typical northern maritime assemblages of belugas, seals, polar bears, and arctic foxes are present and immediately adjacent to typical northern boreal assemblages of moose, beaver, hare, and marten. The seasonal arrival of migratory caribou herds reveals the proximity of tundra within caribou walking distance. Encounters with striped skunks (occasionally in the bush and frequently in stories, as told in Chapters 6 and 10 of this volume) reveal a proximity to southern forests and agricultural lands. Indeed, we believe this is the only region in North America where migratory caribou and striped skunks co-occur (Wilson and Ruff 1999). Wemindji territory lies at the southern boundary of the expansive ecotone between boreal forest and arctic tundra, referred to as the forest to tundra transition (Payette, Fortin, and Gamache 2001). Thus, the territory is boreal in nature, but it contains species that show its proximity to the northern forest fringe and tundra. On the marine side, Hudson and James Bays are large, cold bodies of water that penetrate deep into the North American continent, bringing northern maritime species with them. As a result, James Bay is one of only two places in the world (southern Labrador is the other) where polar bears and arctic foxes occur below the fifty-fifth parallel (DeMaster and Stirling 1981; Audet, Robbins, and Larivière 2002).

The terrestrial mammals of Wemindji occupy a limited array of terrestrial and aquatic environments in the inland portion of the territory. Most upland terrestrial habitats in Wemindji consist of spruce-lichen forest. Spruce, like other conifers, is not a primary food source of most mammalian herbivores, because mature conifer needles are higher in fibre, lower in nitrogen, and higher in phenolics than most other types of vegetation (Hatcher 1990). Red squirrels are one of the few mammals that specialize on conifers, but they primarily consume seed in cones and vegetative or reproductive buds (Steele 1998), all of which are much higher in nutritional value than mature needles (Hatcher 1990). Caribou are one of the few mammals that specialize on lichens, which are highly digestible and rich in energy but low in nitrogen (Storeheier et al. 2002).

For all remaining mammalian herbivores in Wemindji, the stems, leaves, roots, berries, flowers, and seeds of flowering plants, including broadleaf deciduous trees, shrubs, herbaceous plants, and aquatic vegetation, compose the major portion of their diet and dictate the habitats in which they live in greatest abundance. Particularly important to key herbivores such as moose, beaver, hare, and porcupine are birch and aspen trees, willow and alder shrubs, and, for beaver and moose, aquatic plants including water lilies and water sedges. Preferred aquatic plants tend to grow in shallow, stationary waters along the shorelines of small lakes and beaver ponds. Preferred deciduous shrubs are also concentrated along the shorelines of lakes and streams. Thus, the aquatic and terrestrial components of riparian corridors are an important exception to the predominance of coniferous vegetation in the territory, and many mammals concentrate their activity in these riparian zones. Deciduous trees and shrubs also grow well in upland sites characterized by topographical relief, often but not always associated with southern exposures and/or well-drained soils. These sites provide important winter habitat for moose because an abundance of preferred forage occurs in combination with reduced snow depth. Early to intermediate successional stands following fire are also frequently characterized by a richer diversity and greater abundance of mammals than fully mature climax communities, particularly if post-fire successional patterns favour the establishment of aspen rather than jack pine (see Chapter 6, this volume).

The spatial compartments affecting Wemindji mammals can be envisioned as a multi-scale process, from large-scale biome compartments, to meso-scale coastal and watershed compartments, to fine-scale aquatic and terrestrial compartments, which are themselves characterized by a mosaic of local characteristics (Figure 7.4). All but the largest scale of these spatial compartments and integrations are protected by the Paakumshumwaau-Maatuskaau Biodiversity Reserve.

Temporal Compartments and Integrations Affecting Wemindji Mammals

Climate has fundamental influence on mammals (via direct effects on physiology and behaviour, as well as indirect effects operating through their food, competitors, and predators). Wemindji mammals are exposed to extreme climate variation over a wide range of temporal scales. Because Wemindji territory sits at the transition between temperate and polar latitudes, it has experienced repeated glaciations over the last 2.4 million years (Raymo 1994), with extensive glaciations occurring at roughly 100,000-year intervals (Hays, Imbrie, and Shackleton 1976).

Wemindji territory, like other high-latitude regions on Earth, is in the midst of an interglacial period, with the Labrador portion of the Laurentide Ice Sheet estimated to have retreated from the territory eight to nine thousand years ago (Dyke and Prest 1987; Chapter 5, this volume). Following this retreat, the postglacial Tyrrell Sea encompassed large portions of

FIGURE 7.4
Spatial compartments and integration affecting Wemindji mammals and the compartments and interactions included in the Paakumshumwaau-Maatuskaau Biodiversity Reserve (dashed line)

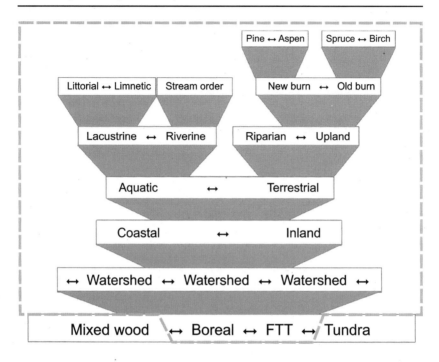

modern Wemindji territory (Hardy 1977). The long-term warming trends of the current postglacial period have been strongly amplified by recent warming produced by anthropogenic greenhouse gas emissions. Since the 1970s, average annual temperature near Wemindji has increased by about three degrees Celsius, which is more than in areas south of James Bay but similar to what has been observed on the Hudson Bay coast (Environment and Climate Change Canada 1961–2010; Ouranos 2015). Thus, Wemindji's climate has been somewhat warmer over the last century than in previous centuries and much warmer during the last few decades.

Over intra-annual time scales, the boreal forest is one of the world's most seasonal environments. Between 1976 and 2006, climate normals near Wemindji could vary from daily minimums of –37°C in January to daily maximums of 35°C in July. Annual precipitation averages 683.9

millimetres, 39 percent of which falls as snow. The growing season (frost-free) lasts, on average, 125 days. Finally, at a daily scale of variation, Wemindji experiences 17.0 hours of daylight on June 21 and 7.5 on December 21. On average, daytime temperature maximums are 7.3°C and 13.0°C warmer than nightly minimums on the winter and summer solstice, respectively (Environment and Climate Change Canada 1961–2010).

Furthermore, Wemindji mammals are also heavily influenced by biotic cycles in forest fire frequency, the presence of migratory caribou, and the abundance of lynx and snowshoe hare.

For the last 4,200 years, black spruce *(Picea mariana)* and jack pine *(Pinus banksiana)* have dominated the boreal forest of northern Quebec, but the relative abundance of the two species changes with variations in the frequency of fires (Arseneault and Sirois 2004). A forest fire cycle of approximately one hundred years prevails in northern Quebec (Payette et al. 1989) except in white-spruce-dominated coastal forest, where fires are extremely rare (Parisien and Sirois 2003). Model simulations predict that fire intervals of less than 60 years and more than 220 years would lead to the extirpation of black spruce and jack pine, respectively (Le Goff and Sirois 2004). In 1989, major fires affected 20,870 square kilometres of Wemindji territory (Couturier and Saint-Martin 1990); most of the Paakumshumwaau-Maatuskaau Biodiversity Reserve falls within this area. A majority of sites previously dominated by black spruce were converted to jack-pine-dominated forests (Lavoie and Sirois 1998), probably because of the short inter-fire interval prior to this burn (Parisien and Sirois 2003; Chapter 6, this volume).

Forest fires have a complex impact on mammals, which varies widely according to the characteristics of the fire, the subsequent pattern of vegetation succession, and the ecology of each mammal species (Fisher and Wilkinson 2005). Immediately after a fire, deer mice and meadow voles generally dominate boreal small-mammal communities (Hooven 1969), whereas red-backed voles are often associated with old growth stands (Nordyke and Buskirk 1991; Fisher and Wilkinson 2005). Martens, red squirrels, and bats are expected to avoid early succession habitats opting instead for mature, closed-crown forests whose abundant snags and coarse woody debris provide denning sites, resources, and travel routes.

Lynx and hare usually select mid-successional stands rather than very young or mature stands (Fisher and Wilkinson 2005). Wildlife biologists generally believe that moose benefit from fire because abundant deciduous forage occurs in early successional stands, whereas Cree hunters tend to focus more on the importance of shelter and vegetation recovery following fire (Jacqmain et al. 2008). Caribou are likely to be negatively affected by fire because lichen, their staple food source, has a slow post-fire recovery (Couturier and Saint-Martin 1990). Black bears at the La Grande River were more abundant in young stands than in old growth stands (Crête et al. 1995), probably due to the presence of wild berries.

Several biotic cycles and variations have additional influence on Wemindji mammals. Long-term fluctuations in the abundance and range use of migratory caribou are well documented in the traditional knowledge of northern Indigenous communities (Ferguson and Messier 1997; Kendrick and Lyver 2005; Berkes 2008) and by dendroecological measures of trampling scars on tree roots along caribou trails (Morneau and Payette 2000; Zalatan, Gunn, and Henry 2006). Migratory caribou were present in Wemindji territory prior to the twentieth century, left it during the early twentieth century, and returned in the 1970s; through the 1980s and until recently, they rapidly increased in abundance (Berkes 2008). Their return during recent decades also coincided with an increased abundance of wolves and martens (Péloquin 2007). Historically absent from Wemindji territory, moose arrived in the 1920s. Why they moved northward is unknown, but many people in the community speculate that climate change, forest fire succession, or logging practices farther south were responsible (see also Darimont, Paquet, and Reimchen 2005; Jacqmain et al. 2008). Beaver drastically declined in Wemindji territory during the 1920s and 1930s due to over-trapping driven by high fur prices and improved access for outside trappers, but they began to recover in the 1940s following the re-establishment of beaver reserves and local stewardship by Cree tallymen (Feit 2005; Chapter 2, this volume). Snowshoe hare and lynx are characterized by synchronized but lagged ten-year cycles, such that hare populations tend to peak once every ten years throughout North America, followed by lynx populations one to two years later (Elton and Nicholson 1942; Vik et al. 2008). In parts of the boreal forest, where the hare is the

FIGURE 7.5

Temporal compartments and integrations affecting Wemindji mammals and the compartments and interactions included in the Paakumshumwaau-Maatuskaau Biodiversity Reserve (dashed line)

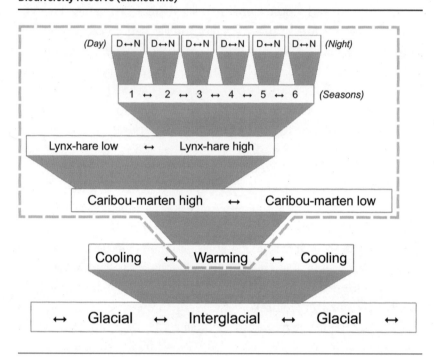

predominant herbivore, the influence of its cycle spreads beyond lynx to drive ten-year cycles in the abundance of much of the vertebrate community (Krebs, Boutin, and Boonstra 2001).

Thus, the temporal compartments affecting Wemindji mammals can be envisioned as a multi-scale process, from long-term glacial and climate cycles, to medium-term cycles of caribou occurrence and fire disturbance/ succession, to decadal cycles in lynx and hare, to annual seasonal cycles, and finally to daily cycles (Figure 7.5). Mammals often respond to these long-term cycles via extinction, extirpation, and colonization, medium-term cycles via changes in abundance, and short-term cycles via changes in behaviour. All but the longest scale of these temporal compartments and integrations are protected by the Paakumshumwaau-Maatuskaau Biodiversity Reserve.

Spatial and Temporal Integration by Wemindji Mammals and Cree Hunters

The relationship between spatial and temporal compartments is complex because integration in one dimension frequently contributes to compartmentalization in the other. For example, long-distance seasonal movements of caribou between summer and winter ranges spatially integrate boreal and tundra regions while contributing to the temporal compartmentalization of winter and summer habitats in both regions. In other words, the absence of caribou during summer and their presence in winter contribute to seasonal differences in Wemindji territory. Two finer-scale examples in which spatial integration contributes to temporal compartments are provided by the behaviour of moose. For instance, moose may spend their summers in riparian lowland habitat then move to upland deciduous habitat for the winter. They can also forage in a regenerating burn at dawn and then bed down in a closed-canopy forest for the rest of the day. In contrast, the year-round presence of moose in boreal regions and of musk-oxen much farther north in tundra regions (including a small introduced herd in Nunavik) contributes to spatial compartments (of the two regions) while also contributing to temporal integration. That is, the year-round presence of moose in boreal and muskox in tundra diminishes seasonal differences in both regions. Finer-scale examples in which temporal integration contributes to spatial compartments are provided by the year-round presence of mink in riparian lowland habitat versus marten in upland forests, and by the constant presence of meadow voles in regenerating burns and of red-backed voles in closed-canopy forests.

As an aquatic mammal that feeds on terrestrial vegetation, beaver clearly integrate across the aquatic and terrestrial food web compartments, but the temporal nature of this integration is paradoxical. Due to the nature of their environment, beaver could be expected to specialize on shrubs and trees during the ice-free season when they can travel between water and land, but then to be frozen in as aquatic specialists during winter when ice cover prevents them from accessing terrestrial habitats. However, beaver construct a floating feeder pile of terrestrial vegetation near their lodge in summer and autumn, which provides an accessible, nutritious winter food source when their capacity to forage under the ice for aquatic vegetation may be limited by its seasonal senescence and their dive endurance and tolerance for immersion hypothermia. Therefore, beaver could be expected to specialize on terrestrial vegetation during winter and on aquatic vegetation in summer. The analysis of naturally occurring stable isotopes in vegetation and beaver fur and teeth shows that both forms of seasonal specialization can occur (Milligan and Humphries 2010). During ice-covered months, beaver that live in streams appear to shift to a terrestrial diet of cached willow (*Salix* spp.) and alder (*Alnus* sp.). Conversely, beaver that inhabit ponds where water lilies *(Nuphar variegatum)* are abundant shift to an aquatic diet. Our results indicate that beaver integrate across aquatic and terrestrial food webs, that the temporal nature of this integration can be complex, and that it is dependent on the composition of forage plants and their seasonal accessibility.

The Crees' hunting traditions respect, benefit from, and, in some cases, amplify the temporal and spatial compartments and integrations that occur naturally in their territory. As explained in Chapter 2 of this volume, Cree land tenure involves traditional family hunting territories, typically defined by watershed boundaries and managed by a local tallyman (*uchimaau*, or hunting boss; see also Feit 1986; Berkes 2008). Thus, locally managed hunting territories represent spatial compartments that conform to natural watershed boundaries, and these boundaries shape the design of the Paakumshumwaau-Maatuskaau Biodiversity Reserve. Differences in the stewardship of individual hunting territories could amplify natural differences in watershed compartments, but frequent exchange, co-operation, and collaboration between families that occupy adjacent territories serve as a form of spatial integration.

Cree hunters use temporal compartments of intensified harvest on particular species, adhering to an annual calendar to maintain efficiency and sustainability (Feit 1988). For example, moose are hunted during the autumn, when bulls travel widely and are receptive to imitated mating calls, and during mid- and late winter, when the mobility of both sexes is limited by deep and crusted snow. Waterfowl and small-game hunting, trapping of beaver and other furbearers, and fishing are also concentrated at times when environmental conditions are most suitable (though synergistic activities may be combined, such as fall moose hunting and beaver trapping) and the fur and/or meat of the target species is at or near its best.

A strategy that combines spatial and temporal compartments is the rotational harvest of beaver and moose (Berkes 2008). This involves dividing a hunting territory into three or four areas, one of which is harvested intensively each year while the others are allowed to rest and recover. The timing and intensity of scheduled rotations are adjusted according to continual assessment of the abundance and condition of the target species and their preferred food in each subdivided area. Thus, the rotational strategy involves spatial compartments of intensified harvest, temporally integrated at a period of three to four years in a manner that ensures long-term success and efficiency while minimizing year-to-year variation.

The Crees demonstrate sensitivity to the spatial scale of compartments dictated by the spatial ecology of the target species by applying local adaptive management to beaver and moose but not to caribou, grouse, or snowshoe hare. Although beaver and moose move between separately managed compartments, which is recognized as a critical contributor to population recovery following intense harvest, the frequency and scale of their movements are not large enough to swamp the maintenance of the compartments. The fact that Cree hunters kill more beaver and moose in compartments that have been rested for several years than in compartments harvested more recently demonstrates the resonance of the spatial scale of local management and the spatial ecology of beaver and moose (Feit 1986). The future abundance of beaver and moose in the Paakumshumwaau-Maatuskaau Biodiversity Reserve will depend mostly on its rates of production, harvest, and habitat transformation. In contrast, the abundance of caribou, hare, and grouse integrates across such large spatial scales that their future numbers in the reserve are most likely to be shaped by non-local processes. Subdividing and locally managing compartments within a hunting territory would be ineffective for these species; spillover from one compartment to the next would be too high, and abundance in any one compartment would be too strongly dictated by factors occurring outside the compartment and hunting territory. Although these species cannot be managed according to spatial compartments, other approaches are used. For example, some Indigenous communities that shoot caribou on their migration routes between calving and wintering grounds avoid killing the leaders, so that the bulk of the herd will continue undisturbed to its destination. This strategy is similar to Cree management of the spring and fall goose hunt in terms of efforts to minimize disturbance and to avoid hastening the departure of migrating geese from the territory (see Chapter 9, this volume).

POTENTIAL THREATS TO WEMINDJI MAMMALS

Major foreseeable threats to Wemindji mammals include an acceleration of industrial development, over-harvesting of ungulates and furbearers, and climate change.

Industrial Development

At present, the most likely forms of industrial development include additional hydro-electric dams and diversion projects, mining activities, and forestry. Even if they do not occur in the biodiversity reserve, they may affect regional wildlife populations in a manner that affects the mammals in the reserve.

Over-Harvesting

Over-harvesting of wildlife species is a concern, particularly when road and trail construction associated with industrial development leads to increased access by non-traditional users. Low fur prices and declining trapper numbers mean that over-harvesting of furbearers is presently less of a concern than over-hunting of ungulates, such as moose and caribou. Migratory caribou were very abundant in Wemindji territory during recent winters, coincident with the continued growth and range expansion of the Leaf River caribou herd (Couturier 2007). Nevertheless, the incursion of large numbers of southern recreational caribou hunters into Wemindji territory (via a highway originally constructed in support of hydro-electrical development) generated conflicts related to access, safety, and local authority. An even more pressing concern is the over-hunting of moose by outsiders who use hydro roads, transmission lines, and reservoirs to enter Wemindji territory. Because moose occur at low densities in forested habitats, increased harvester access facilitated by corridors, roads, and trails associated with industrial development is a major contributor to their over-exploitation (Rempel et al. 1997).

Climate Change

Of course, the potential impacts of climate change on Wemindji mammals are a major concern. However, because the problem is global in nature, addressing it with local protected area measures is difficult. How much Wemindji territory will warm in the coming decades is unknown, but a four-degree Celsius increase in average annual temperature by 2050 is plausible, based on a mid-range greenhouse gas emission scenario and an averaging of three global climate models (prediction range 3.1 to 4.7 °C; Hijmans et al. 2005).

Latitudinal gradients in climate and biodiversity provide valuable null models for predicting the future impacts of a warmer climate (Humphries 2009). According to this latitudinal shift or climate envelope approach, to the extent that species range limits are imposed by climatic tolerance, climate warming will cause the plants and animals in a given region to more closely resemble those in warmer locations, which are typically at lower latitudes (Araújo et al. 2005). The climate conditions that are predicted to occur in Wemindji territory in 2050 currently prevail about five hundred kilometres to the south, between the communities of Amos and Matagami, Quebec. Thus, a simplistic but potentially robust prediction is that the plants and animals of Wemindji territory in 2050 might resemble those currently found in the vicinity of Amos and Matagami. Because the terrestrial mammals of Wemindji are mainly boreal species approaching the northern extent of their continental distributions, there is almost complete overlap between the terrestrial mammals in both locales (Wilson and Ruff 1999). Notable exceptions are white-tailed deer, which live in Amos/Matagami but not in Wemindji, and migratory caribou, which are largely absent from Amos/Matagami but routinely present in Wemindji (Couturier 2007). In addition, four small mammal species (one bat, one insectivore, and two rodents) live near Amos/Matagami but have not been confirmed in Wemindji territory.

In contrast, the marine mammals in Wemindji territory are mainly arctic species that are near the southern limit of their ranges, with the present southern distribution of most species geographically constrained by the southern terminus of James Bay (Wilson and Ruff 1999). Continued warming in the coming decades may introduce additional climate constraints that supersede this geographic limitation, causing some marine mammal species to retreat north out of James Bay and away from Wemindji territory. Two key local features make this region a particularly interesting case study for understanding the impacts of climate change on marine mammals. First, the geographic barriers that currently prevent the Hudson/James Bay marine assemblage from extending farther south also prevent more southerly marine assemblages from entering the ecosystem, unless they do so from the far north via Foxe Basin or Hudson Strait. Thus, whereas the impacts of climate change on most ecosystems are dictated by how the original community assemblage copes with new species assemblages under the new environmental conditions, impacts in James Bay may be weighted more toward direct environmental impacts on the original community. Second, though warming potentially decreases the extent of climatically favourable habitat in James Bay, the bay itself could also contract or expand, depending on the relative magnitudes of isostatic rebound (Mitrovica, Forte, and Simons 2000) and sea level rise (Church et al. 2001). The scale of coastline shifts may be much smaller than climate-driven changes in environmental suitability, but they may have a substantial influence on coastal and submerged habitat features of particular importance to marine species.

MAMMALS AND THE PAAKUMSHUMWAAU-MAATUSKAAU BIODIVERSITY RESERVE

The large Paakumshumwaau-Maatuskaau Biodiversity Reserve represents a long-term vision of protection. An incredible wealth of mammal diversity flourishes within its borders, including a unique convergence of marine, terrestrial, boreal, and arctic species. However, the reserve's richest and most unique mammalian feature is the relationship between people and wild mammals that persists today as it has since time immemorial. The harvest of wild animals forms the foundation of this relationship; engagement, action, respect, and stewardship partly define its nature. Protecting this relationship is as important as protecting the area itself and the biodiversity present there. In fact, one cannot be accomplished without the others, for the land, animals, and people are all part of each other. There is grandeur in this view of life (Darwin 1859; Berkes 2008).

Critically, many contributors to the mammal diversity and the traditional relationships between people and wild mammals in the biodiversity reserve operate at spatial and temporal scales beyond the reserve's boundaries. Spatially, these include the seasonal migration of tundra caribou into and out of Wemindji territory, the interaction of boreal and marine species along the James Bay coast, and the continental-scale synchrony of snowshoe hare and lynx cycles. Temporally, coastal uplift, cycles of fire frequency and forest succession, climate change, and the periodic absence and presence of wintering caribou in the territory are long-term processes of change that are likely to continue with or without local biodiversity protection. This issue of restricted spatial and temporal scale is certainly not unique to mammal protection in the reserve, but rather is a universal limitation of protected areas (Joppa, Loarie, and Pimm 2008). Thus, rather than constituting a weakness of the reserve itself, these limitations emphasize the importance of the context in which it exists. Spatially, this means situating the reserve within an appropriate regional context of marine and terrestrial biodiversity protection, sustainable development, and traditional use. Temporally, this means situating the reserve within an appropriate context of change, not attempting to maintain Wemindji mammals as they are now, frozen in time, but ensuring a certain continuity of change that is inherent to both Wemindji mammals and, not coincidentally, Cree worldviews.

WORKS CITED

Araújo, Miguel B., Richard G. Pearson, Wilfried Thuiller, and Markus Erhard. 2005. "Validation of Species-Climate Models under Climate Change." *Global Change Biology* 11 (9): 1504–13.

Arseneault, Dominique, and Luc Sirois. 2004. "The Millennial Dynamics of a Boreal Forest Stand from Buried Trees." *Journal of Ecology* 92 (3): 490–504.

Audet, Alexandra M., C. Brian Robbins, and Serge Larivière. 2002. "Alopex Lagopus." *Mammalian Species* 713: 1–10.

Berkes, Fikret. 2008. *Sacred Ecology.* New York: Routledge. Originally published 1999.

Church, John A., Jonathon M. Gregory, Philippe Huybrechts, Michael Kuhn, Kurt Lambeck, Mai T. Nhuan, Dahe Qin, and Phil L. Woodworth. 2001. "Changes in Sea Level." In *Climate Change 2001: The Scientific Basis: Contribution of Working Group I to the Third Assessment Report of the Intergovernmental Panel on Climate*

Change, ed. John T. Houghton, Yanni Ding, David J. Griggs, Maria Noguer, Paul J. Van der Linden, X. Dai, Kathy Maskell, and C.A. Johnson, 639–94. New York: Cambridge University Press.

Couturier, Serge. 2007. "Génétique et condition physique des trois écotypes de caribou du Québec-Labrador." PhD thesis, Université Laval.

Couturier, Serge, and Guy Saint-Martin. 1990. "Effets des feux de forêts sur les caribous migrateurs." Ministère du Loisir, de la Chasse et de la Pêche du Québec, Direction générale de Nouveau-Québec, Ste-Foy, Quebec.

Cree Trappers Association. 1989. *Cree Trappers Speak.* Chisasibi, QC: James Bay Cree Cultural Education Centre.

Crête, Michel, Bruno Drolet, Jean Huot, Marie-Josée Fortin, and G. Jean Doucet. 1995. "Postfire Sequence of Emerging Diversity among Mammals and Birds in the North of the Boreal Forest in Québec." *Canadian Journal of Forest Research/Revue canadienne de recherche forestière* 25 (9): 1509–18.

Darimont, Chris T., Paul C. Paquet, and Thomas R. Reimchen. 2005. "Range Expansion by Moose into Coastal Temperate Rainforests of British Columbia, Canada." *Diversity and Distribution* 11 (3): 235–39.

Darwin, Charles. 1859. *On the Origin of Species by Means of Natural Selection: Or the Preservation of Favored Races in the Struggle for Life.* London: John Murray.

DeMaster, Douglas P., and Ian Stirling. 1981. "Ursus Maritimus." *Mammalian Species* 145: 1–7.

Dyke, Arthur S., and Victor K. Prest. 1987. *Paleogeography of Northern North America, 18,000–5,000 Years Ago.* Geological Survey of Canada, "A" Series Map 1703A. https://doi.org/10.4095/133927.

Elton, Charles, and Mary Nicholson. 1942. "The Ten-Year Cycle in Numbers of the Lynx in Canada." *Journal of Animal Ecology* 11: 215–44.

Environment and Climate Change Canada. 1961–2010. Canadian Climate Normals and Averages. Online database. http://climate.weather.gc.ca/climate_normals/.

Feit, Harvey A. 1986. "Hunting and the Quest for Power: The James Bay Cree and Whitemen in the Twentieth Century." In *Native Peoples: The Canadian Experience,* ed. R. Bruce Morrison and C. Roderick Wilson, 171–207. Toronto: McClelland and Stewart.

–. 1988. "Waswanipi Cree Management of Land and Wildlife: Cree Cultural Ecology Revisited." In *Native Peoples: Native Lands,* ed. Bruce Cox, 75–91. Ottawa: Carleton University Press.

–. 2005. "Recognizing Co-management as Co-governance: Histories and Visions of Conservation at James Bay." *Anthropologica* 47 (2): 267–88.

Ferguson, Michael A.D., and Francois Messier. 1997. "Collection and Analysis of Traditional Ecological Knowledge about a Population of Arctic Tundra Caribou." *Arctic* 50 (1): 17–28.

Fisher, Jason T., and Lisa Wilkinson. 2005. "The Response of Mammals to Forest Fire and Timber Harvest in the North American Boreal Forest." *Mammal Review* 35 (1): 51–81.

Hardy, Léon. 1977. "La déglaciation et les épisodes lacustre et marin sur le versant québécois des basses terres de la baie de James." *Géographie Physique et Quaternaire* 31 (3–4): 261–73.

Hatcher, Paul E. 1990. "Seasonal and Age-Related Variation in the Needle Quality of Five Conifer Species." *Oecologia* 85: 200–12.

Hays, James D., John Imbrie, and Nicholas J. Shackleton. 1976. "Variations in the Earth's Orbit: Pacemaker of the Ice Ages." *Science* 194 (4270): 1121–32.

Henry, J. David. 2002. *Canada's Boreal Forest.* Smithsonian Institute Natural History Series. Washington, DC: Smithsonian Institution Press.

Hijmans, Robert J., Susan E. Cameron, Juan L. Parra, Peter G. Jones, and Andy Jarvis. 2005. "Very High Resolution Interpolated Climate Surfaces for Global Land Areas." *International Journal of Climatology* 25: 1965–78.

Hooven, Edward F. 1969. "The Influence of Forest Succession on Populations of Small Animals in Western Oregon." In *Wildlife and Reforestation in the Pacific Northwest,* ed. H.C. Black, 30–34. Corvallis: School of Forestry, Oregon State University.

Humphries, Murray M. 2009. "Mammal Ecology as an Indicator of Climate Change." In *Climate and Global Change: Observed Impacts on Planet Earth,* ed. Trevor Letcher, 197–214. New York: Elsevier.

Jacqmain, Hugo, Christian Dussault, C. Réhaume Courtois, and Louis Bélanger. 2008. "Moose-Habitat Relationships: Integrating Local Cree Native Knowledge and Scientific Findings in Northern Québec." *Canadian Journal of Forest Research* 38 (12): 3120–32.

Joppa, Lucas N., Scott R. Loarie, and Stuart L. Pimm. 2008. "On the Protection of 'Protected Areas.'" *Proceedings of the National Academy of Sciences of the United States of America* 105 (18): 6673–78.

Kendrick, Anne, and Phil O.B. Lyver. 2005. "Denesoline (Chipewyan) Knowledge of Barren-Ground Caribou *(Rangifer tarandus groenlandicus)* Movements." *Arctic* 58 (2): 175–91.

Krebs, Charles J., S. Boutin, and R. Boonstra, eds. 2001. *Ecosystem Dynamics of the Boreal Forest: The Kluane Project.* New York: Oxford University Press.

Lavoie, Luc, and Luc Sirois. 1998. "Vegetation Changes Caused by Recent Fires in the Northern Boreal Forest of Eastern Canada." *Journal of Vegetation Science* 9: 483–92.

Le Goff, Héloïse, and Luc Sirois. 2004. "Black Spruce and Jack Pine Dynamics Simulated under Varying Fire Cycles in the Northern Boreal Forest of Québec, Canada." *Canadian Journal of Forest Research* 34: 2399–409.

Levin, Simon A. 1992. "The Problem of Pattern and Scale in Ecology: The Robert H. MacArthur Award Lecture." *Ecology* 73 (6): 1943–67.

McCann, Kevin, James Umbanhowar, Murray M. Humphries, and Joseph S. Rasmussen. 2006. "Role of Space, Time, and Variability in Food Web Dynamics." In *Dynamic Food Webs: Multispecies Assemblages, Ecosystem Development, and Environmental Change – A Volume of Theoretical Ecology,* ed. Peter C. de Ruiter, Volkmar Wolters, and John C. Moore, 56–70. New York: Elsevier.

Milligan, Heather E., and Murray M. Humphries. 2010. "The Importance of Aquatic Vegetation in Beaver Diets and the Seasonal and Habitat Specificity of Aquatic-Terrestrial Ecosystem Linkages in a Subarctic Environment." *Oikos* 119 (12): 1877–86.

Mitrovica, Jerry X., Alessandro M. Forte, and Mark Simons. 2000. "A Reappraisal of Postglacial Decay Times from Richmond Gulf and James Bay, Canada." *Geophysical Journal International* 142 (3): 783–800.

Morneau, Claude, and Serge Payette. 2000. "Long-Term Fluctuations of a Caribou Population Revealed by Tree-Ring Data." *Canadian Journal of Zoology* 78: 1784–90.

Naiman, Robert J., and Henri Décamps. 1997. "The Ecology of Interfaces: Riparian Zones." *Annual Review of Ecology and Systematics* 28: 621–58.

Nordyke, Kirk A., and Steven W. Buskirk. 1991. "Southern Red-Backed Vole (*Clethrionomys gapperi*) Populations in Relation to Stand Succession and Old-Growth Character in the Central Rocky Mountains." *Canadian Field Naturalist* 105 (3): 330–34.

Ouranos. 2015. *Vers l'adaptation: Synthèse des connaissances sur les changements climatiques au Québec*. Montreal: Ouranos. https://www.ouranos.ca/publication-scientifique/SyntheseRapportfinal.pdf.

Parisien, Marc-André, and Luc Sirois. 2003. "Distribution and Dynamics of Tree Species across a Fire Frequency Gradient in the James Bay Region of Québec." *Canadian Journal of Forest Research* 33: 243–56.

Payette, Serge, Marie-Jose Fortin, and Isabelle Gamache. 2001. "The Subarctic Forest Tundra: The Structure of a Biome in a Changing Climate." *BioScience* 51 (9): 709–18.

Payette, Serge, Claude Morneau, Luc Sirois, and Mireille Desponts. 1989. "Recent Fire History of the Northern Québec Biomes." *Ecology* 70: 656–73.

Pellikka, Jani, Hannu Rita, and Harto Lindén. 2005. "Monitoring Wildlife Richness – Finnish Applications Based on Wildlife Triangle Censuses." *Annales Zoologici Fennici* 42: 123–34.

Péloquin, Claude 2007. "Variability, Change and Continuity in Social–Ecological Systems: Insights from James Bay Cree Cultural Ecology." Master's thesis, University of Manitoba.

Pruitt, William O. 1978. *Boreal Ecology*. Institute of Biology Studies No. 91. London: Edward Arnold.

Raymo, Maurenn E. 1994. "The Initiation of Northern Hemisphere Glaciation." *Annual Review of Earth and Planetary Sciences* 22: 353–83.

Rempel, Robert S., Philip C. Elkie, Arthur R. Rodgers, and Michael J. Gluck. 1997. "Timber-Management and Natural-Disturbance Effects on Moose Habitat: Landscape Evaluation." *Journal of Wildlife Management* 61 (2): 517–24.

Silva, Marina, and John A. Downing. 1995. *CRC Handbook of Mammalian Body Masses*. Boca Raton, FL: CRC Press.

Steele, Michael A. 1998. "Tamiasciurus hudsonicus." *Mammalian Species* 586: 1–9.

Storeheier, Pal V., Svein D. Mathiesen, Nicholas J.C. Tyler, I. Schjelderup, and Monica A. Olsen. 2002. "Utilization of Nitrogen- and Mineral-Rich Vascular Forage Plants by Reindeer in Winter." *Journal of Agricultural Science* 139 (2): 151–60.

Vik, Jon Olav, Christian N. Brinch, Stan Boutin, and Nils C. Stenseth. 2008. "Interlinking Hare and Lynx Dynamics Using a Century's Worth of Annual Data." *Population Ecology* 50: 267–74.

Wilson, Don E., and Sue Ruff. 1999. *The Smithsonian Book of North American Mammals*. Washington, DC: Smithsonian Institution Press.

Zalatan, Rebecca, Anne Gunn, and Gregory H.R. Henry. 2006. "Long-Term Abundance Patterns of Barren-Ground Caribou Using Trampling Scars on Roots of *Picea mariana* in the Northwest Territories, Canada." *Arctic, Antarctic, and Alpine Research* 38 (4): 624–30.

8

Coastal Goose Hunt of the Wemindji Cree
Adaptations to Social and Ecological Change

CLAUDE PÉLOQUIN and FIKRET BERKES

Wemindji hunters interact with the land through relationships of reciprocity and respect that entail attentiveness, care, and humility. These relationships involve humans, animals, and many other members of the "community of beings" that constitutes the Cree world. Humans, as well as all others, are part of an integrated, interdependent social-ecological system in which the natural world and the Cree world sustain one another over time.

The subarctic or boreal ecosystem in which the Crees live is characterized by a high degree of variability and unpredictability (Berkes 1998; Elmqvist et al. 2004). In this chapter, we examine the ways in which Wemindji hunters deal with variability and make sense of complexity as it shapes their subsistence activities. The Canada goose *(Branta canadensis)* hunt, as practised by coastal Crees, has been described as particularly adaptive to shifts and changes in conditions. At present, this hunt appears to be undergoing transformations of greater magnitude than the usual year-to-year variations. How do Wemindji hunters talk about and deal with the variability, unpredictability, and change characterizing the hunt? What factors of change do they identify? How are they responding to these changes?

To answer these questions, we conducted research that was participatory (Mulrennan, Mark, and Scott 2012) and largely qualitative. We examined variables and factors related to change, not quantitatively but qualitatively, as verbal categories in the fuzzy logic sense (Berkes and Berkes 2009). Favouring an ethnographic approach, we interviewed Wemindji hunters and accompanied them on their hunting territories, learning from their knowledge and practice.[1] We looked at ecological knowledge as a process, not so much about "what is known" but rather as "ways of knowing" (Péloquin and Berkes 2009; Berkes 2018). We learned from Wemindji Crees about ways of making observations and making sense of these observations, dealing with a large number of variables qualitatively (fuzzy logic), and how this in turn influenced human-environment relations. We explored the two-way relationship between people and their environment as a set of processes occurring at multiple scales, involving both biophysical and social-cultural factors, focusing on social-ecological systems – rather than social systems or ecological systems per se.

Following Colin Scott (1983, 1986, 1996) and Fikret Berkes (1982), we present the Cree coastal goose hunt as a set of practices that involves constant observing, monitoring, and adjusting of hunting, in tune with ecosystem variability. We discuss hunters' observations of environmental change processes, and their account of the many variables involved. We examine how their explanations for changes reveal the synergistic effects of both biophysical and social-cultural factors that have necessitated a reorganization of the hunt. This transformation highlights the adaptability of traditional knowledge and resource management systems in responding to very real contemporary challenges.

THE GOOSE HUNT

Canada geese, or *niskw* in the Cree language, use a wide range of habitats along the James Bay coast for feeding and staging as they travel to, and from, their northern breeding grounds. These habitats include salt- and freshwater marshes, bays, lakes and rivers, and islands and headlands covered with lichen and shrub heaths (Reed et al. 1996; Chapter 6, this volume). Their use of these habitats fluctuates greatly, according to local geography, season, current weather patterns, and other factors.

The goose hunt, which is one of the most important subsistence activities for the coastal Crees, entails a high level of social co-ordination (Scott 1986). The Wemindji Nation is divided into hunting territories; the coastal territories also serve as management units for goose hunting, each under the supervision of a *paaschimaau uchimaau*, or goose boss. Territories contain multiple sites that are suitable for the hunt, some of which have been engineered through small-scale landscape modifications to enhance their effectiveness (see Chapter 9, this volume). These modifications include the creation of small dikes, new ponds, and *tuuhiikaan,* corridors made in the forest by cutting trees along given lines, which the Crees describe as "flyways" in English (Scott 1983).

As geese are migratory and not bound to any particular area, it is important for hunters to ensure that they do not associate a territory with danger (Scott 1983; Bearskin et al. 1989). Geese are highly intelligent and observant, able to detect even subtle signs of hunters' presence (Scott 1996). Any material of bright colour, such as red or orange, must be hidden, hunting blinds must be well camouflaged, and garbage must not be left around. The latter is part of a broad commitment to keeping camps clean, thus manifesting respect to the animals (Tanner 2014). Fires and others sources of light are minimized, and hunters avoid making noise, through superfluous shooting, for example. It is also important to avoid shooting after dusk so that the geese will not see the flame at the muzzle of the shotgun.

Using a combination of stones, branches, snow, and canvas, hunters build individual cone-shaped blinds on each site, 1.2 to 1.5 metres high. They kneel in these and wait for the time to shoot. Early each morning, the goose boss evaluates the hunting potential of the day. The key factor is weather, especially wind, which influences the foraging patterns of geese. The birds fly more and they fly lower on windy days than they do on calm days, and a strong wind muffles the sound of shotguns: "We go to places when it's not that windy, know where to go by the weather, and where the wind comes from" (R. Atsynia, pers. comm., August 7, 2007).[2] No hunting should occur until the wind is strong, which could take from a couple of hours to a few days. When the wind is sufficiently forceful, its direction largely determines the choice of hunting location so as to prevent incoming geese from hearing the shots before they arrive and to increase their likelihood of visiting the site.

The conditions in which geese remain for a long season are ones in which periods of southerly wind alternate with periods of northerly wind. Implicit in the following comment is the idea that rotation between sites, attuned to wind patterns, encourages the birds to stay: "When the wind is southwest, south, we go around in a bay that is good for that ... On the north side, then when wind is north, we go for a place that is good for that. This is the best way, it keeps them around" (L.U., pers. comm., July 10, 2006).

Other key factors that influence the success of the day include tide levels, temperature, and recent hunting pressure. The goose boss, called the tally-man in the passage below, observes the birds from a distance, often using binoculars, to evaluate their numbers and behaviour:

> It is the tallyman who decides where we hunt and don't hunt. There may be two or three places on the trapline where we could go, but he says only one, at least one or two in one day, you just go to one, that's it. You have to listen to the tallyman. Maybe next day we'll go to that other one. (A. Gilpin, pers. comm., July 12, 2006)

A new site is chosen almost every day. This rotation diffuses hunting efforts in time and space through judicious site selection that provides the best fit to given weather and goose foraging patterns (Scott 1983). While one site is in use, others are rested, and on some days the goose boss suggests that no hunting should occur at a specific site, even if conditions seem favourable. In addition, there are days when the entire territory is rested and hunters stay at camp. This usually occurs when winds are too calm but may also apply when the hunt of the previous day was especially large and successful, as a continued hunt might become too much if a break is not taken. The result is a system of "no-take zones" that fluctuate according to ecological dynamics. It is comparable to the fallow system used in agriculture, though on a much shorter time scale, as it encompasses days rather than months or years (Scott 1983, 54).

Two main strategies determined by environmental conditions and goose numbers are applied to the hunt. In the first, hunters wait for the birds to fly over on their way from one site to another. Next, they choose a suitable site that is ice-free. Alternatively, if there are no ice-free areas of open water, they dig a depression in the snow that is suggestive of a pond and place decoys there. Then, they wait in their blinds to intercept the incoming geese as they land. The second strategy involves walking slowly toward a flock of feeding geese to chase them away. When the birds fly up, no one shoots, because this would risk scaring them off permanently. Instead, the hunters wait for the birds to circle back to the feeding area, as they commonly do:

> When they hang around in one area, if it's a north wind or something, we chase them out, and then they fly back, that's when we shoot them. We don't slaughter them. We just walk in, don't shoot, just chase them away, they come back after, later on. They won't come back if we shoot them right away. (A. Gilpin, pers. comm., July 12, 2006)

This strategy usually brings higher returns than attempting to shoot a larger flock at the outset, and is reliable under the correct conditions: when strong winds are blowing in the right direction and large numbers of geese are present.

There are consequences when people do not adhere to these practices. In the following example, lack of care for an especially sensitive congregation site spoiled the hunt:

> There is this one site, my grandfather used to go there late in the fall, just to check on the geese, and would not allow others to come with him. He would go by himself, and climb up a tree to see them without disturbing them. This is how sensitive it was. The geese do not congregate there anymore. Once he got there and there were people. He asked where they parked their canoe, and it was where the geese congregate, so that was the end of that hunt. (R. Atsynia, pers. comm., August 4, 2006)

Elders and hunters mentioned a specific event in the mid-1950s when a careless hunter shot geese at night, firing multiple rounds that scared the birds (S. Hughboy, pers. comm., 2006; S. Mistacheesick, pers. comm., 2006). After that, they avoided the area. The incident was also reported to Colin Scott (1983, 47). Some Crees cite this sort of neglect as partly responsible for the shift in snow geese migration patterns. Once present in high numbers, these geese began to fly elsewhere during the 1950s and 1960s. Others explain the change in snow geese abundance in terms of "animals come and go." Before, the snow geese had been abundant and the

Canada geese relatively rare. Then, as snow geese declined, Canada geese became abundant.

Living with Variability and Unpredictability to Maintain the Resource

> Maybe the year before was not too good, if no geese one year maybe you expect more the following year, so you don't know, but it could be the same again. People used to live like that: we go there and we expect to have a good season but we don't know. (J. Blackned, pers. comm., August 6, 2006)

Variability and unpredictability are key aspects of life in the bush. Cree ecology does not explain this reality in terms of mechanical laws but rather in terms of relationships within a community of beings (Preston 2002). In the Cree world, both humans and non-humans, including the geese, are sentient beings who have agency and who interact in intricate webs of relationships. As Scott (1983, 196) puts it, "Hunters cannot always be sure why an animal has gone away or why it returns, but one important factor they can control is their own manner of hunting, and respect shown to the animal in general terms."

The notion of respect entails humility rather than boasting, using only as much as is needed, and ensuring reciprocity in relationships through gifts and other exchanges (Bearskin et al. 1989; Feit 1995; Tanner 2014). In one way, the obligation to show self-restraint and respect may be seen as part of the Cree side of the relationships with animals. At the same time, one might say that animals show respect toward hunters in providing for human needs, but they also manifest restraint, in that their gifts are measured and may be withheld (Scott 1983).

In the context of the goose harvest, this outlook informs a management approach that is highly flexible, responsive to the behaviour of the birds, to the wind, and to many other factors. The Crees do not necessarily seek to maximize short-term returns but instead favour active restraint on hunts to ensure long-term viability (Feit 1987). Restraint is informed by, and enacted through, specific elements of knowledge, practices, and institutional arrangements (Berkes 2018). This differs from more static arrangements in which access and use are allocated by spatial boundaries that are defined a priori and rigidly, as is often the case in conventional resource management. The application of rules, or codes of practice, concerning the goose hunt is determined by where the birds are, how many there are, and what they are doing, factors that constantly change. This is consistent with the notion of "rules-of-thumb," or the relationship between general principles or propositions that are flexible and situation-specific rules that are compatible with these general principles but also tailored to particular circumstances.

Concerns over Declining Goose Availability

Cycles are a recurring theme in the Cree view of life (Bearskin et al. 1989). Animals "leave" and "come back"; fires are followed by regrowth and renewal (see Chapter 6, this volume). In recent years, however, conversations about the goose hunt in Wemindji point to change of higher than usual magnitude. Below are some comments about the current state of the hunt:

> It's been getting worse every year, bad goose hunt last two years. I did not catch any goose this spring [2006]. Many others also did not catch any. It used to be a hundred in a season. (F. Stewart, pers. comm., June 15, 2006)

> Before, I would get 130 geese or so in a spring, now usually 10, this year, 6. (S. Georgekish, pers. comm., September 25, 2006)

> Hardly any geese anymore. In 1984, got fifty geese a day, now you get ten and return home because you know you won't get any more. (J. Blackned, pers. comm., August 6, 2006)

> This year I killed three only. We were there about three weeks I guess. There was hardly any, we would see them flying. That's it. They wouldn't stop. (A. Gilpin, pers. comm., July 12, 2006)

> This year and last year, the hunting wasn't too good, spring hunting. That's why, I bought three boxes of shell, and I still have them all. There's hardly any geese ... Some men didn't even kill a goose in the spring. (S. Mistacheesick, pers. comm., June 27, 2006)

Almost all the interviewed hunters mentioned that the goose hunt in the coastal portion of Wemindji territory was becoming increasingly unrewarding.[3] Cree participants in the Voices from the Bay project mentioned that geese began to decline in eastern James Bay in 1988 (McDonald, Arragutainaq, and Novalinga 1997), but this was part of a decades-long trend, as Scott (1983) reported. However, though the current changes are similar to the earlier ones, it appears that the rate of change has dramatically increased.

This is surprising, given that Canada goose populations reached an all-time high during the 2000s (Hass 2002; Harvey and Rodrigue 2006). Throughout North America, the numbers of most Canada goose subspecies have increased since the 1940s, when they were endangered due to overhunting and habitat loss (Hass 2002). Although population estimates at this level are imprecise and difficult to calculate, it is clear that the Atlantic Flyway sub-populations expanded tremendously from approximately 30,000 breeding pairs in 1995 to over 160,000 in 2005 (Harvey and Rodrigue 2006). This five-fold increase coincides with the period when the Wemindji goose hunt went from bad to worse. Obviously, given the statistics cited above, a drop in the number of birds is not responsible for this decline. Other factors must be at play, including changes in goose behaviour, which are in turn partly related to biophysical and social-cultural changes in James Bay.

It is important to point out that Crees often see many changes as manifestations of cyclical patterns. Elders Samuel Hughboy and Sinclair Mistacheesick noted that the ongoing trends of "not many geese" had happened before, most recently around fifty years ago. Freddie Atsynia (pers. comm., July 16, 2006) referred to an even earlier period: "A long time ago, like my mother used to say a long time ago, before I was born, her walking out ceremony, she only had eight geese to celebrate with, that spring. That was long time ago. I'm sixty-four years old, and that was before I was born."

These accounts suggest that periods of goose scarcity and abundance are part of the normal course of events and part of the larger web of processes in which "fire, things burn, grow back, moose and beaver leave, then they come back" (F. Atsynia, pers. comm., July 16, 2006). Yet, though goose behaviour and abundance may be cyclical, hunter-goose interactions are embedded in a broader context that is also changing drastically.

HUNTERS' VIEWS OF CHANGES IN THE GOOSE HUNT

Wemindji hunters offered a wide range of observations and interpretations regarding the changes in the goose hunt. They implicated many factors as causes, as symptoms, or both. The factors reported, and the suggested connections between them, are summarized in Figure 8.1. The items first reported usually involved shifts in goose behaviour, but others included changes in Cree values and practices, biophysical modifications such as climate-related trends, and alterations in the vegetative cover of key areas. Also of interest are differences in abundance and behaviour of other animal species, many of which were assumed to have direct or indirect impacts on goose behaviour. Figure 8.2 presents a simplified schema of the principal categories.

Goose Behaviour and Migratory Patterns

Goose behaviour and migratory patterns are key determinants of goose availability to hunters, and consequently, of the viability of the harvest. For example, hunters reported that the birds increasingly fly inland instead of along the coast. This is part of a trend that some hunters identified as dating from about the 1960s, which gradually transitioned from exceptional occurrences to the dominant pattern: "In 1953, on the inland portion of my trapline, I seldom saw geese over there. I first saw geese inland in the fall of 1959, it was unusual. By 1969, I realized that it was usual that there were geese inland in the springtime" (R. Atsynia, pers. comm., August 4, 2006).

This has important implications. First, geese can no longer be relied upon to visit the coast, where hunters wait for them according to the practices discussed earlier. Second, the birds land in scattered patterns throughout the territory, instead of congregating in large numbers, as they did in coastal bays. In addition, they are less likely to return after being chased. If they fail to come back, the "drive" hunting technique will yield no results.

A related trend is that geese now tend to fly high and are less likely to land in the spring, this is associated with fast ice breakup and snow melt:

> This year and last year, we had an early spring, early open water; the ice went really fast, so there is less geese. Because the snow is really going

FIGURE 8.1
Web of factors of change affecting the goose hunt

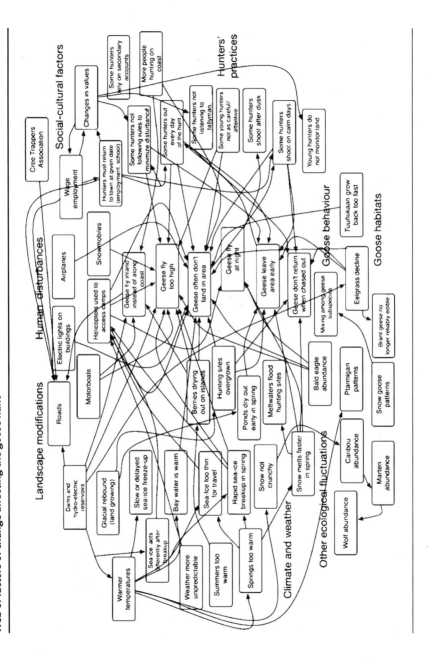

FIGURE 8.2
Simplified summary of factors of change

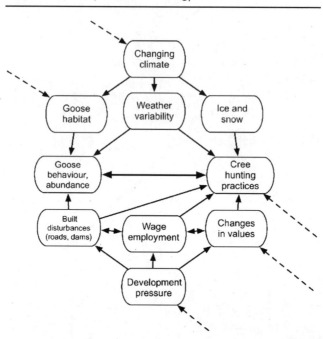

fast, and there is hardly any water in the swampy areas, the geese don't land and don't stick around. This spring and last spring, I noticed it's early spring, and there's hardly any geese. (S. Mistacheesick, pers. comm., June 27, 2006)

Sometimes, it all depends on the snow, the weather. If the snow melts fast they don't stop, if the snow hangs around they're gonna stop. It was fast spring in the last two years, and they didn't stay. (A. Gilpin, pers. comm., July 12, 2006)

As we will see later, such weather patterns are becoming more frequent and are increasingly perceived as local manifestations of global climate change. Some hunters suggested that this may partly explain why geese are now more prone to fly high and to refrain from landing in the area. But this explanation seems only partly valid, since the birds sometimes decide not to land even when conditions do appear suitable.

Another important behavioural change is that the geese have started flying at night instead of during the day, which makes hunting impossible: "They fly at 2:00, 3:00, 4:00 a.m., we see them but we don't shoot" (J. Blackned, pers. comm., August 6, 2006). In the late 1970s, this behaviour was regarded as exceptional (Scott 1983, 200). But nocturnal flights have become increasingly common since the early 2000s, and are especially frequent "between quarter and full moon [when] they use moonlight for navigation" (A.V., pers. comm., August 8, 2006). Hunters did not single out a specific cause as solely responsible for all these behavioural changes. Rather, they were linked with myriad other social-ecological changes. These were understood within a broader web of fluctuations, in which events may or may not be correlated. During our conversations about change on the land, hunters often referred to certain events and observations, which they juxtaposed. These associations between events first appeared to be mere temporal reference points, but they were also mentioned in a way that suggested linkages between them, links that may or may not be causal.

Weather- and Climate-Related Factors of Change

Climate and weather in Wemindji have changed drastically and at accelerating rates over the past decades (Syvänen 2011). The weather has reportedly become warmer in the Wemindji area since the late 1970s (J. Blackned, pers. comm., July 27, 2006); some even say it is "too hot" (F. Stewart, pers. comm., June 24, 2006). This is consistent with reports from elsewhere in the Subarctic and the Arctic (McDonald, Arragutainaq, and Novalinga 1997; Nichols et al. 2004; Ford et al. 2008; Ford, McDowell, and Pearce 2015). Similarly, these climate changes are manifest through multiple signs on the land (Krupnik and Jolly 2002). In Wemindji, for example, with warmer temperatures, "snow is not crunchy ... It doesn't scare the caribou when we walk toward them" (A. Mayappo, pers. comm., July 16, 2006). This warmer temperature also thins the ice on the bay:

> The weather has been changing a lot since the late 1970s. It's not as cold in the wintertime, and after freeze-up you have to wait a long time before you can travel on the ice. And people say the ice is not as thick as it used to be, even out in the bay. In late February I put out my fish nets,

five kilometres from here. I was surprised that the ice was very thin, it was about this thin [around thirty centimetres]. It used to be about one metre thick. It makes it easier for digging a hole in the ice. (J.M., pers. comm., July 10, 2006)

The decrease in sea-ice thickness makes it dangerous to travel on the frozen bay in the fall and spring, which has implications for harvesting activities:

Freeze-up takes longer, we must wait a long time before going on ice [in the fall], and then in the spring, ice goes out really fast, too fast. (L.U., pers. comm., July 3, 2006)

In the past, say twenty-five to thirty years, it would not be unusual to travel by snowmobile as late as the fifteenth of May, whereas nowadays the ice is often too thin for such travel as early as mid-April. (J. Blackned, pers. comm., August 6, 2006)

In the fall, we used to travel on the ice in November. Nowadays, it's often only safe after the New Year. (S. Georgekish, pers. comm., September 25, 2006)

Hunters reported that spring came earlier and was shorter, the snow and ice melting earlier and faster. During the spring of 2005, on at least one coastal hunting territory, "when the geese arrived there was no snow left, only ice" (S. Georgekish, pers. comm., August 1, 2006). A hunter from the neighbouring territory to the north reported a similar trend: "Over the last thirty years hunting at Moar Bay, I noticed that there is no more snow nowadays [in the spring when the geese arrive]" (A.V., pers. comm., August 8, 2006). In addition to this warming trend, weather is more chaotic, more difficult to predict: "In the summer, sometimes it's very hot for a few days, and then it can change very rapidly. It changes faster than it used to" (J.M., pers. comm., July 10, 2006).

Some key climate-related factors of change, and their interactions, are summarized in Figure 8.3. These factors diminish the reliability of the ecological indicators on which the Crees normally base their decisions, making subsistence activities on the land more hazardous. Moreover, they directly affect the availability of geese. Early and fast springs and ice

FIGURE 8.3

Climate-related factors of change affecting the goose hunt

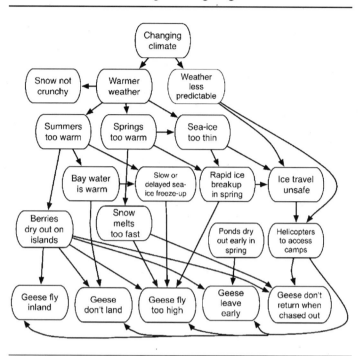

breakup, as well as warm weather, are key factors influencing the birds: "It's too warm, it's not good for the geese, they fly right through, it's probably why the geese change their patterns" (A.V., pers. comm., August 8, 2006).

One of these factors involves the crop of black crowberries, which the Crees call "blackberries" in English and *aschiimin* (earth or moss-berries) in Cree, on which geese feed as they fly south in the fall. When the summer is too warm and dry, the berry crop fails on coastal islands and peninsulas. Under the heat of the sun, the berries dry out before they can ripen and are unpalatable when geese start coming back in August-September. As a result, the birds do not stop and are not locally available for harvesting. Whereas such crop failure occurred sporadically in the past (Scott 1983, 50), it seems to be more frequent now: "Last few years, not very many berries, so geese don't stop in the fall" (F. Stewart, pers. comm., June 13, 2006). "The berries don't grow as much because it's too hot in the summer" (I. Mistacheesick, pers. comm., June 27, 2006). However, though berries have become less reliable on headlands and islands in Wemindji territory, hunters reported that they have become more abundant during the last twenty years in Whapmagoostui, the northernmost Cree community some 250 kilometres north of Wemindji (I. Mistacheesick, pers. comm., June 27, 2006). This appears to be consistent with the northward expansion of certain species as average temperatures rise, as part of climate change.

At the same time, hunters' behaviour is also influenced by these manifestations of climate change. For example, their ability to move from one site to another is determined by the thickness and reliability of ice on the bay. In many instances, they prefer staying close to the camp to avoid unnecessary risk. Hunting pressure is more concentrated as a result. Also, ongoing land uplift in the area (see Chapters 5 and 9, this volume) means that many hunting sites are "drying out." Portions of the bay are now part of the mainland, and "what the geese eat is buried" (J. Blackned, pers. comm., July 26, 2006), as woody plant species replace the marshy species that constitute "goose food."

Large-Scale Human Modifications of the Landscape

In addition to these climatic and other biophysical factors of change, a series of large-scale disturbances – anthropogenic this time – has occurred, associated with the massive projects of Hydro-Québec. They are central to Cree perception of a changing world (Carlson 2008). Their magnitude is so great that even if they were not directly implicated as causal factors, it is possible that they contributed indirectly to social-ecological change. Most Crees linked the majority of changes in the goose hunt to the construction of dams and reservoirs as part of the James Bay hydro-electric project:

> I think since Hydro-Québec made the reservoirs, the geese changed their patterns. If you look at the maps all the way to Eastmain River, there is a lot of water, just like James Bay. That's why I think that's one thing that they follow. And along the bay, there used to be grass. How do you call that? We call it in the Cree *sishkabash* [eelgrass, *Zostera marina*]. Over ten years now, there used to be lot of sishkabash, so I noticed when I set the nets in the water there is just a little bit of that now ... They say

it came from La Grande, I think it changed the water. I don't know. (F. Atsynia, pers. comm., July 16, 2006)[4]

Ever since they built the dams, geese are flying more inland. A person from Labrador, Schefferville, he didn't see geese back then. But ever since they built the dams they can see geese, even ptarmigans, they didn't have before. They used to fly along the coast, and now they fly inland. (S. Hughboy, pers. comm., June 23, 2006)

In the early days, geese would rarely fly inland. They always went along the coast. Since they built the dams, the geese go more inland, there's water up there. (S. Hughboy, pers. comm., June 23, 2006)

Many hunters saw the reservoirs as contributing to the changing pattern of geese flying inland. Some suggested that the most recent phase of hydro-electric development – the Eastmain-1-A and Rupert Diversion Project – will exacerbate the situation. Hunters also related climate change to these projects: "The weather is warmer, ever since the reservoirs" (F. Atsynia, pers. comm., July 16, 2006); "Ever since the dams, since the late 1970s, it has been warmer" (J.M., pers. comm., July, 10 2006).

"The dams," and everything that came with them, certainly produced the most rapid, catastrophic, and overwhelming changes that the Crees have had to face, at least in living memory. Since the late 1970s, hunting territories have been flooded, rivers dammed and diverted, and the hydrology of a very large basin has been modified. Changes have included spatial and temporal shifts in the freshwater discharge into James Bay, with implications for the salinity and temperature in various parts of it (Rosenberg et al. 1997), as well as for the distribution and health of eelgrass, a primary food source of Canada geese (Short 2008).

Informants were quick to offer caveats on the impact of hydro development, however, and corrected us whenever our interpretation of their account seemed overly simplistic. One hunter made the case that most changes had to be understood in the context of "normal" fluctuations and that knowledge and connectedness with the land and animals allowed the Crees to appropriately respond and adapt to such changes. As he put it, "Since the dams, it changed, but it's not that bad, it didn't harm the

FIGURE 8.4
Social-cultural factors of change

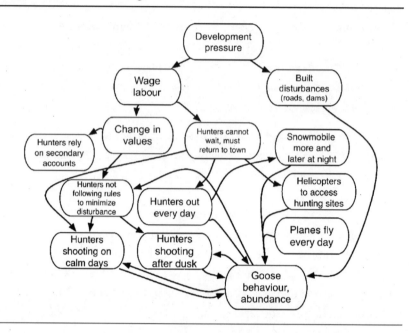

environment that much. The caribou are here, they were not here before, they came after the Hydro project, now you have to scare them away out of the cabin, it's like dogs" (O.V., pers. comm., August 3, 2006).

Social-Cultural Change

Figure 8.4 presents some of the social-cultural factors that pertain to changes in the Wemindji goose hunt. Hunters now spend more time in the village than out on the land, in both their numbers and the duration of their stay. The Cree population has increased substantially over the last decades, and the bush economy alone cannot support these growing numbers in the long run. At the same time, employment opportunities exist in town as well as elsewhere in the James Bay area, which keep some would-be hunters away from the bush: "Jobs are in town; people don't hunt as much" (L.U., pers. comm., July 10, 2006). Cree youth attend school in Wemindji, and many elders also stay there to receive medical support. This is seen as both resulting from and causing a measure of change in

Cree values and lifestyle. For example, the participation of Wemindji residents in the Cree Hunters and Trappers Income Security Program began to drop off in 2000.[5] Many elders and occupational hunters noted that some young Crees were not as well connected with the land as in the past:

> Before they would know when caribou are coming, could predict a storm. Now with technology, people are disconnected from the land. (S.B., pers. comm., August 2, 2006)

> Nowadays people don't observe like in the past, for them a rabbit is a rabbit. Back then in the old days, people observed a lot, they would find things, observe change. (S. Georgekish, pers. comm., July 19, 2006)

When deciding when and where to hunt, some younger hunters rely more on second-hand information, supplied by the community radio or other hunters, than on their own knowledge. According to elders, this was due to a lack of connectedness with their environment, which prevented them from "reading the land" themselves. Spiritual connections had been replaced by the mundane: "Nowadays people only believe in bingos, even the old ladies" (S. Georgekish, pers. comm., July 19, 2006). Customary rules were followed less closely: "People don't monitor anymore, and they take off whenever they want" (R. Atsynia, pers. comm., August 4, 2006). It is important to point out that this trend should not be overstated. The same respondents quickly added that youth were respectful and attentive when they did go out on the land to learn the proper ways of interacting with animals (A. Gilpin, pers. comm., July 12, 2006; J. Blackned, pers. comm., August 6, 2006). Similarly, not everyone agreed with the suggestion that the Crees were changing their approach to resource harvesting, maintaining instead that biophysical factors, such as habitat change, were to blame (J. Blackned, pers. comm., August 6, 2006).

Yet some trends in social change could be linked with changes in the goose hunt. These are related to the synergistic effect of three factors: less flexible schedules for many hunters, change in the relationship between the Crees and the land, and reduction in goose availability. As one hunter explained, "I work, so I get three weeks for goose break. With jobs in town, I can't wait around" (J. Blackned, pers. comm., August 6, 2006). Once at the hunting site, most hunters wait for the right conditions, staying in camp, or not shooting when restraint is the appropriate course. But our interviewees reported that some became impatient: "Not enough days because work, people just shoot, don't wait" (J. Blackned, pers. comm., August 6, 2006). Some hunted every day – except Sundays – which meant that they did not rest the sites, or they could end up shooting on calm days (A. Gilpin, pers. comm., July 12, 2006). In some instances, "young hunters run at incoming geese instead of waiting in the blinds" (R. Atsynia, pers. comm., August 5, 2006). Also, some younger hunters had shot geese later into the evening, not so much due to wilful neglect of customary rules but because geese were sometimes available only at that point.

Greater reliance on wage employment, hospital medicine, and store-bought food – "the whiteman's food" as the Crees call it – has reduced the risks, uncertainty, and hardships that once characterized bush life (Tanner 2014). Elders highlighted that things were often difficult "in the old days," when hunger and starvation were always possible outcomes of "bad luck" in harvesting, animal scarcity, or bad winters (S. Hughboy, pers. comm., June 23, 2006; R. Atsynia pers. comm., August 4, 2006; W. Asquabaneskum, pers. comm., September 23, 2006). In the past, hunters were more vulnerable to the effects of fluctuations in animal availability. The combination of uncertainty and vulnerability, and their influence on contingency in the Cree world, probably played a key role in informing an ethos of respect and flexibility. Whereas the Crees largely welcomed the new reality, they also associated it with an erosion of connectedness to their environment.

RESPONDING TO CHANGE

Continuity in social-ecological systems involves both adapting to and shaping change (Berkes, Colding, and Folke 2003). Wemindji hunters are responding to change in the goose hunt by modifying their practices. Their reactions provide insights on how social arrangements for resource use can navigate turbulence.

In some cases, the customary practice of site selection and rotation on coastal territories remains flexible enough to meet these changes. To illustrate how this plays out, Figure 8.5 maps the sites used by a group of hunters during the spring hunts of 1981 and 2006 (see Scott 1983, 66).

FIGURE 8.5
Site selection during the spring goose hunt, 1981 and 2006

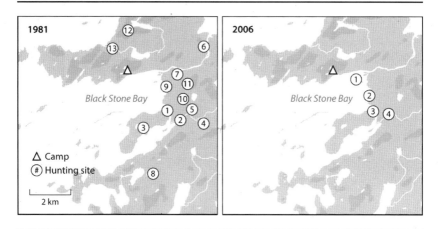

The left-hand portion of the map displays the territory and the various sites used over a four-week period in Blackstone Bay, about ten kilometers to the South of Wemindji, 1981. An extensive rotation and switching of sites was used in that year, and there were many days when no hunting took place, to rest the territory. The right-hand portion of the map shows the May 2006 hunt, which lasted for about three weeks. During that season, only four sites were used, down from the thirteen of 1981, all clustered at the centre of the territory. Hunters' accounts of the 2006 hunt indicate that biophysical changes in the territory limited the number of suitable sites. Land uplift had dried out important parts. This was associated with vegetative succession on these sites, in which woody plant species replaced the marshy species that geese prefer. The ice had also gotten thinner, which further reduced access to certain sites (see Chapter 9, this volume, for an account of how Crees modify the landscape to counter some changes and attract geese to certain locations).

Yet for many hunters, the various factors that simultaneously affected their practices did invite adaptive responses. Their ability to continue old practices was often constrained not only by biophysical factors beyond their control (the geese flew elsewhere, too high, or simply did not land), but also by socio-cultural factors that impeded their flexibility.

One change in the goose hunt involves the use of helicopters to access the coastal hunting sites when ice conditions are unsafe for travel by boat or snowmobile. This airlift, available to members of the Cree Trappers Association, has been in play since the mid-1980s. At that point, it was mostly used at the end of the hunt, when ice breakup prevented travel by boat or snowmobile. Due to various commitments, hunters and youth must often return to town by a given date at the end of the three-week hunting break, which means that they cannot wait until travel conditions improve. The airlift has been especially important in recent years due to rapid warming in spring that makes the ice unsafe for travel. However, helicopters scare the geese away:

> People are flown back because they have to come back right away. They can't wait, because they are workers or students, so they need the airlift. The spring airlift, using helicopters down the coast, all the communities do that so it must scare them. It is since they've been using that in 1985, there is less geese, there used to be more, now they are scared. (S. Georgekish, pers. comm., August 1, 2006)

A more major transformation of the hunt is the shift away from the coast to the inland territories. This is made possible by the hundred-kilometre all-weather gravel road that runs from Wemindji village to the paved James Bay Highway, which crosses the James Bay region on a north-south axis. These roads are important to Wemindji's economy, but they have also brought an influx of non-Indigenous sport hunters (Scott and Webber 2001). Many Cree winter hunting and trapping camps have been established along these two roads. The Cree Trappers Association provides transportation to these sites for members who do not have a vehicle and helps with gas expenses for those who do.

Hunters are now favouring road-based travel for the goose hunt, in both the spring and the fall. For example, Cree Trappers Association maps of hunting camp radio locations reveal that there are many more camps along the roads than on the coast. Of the thirty-one spring camps that were active in 2006, eight were on the coast and nineteen on roads: ten on the Wemindji road, seven on the James Bay Highway, and two on the trans-taiga road south of the LG3 dam, the remaining four scattered elsewhere in the territory. Road travel allows mobility over inland portions of Wemindji territory, even in the spring and early fall when absence of snow cover renders snowmobile travel impossible. Hunters can reach the inland areas where geese now fly more frequently. Geese are scattered inland, making it difficult to predict their location. Thus, the ability to drive from one area to another is advantageous.

Access to camps via roads is less vulnerable than boat and/or air travel to changing weather patterns and ice conditions. As well, there is no need either to rush to travel before ice breakup or to wait for the ice to clear before heading home. It is safer, less costly, and less complicated than flying by helicopter. Vehicles do not frighten the geese, so driving inland diminishes the disturbance associated with the airlift and diffuses hunting pressure over a broader area. As an experienced Wemindji hunter suggested, "Helicopters are expensive and noisy, let's hunt geese along the road. Leave the coast a chance to rest" (O.V., pers. comm., August 3, 2006). His reference to letting the coast rest is significant: if hunting were to concentrate on the inland sites, the coast might become attractive to the geese once again, and the site rotation customarily practised at the scale of individual territories would extend to the entire community's territory.[6]

Some hunters combine the fall moose hunt with some inland goose hunting. For them, the coastal goose hunt is not productive or reliable enough to justify making separate trips. Instead, they decrease the risk in their investments of energy and time by combining the goose hunt with the moose hunt and beaver trapping season, which are based at inland bush camps. Figure 8.6 shows the role of road travel in the context of adapting to changing conditions.

DISCUSSION AND CONCLUSIONS

Goose harvesting by the Wemindji Crees is grounded in an ethos of care and attention to ecological events and processes. Some features of Cree land stewardship are well known. They include treading lightly in a world of contingency, where actions can have unforeseen consequences (Scott 1983; Tanner 2014); "doing the right thing" in hunting, which entails respect, humility, attentiveness, and flexibility (Feit 1987; Preston 2002);

FIGURE 8.6

Road use as adaptation strategy in the goose hunt

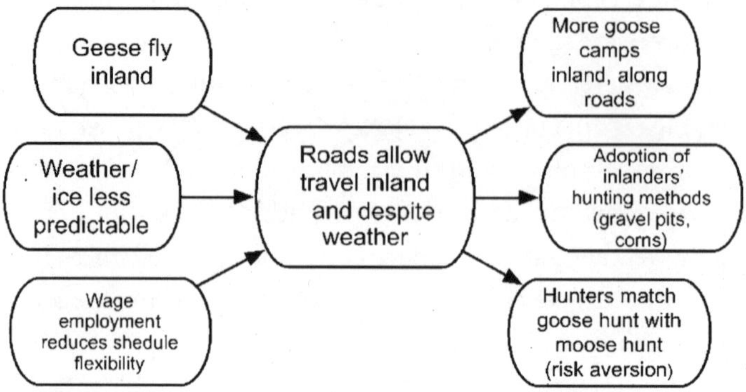

```
┌─────────────┐                          ┌──────────────┐
│ Geese fly   │                          │ More goose   │
│ inland      │                          │ camps        │
│             │                          │ inland, along│
└─────────────┘                          │ roads        │
                  ┌──────────────┐       └──────────────┘
┌─────────────┐   │ Roads allow  │       ┌──────────────┐
│ Weather/    │   │ travel inland│       │ Adoption of  │
│ ice less    │──▶│ and despite  │──▶    │ inlanders'   │
│ predictable │   │ weather      │       │ hunting      │
└─────────────┘   └──────────────┘       │ methods      │
                                         │ (gravel pits,│
┌─────────────┐                          │ corns)       │
│ Wage        │                          └──────────────┘
│ employment  │                          ┌──────────────┐
│ reduces     │                          │ Hunters match│
│ shedule     │                          │ goose hunt   │
│ flexibility │                          │ with moose   │
└─────────────┘                          │ hunt (risk   │
                                         │ aversion)    │
                                         └──────────────┘
```

and recognizing that the world cannot be fully understood or controlled. (Scott 1996). We argue that these three points are supplemented by a fourth: the capacity to deal with environmental variability and complexity".

This chapter has documented the detailed ways in which the Crees account for the complexity and irreducible uncertainty of their subarctic ecosystems (Berkes 1998; Elmqvist et al. 2004; Sayles and Mulrennan 2010). They observe and monitor change in great detail and suggest possible connections among various factors. For example, the goose hunt can be represented as sets of relations linking hunters and birds, showing that the former recognize a bewildering diversity of variables in the complexity of the hunt (Figures 8.1 and 8.2). Hunters communicate, exchange observations, and, as appropriate, attempt to adjust their behaviour according to their interpretations of ongoing changes. This flexible monitoring of change relies on opportunistic observations of unusual occurrences. Cree ways of knowing, in this context, appear to be largely – but not exclusively – qualitative and probabilistic. The collective understanding of this web of factors suggests that resource users rely on a deep understanding and recognition of what we might call complexity, for which they constantly monitor a wide range of environmental signs and signals (Berkes and Berkes 2009; Péloquin and Berkes 2009).

They note unusual events but do not seek to measure trends or observations of change as scientists might. Their understanding does not rely on simple cause-effect relationships; propositions about the causal relations between things are subjected to the "proof" of experience. For instance, when eelgrass is absent, geese seek other sources of food, but other factors are also at play such as strong winds that cause them to fly more often and at lower altitudes, which in turn affects their feeding patterns. Crees make their observations in a relational context; causality itself remains uncertain. They perceive changes in goose behaviour and availability within a view of their social-ecological system that could be described as a complex and dynamic web of interactions. Given that so many variables are involved, treating them quantitatively is, in any case, not feasible because an inverse relationship exists between the complexity of a system and the degree of precision in descriptions of it (Zadeh 1973).

Science approaches problems such as climate change by quantifying a relatively small number of variables, such as mean temperature. By contrast, many Indigenous ways of knowing, including that of the Crees, seem to approach these problems with a different strategy – by qualitatively scanning a very large number of variables. In this sense, the ability of Indigenous knowledge systems to deal qualitatively with a large number of variables may be analogous to the use of fuzzy logic in Western science (Berkes and Berkes 2009; Péloquin and Berkes 2009). This way, Cree hunters seem able to process an astonishingly large amount of information to account for the high degree of complexity in processes of change.

Such an analysis would be in agreement with Richard Preston's (2002, 152) suggestion that "precise answers are not easily given by ... the Cree, for the contingencies of the Cree world are not predictably patterned and directly apprehended in all their complexity." The Crees do not seek to diminish complexity or uncertainty. Instead, they embrace it. They act upon a relational model of their environment in which events and patterns are understood in probabilistic terms, an approach that allows for the treatment of large numbers of variables, especially at the collective level, when hunters and elders deliberate over the meanings of their observations.

In the world of the Wemindji Crees, large-scale changes are in some instances overwhelming local processes, inviting adaptation (Royer and Herrmann 2013). The impact of combined multiple external factors makes variability management strategies at the local level, such as hunting area rotation, increasingly less practical. One response involves using air travel to access coastal hunting sites, but this is a costly solution that contributes to disturbance and decline in goose availability. Another strategy – using roads to hunt geese inland – has replaced coastal air travel.

Maintaining options, or resilience in a social-ecological system (Berkes and Ross 2013), involves opportunistic use of whichever capital is available for reorganization when initial conditions are no longer suitable due to changes in the overall system. Cree livelihood shows evidence of cultural continuity. This entails the use of social memory in the form of knowledge, institutions, and practices, as well as technologies, flexibility, and autonomy. These allow hunters to maintain the relevance of customary practices and develop new ones in the face of contemporary problems.

NOTES

1   For more detail on methods, see Claude Péloquin (2007). We did our interviewing during 2006 and 2007, and verified and confirmed the results and interpretations in 2007–08.
2   For interviewees who requested anonymity, we use randomly selected initials.
3   Interestingly, even as the productivity of the coastal hunt diminished during the last dozen years, people camping inland bagged increasing numbers of geese. Their success was not enough to compensate for the decline of the coastal hunt, but it does suggest that the growing numbers of geese that use the eastern James Bay flyway may be spreading out as they migrate.
4   The decline of eelgrass in Hudson and James Bays remains puzzling and is associated with changes in water temperature, salinity, and turbidity, with impacts on the ecology of waterfowl, especially Brant geese but also Canada geese and ducks (Short 2008).
5   In that year, 310 (or 29.0 percent) of Wemindji residents participated in the program. By 2005–06, the number had declined to 173 (or 14.1 percent of residents). During the same period, the total number of days in the bush claimed by participants (full-time subsistence hunters) dropped from 39,046 to 25,605 (see Péloquin 2007, 153–54, for further details).
6   See Jesse Sayles (2008) for a discussion of the positive feedback associated with fewer geese on the coast – less foraging and changes to wetland vegetation as a result. His argument suggests that as fewer geese stop on the coast, the landscape changes to become even less attractive to them.

WORKS CITED

Bearskin, Joe, George Lameboy, R. Matthew, J.S. Pepabano, A. Pisinaquan, W. Ratt, and D. Rupert. 1989. *Cree Trappers Speak*. Comp. and ed. Fikret Berkes. Chisasibi, QC: James Bay Cree Cultural Education Centre.

Berkes, Fikret. 1982. "Preliminary Impacts of the James Bay Hydroelectric Project, Québec, on Estuarine Fish and Fisheries." *Arctic* 35 (4): 524–30.

–. 1998. "Indigenous Knowledge and Resource Management Systems in the Canadian Subarctic." In *Linking Social and Ecological Systems,* ed. Fikret Berkes and C. Folke, 98–128. Cambridge: Cambridge University Press.

–. 2018. *Sacred Ecology.* 4th ed. New York: Routledge.

Berkes, Fikret, and M. Kislalioglu Berkes. 2009. "Ecological Complexity, Fuzzy Logic and Holism in Indigenous Knowledge." *Futures* 41: 6–12.

Berkes, Fikret, Joan Colding, and Carl Folke, eds. 2003. *Navigating Social-Ecological Systems: Building Resilience for Complexity and Change.* Cambridge: Cambridge University Press.

Berkes, Fikret, and Helen Ross. 2013. "Community Resilience: Toward an Integrated Approach." *Society and Natural Resources* 26: 5–20.

Carlson, Hans M. 2008. *Home Is the Hunter: The James Bay Cree and Their Land.* Vancouver: UBC Press.

Elmqvist, Thomas, Fikret Berkes, Carl Folke, Per Angelstam, Anne-Sophie Crépin, and Jari Niemelä. 2004. "The Dynamics of Ecosystems, Biodiversity Management and Social Institutions at High Northern Latitudes." *Ambio* 33 (6): 350–55.

Feit, Harvey A. 1987. "Waswanipi Cree Management of Land and Wildlife: Cree Cultural Ecology Revisited." In *Native People, Native Lands: Canadian Indians, Inuit and Metis,* ed. B. Cox, 75–91. Ottawa: Carleton University Press.

–. 1995. "Hunting and the Quest for Power: The James Bay Cree and Whitemen in the Twentieth Century." In *Native Peoples: The Canadian Experience,* ed. R.B. Morrison and C.R. Wilson, 181–223. Toronto: McClelland and Stewart.

Ford, James D., Graham McDowell, and Tristan Pearce. 2015. "The Adaptation Challenge in the Arctic." *Nature Climate Change* 5: 1046–53.

Ford, James D., Barry Smit, Johanna Wandel, Mishak Allurut, Kik Shappa, Harry Ittusarjuat, and Kevin Qrunnut. 2008. "Climate Change in the Arctic: Current and Future Vulnerability in Two Inuit Communities in Canada." *Geographical Journal* 174 (1): 45–62.

Harvey, William F., and Jean Rodrigue. 2006. *A Breeding Pair Survey of Canada Geese in Northern Québec 2006.* Baltimore: Maryland Department of Natural Resources and Canadian Wildlife Service.

Hass, G. 2002. *The Canada Goose: Branta Canadensis Atlantic Flyway Resident Population.* Hadley, MA: United States Fish and Wildlife Service, Migratory Birds.

Krupnik, Igor, and Dyanna Jolly. 2002. *The Earth Is Faster Now: Indigenous Observations of Arctic Environmental Change.* Washington, DC: Arctic Research Consortium of U.S.

McDonald, Miriam, Lucassie Arragutainaq, and Zack Novalinga. 1997. *Voices from the Bay: Traditional Ecological Knowledge of Inuit and Cree in the Hudson Bay Bioregion.* Ottawa: Canadian Arctic Resources Committee and Municipality of Sanikiluaq.

Mulrennan, Monica E., Rodney Mark, and Colin H. Scott. 2012. "Revamping Community-Based Conservation through Participatory Research." *Canadian Geographer* 56 (2): 243–59.

Nichols, Theresa, Fikret Berkes, Dyanna Jolly, Norman B. Snow, and the Community of Sachs Harbour. 2004. "Climate Change and Sea Ice: Local Observations from the Canadian Western Arctic." *Arctic 57*: 68–79.

Péloquin, Claude. 2007. "Variability, Change and Continuity in Social-Ecological Systems: Insights from James Bay Cree Cultural Ecology." Master's thesis, University of Manitoba.

Péloquin, Claude, and Fikret Berkes. 2009. "Local Knowledge, Subsistence Harvests, and Social-Ecological Complexity in James Bay." *Human Ecology* 37: 533–45.

Preston, Richard J. 2002. *Cree Narrative: Expressing the Personal Meanings of Events.* 2nd ed. Montreal and Kingston: McGill-Queen's University Press. Originally published 1975.

Reed, Austin, Réjean Benoit, Michel Julien, and Richard Lalumière. 1996. *Goose Use of the Coastal Habitats of Northeastern James Bay.* Ottawa: Canadian Wildlife Service.

Rosenberg, D.M., Fikret Berkes, R.A. Bodaly, R.E. Hecky, C.A. Kelly, and J.W.M. Rudd. 1997. "Large-Scale Impacts of Hydroelectric Development." *Environmental Reviews* 5: 27–54.

Royer, Marie-Jeanne S., and Thora M. Herrmann. 2013. "Cree Hunters' Observations on Resources in the Landscape in the Context of Socio-environmental Change in the Eastern James Bay." *Landscape Research* 38 (4): 443–60.

Sayles, Jesse. 2008. "*Tapaiitam:* Human Modifications of the Coast as Adaptations to Environmental Change, Wemindji, Eastern James Bay." Master's thesis, Concordia University.

Sayles, Jesse, and Monica E. Mulrennan. 2010. "Securing a Future: Cree Hunters' Resistance and Flexibility to Environmental Changes, Wemindji, James Bay." *Ecology and Society* 15 (4): art. 22. http://www.ecologyandsociety.org/vol15/iss4/art22/.

Scott, Colin. 1983. "The Semiotics of Material Life among Wemindji Hunters." PhD thesis, McGill University.

–. 1986. "Hunting Territories, Hunting Bosses and Communal Production among Coastal James Bay Cree." *Anthropologica* 28 (1–2): 163–73.

–. 1996. "Science for the West, Myth for the Rest? The Case of James Bay Cree Knowledge Construction." In *Naked Science: Anthropological Inquiry into Boundaries, Power and Knowledge,* ed. Laura Nader, 69–86. New York: Routledge.

Scott, Colin, and Jeremy Webber. 2001. "Conflicts between Cree Hunting and Sport Hunting: Co-management Decision-Making at James Bay." In *Aboriginal Autonomy and Development in Northern Québec and Labrador,* ed. Colin Scott, 149–74. Vancouver: UBC Press.

Short, Frederick T. 2008. *The Status of Eelgrass in James Bay.* Report to the Cree Nation of Chisasibi. Durham: University of New Hampshire.

Syvänen, Andra L. 2011. "Wemindji Cree Observations and Interpretations of Climate Change: Documenting Vulnerability and Adaptability in the Sub-Arctic." Master's thesis, Concordia University.

Tanner, Adrian. 2014. Bringing Home Animals: Mistissini Hunters of Northern Quebec. Saint-John's, NL: ISER Books (Institute of Social and Economic Research, Memorial University).

Zadeh, Lotfi A. 1973. "Outline of a New Approach to the Analysis of Complex Systems and Decision Processes." *IEEE Transactions on Systems, Man, and Cybernetics* 1: 28–44.

9

Coastal Landscape Modification by Cree Hunters

JESSE S. SAYLES and MONICA E. MULRENNAN

Coastal James Bay is undergoing dramatic environmental changes, principally in response to land emergence at rates of more than a metre per century (see Chapters 5 and 6, this volume). Associated changes are noticeable within a human lifetime and include the merger of islands with the mainland, the infilling of bays, and shifts in coastal and near-shore vegetative communities, which in turn result in changes to local wildlife. Climate-related changes, hydro-electric development in the broader region, and larger-scale land-use/cover changes in North America contribute to the complexity of this process (see Chapter 8, this volume).

Wemindji Cree hunters and fishers have occupied the eastern shores of James Bay for generations. Crees have traditionally been a hunter-gatherer society, and these activities remain vital today, both economically and culturally (Scott and Feit 1992; Colin Scott 1996). Although only a small percentage of families live as full-time subsistence hunters, bush food and life on the land are thoroughly integrated for many community members. Goose hunting during the spring and fall is of particular significance for coastal Crees and is associated with a set of harvesting practices, overseen by a tallyman or hunting boss, that respond to environmental variability and complexity. Local adaptations to environmental fluctuations include the relocation of hunting and fishing camps, adjustments in travel routes, and revised hunting strategies in response to habitat change and shifts in the abundance and availability of wildlife (see Chapter 8, this volume).

In this chapter, we focus on a largely unacknowledged dimension of the Cree response to environmental change by presenting details of hunters' efforts to resist, retard, delay, and/or redirect landscape alteration. Drawing upon a range of methodologies, including ethnography,[1] aerial photography, and satellite image measurements and interpretations, as well as in-field observations of topography and vegetation, we discuss two Cree landscape modifications whose purpose is to create or enhance resource-rich areas for the goose hunt. These are building dikes to prevent wetland succession and cutting *tuuhiikaan*, or corridors, through the coastal forest to enhance the visibility and predictability of approaching geese (*tuuhiikan* is singular; tuuhiikaan is plural). We also examine more recent practices, such as marsh restoration, laying down corn, and prescribed burning to illustrate continuity with past practices, as well as a willingness among Cree hunters to embrace new technologies and ideas along with landscape change itself.

Our study is intended to improve the understanding of the link between landscape change, resource availability, and human efforts to manage and manipulate change. Modifications to the landscape by Cree hunters demonstrate occupancy and agency, and reflect their informed understanding of the biophysical and ecological dynamics of this environment. We argue that their practices are motivated by a desire to enhance resource predictability. Resource management in such a setting requires striking a balance between resistance to environmental change and investment in place, on the one hand, and an ability to adapt, be flexible, and experiment in response to change, on the other. Our study also has practical implications for management of the biodiversity reserve. Landscape modifications, integral parts of an approach that dates back generations, enhances resource harvesting, and supports local knowledge transmission and cultural continuity, are entirely consistent with Wemindji Cree goals for protected areas.

LANDSCAPE MODIFICATION IN THE CONTEXT OF ENVIRONMENTAL CHANGE

As land rises in James Bay, shorelines retreat and upland vegetation invades seaward at roughly equivalent rates (von Mörs and Bégin 1993), causing habitats and associated resource uses to shift. For example, seaward invading upland species overgrow berry crops, requiring a change in resource harvesting. A major outcome of land emergence is thus to drive ecological changes in coastal wetlands. Because these wetlands are preferred feeding grounds for geese, these changes also affect where the birds are hunted. Such changes are dramatically noticeable within human lifetimes. Older hunters point out numerous places where they hunted in their youth, places that geese no longer visit today. These coastal wetlands, once dominated by the water-loving plants that geese prefer, such as chaffy sedge *(Carex paleacea)*, mare's tail *(Hippuris tertaphylla)*, and spikerush *(Eleocharis* spp.), have changed to willow-grass lands dominated by plants such as *Salix* spp. and *Festuca rubra*, which are unpalatable to the birds (Reed et al. 1996; Handa, Harmsen, and Jefferies 2002).

Geese feed in high marsh pools during spring and in low marshes during fall (Reed et al. 1996). Bulbs and rhizomes form the bulk of their diet. By digging in the low marsh, geese can suppress seaward invasion of the high marsh that follows land emergence, thus increasing the low marsh areas that geese find more attractive (Hik, Jefferies, and Sinclair 1992). Studies of lesser snow geese show that if this negative feedback is maintained, the low marsh can persist for ten to fifty years, but if it is discontinued, the low marsh will be replaced by high marsh in as little as two to five years (Hik, Jefferies, and Sinclair 1992; Handa, Harmsen, and Jefferies 2002). Wemindji Crees have observed such changes in coastal marshes and correlated them with the local decline of lesser snow geese in the mid- to late twentieth century (see Chapter 8, this volume), a decline that is also documented in the scientific literature (Abraham, Jefferies, and Alisaukas 2005). According to Lot Kakabat (pers. comm., June 17, 2007),[2] the loss of lesser snow geese abundance in Wemindji, and their ability to "break the mud" for Canada geese, had a significant impact on local populations of the latter. Once the negative feedback between lesser snow goose foraging and succession broke down, the feeding grounds became unattractive to Canada geese.

When discussing longer time scales of environmental change, hunters pointed to former swan hunting areas that were coastal sites during the eighteenth century, but that were now located 7.5 to 11.0 kilometres inland (based on interviews using 1:30,000 topographic maps). Old stone blinds still stand at some of these places. Hunters also identified channels, used for fishing in the early twentieth century, that are now coastal marshes used for goose hunting. Such transitions from fishing sites to goose hunting grounds illustrate an interesting dynamic between coastal change and resource use; places are not only used, but are often reused in different capacities as the landscape alters. Indeed, hunters mentioned certain young rocky coastal islands, currently visited for egg or driftwood gathering, and noted that they would probably provide favourable locations for fishing camps one day, if and when they had to spatially shift resource harvesting activities due to continued environmental changes.

Land emergence drives many changes that affect the use of land and resources. Because the rates of change are directly linked to the local gradient of the land, the rates of change are not constant everywhere. While rapid alteration is a dominant characteristic of the coastal landscape, some hunting and camping locations have remained relatively constant over recent generations. Embedded in these places are stories of the past: a deceased grandfather's favourite hunting spot, places where the Hudson's Bay Company used to land boats to acquire salted geese from the Crees, a campsite where a grandmother lost her wedding ring, and the gravesite of a loved one. Such history is an important feature of this landscape (see Chapter 10, this volume).

Though land emergence is a major regional driver of change, modifications occurring at distant locations and larger scales also exert important influences on the abundance and behaviour of waterfowl over decades. For example, swans were an important resource in the past but are no longer available today. As previously noted, Cree hunters observed a drastic decrease in the numbers of lesser snow geese from the 1950s to the 1980s. Shifts in snow goose migration during that period have been linked to changes in agricultural and wildlife sanctuary patterns and practices in the Gulf Coast and the mid-continental United States, as well as climate changes in the James and Hudson Bay region (Abraham, Jefferies, and Alisaukas 2005). Since the late 1980s and the 1990s, many Crees have observed that Canada geese are shifting their migration path inland from the coast, in response to a complex range of variables, including regional hydro-electric development (see Chapter 8, this volume). As mentioned above, the birds may also have found the Wemindji coast less attractive once snow geese were no longer breaking up the mudflats (Lot Kakabat, pers. comm., June 17, 2007).

These environmental changes have certainly influenced Cree harvesting success over the years, but the hunters have not been passive recipients of change, choosing to modify and create some parts of the landscape (Sayles and Mulrennan 2010; Sayles 2015). Some of these practices represent centuries of tradition. Others have arisen more recently and illustrate the adaptive capacity and ingenuity of Cree hunters in integrating traditional techniques with the opportunities available to them as a modern globally linked society.

PAST PRACTICES OF LANDSCAPE MODIFICATION

Cree elders recall several historic forms of landscape modification that are no longer employed. Abandoned stone fish weirs on several Wemindji rivers were built during their "grandfather's time" or much earlier: "before the white man came" and "before there were nets" (Billy Gilpin, pers. comm., July 15, 2007, August 17, 2007; William Mistacheesick, pers. comm., August 2, 2007). These stone weirs were bowl shaped and created pools in which fish aggregated and were speared or caught by hand. One family used them until the 1940s, as an addition to gill net fishing. A related practice was the construction of stone deflectors, or rock piles, whose purpose was to "funnel fish into a specific spot," typically a natural pool, where they could be killed with a bow or spear. The deflectors were used historically, "before nets ... before the Hudson's Bay Company" (Lot Kakabat, pers. comm., June 17, 2007). Toby Morantz (1983, 30) mentions the "damming of creeks" as a fishing strategy, but otherwise the literature provides little information about weirs.

Two elders, one a tallyman, stated that marshes were dug up to make them more appealing to geese. As they explained, digging would have mimicked the foraging of the birds and exposed roots and rhizomes, and suppressed the development of plants that geese dislike. Digging maintained or created areas that were attractive to geese and yielded predictable hunting places. However, three elders from other hunting territories, one of whom was a tallyman, had no recollection of people digging to manage marshes. They speculated that the practice would have been limited because digging with a shovel is hard work. They also stated that there would have been little need for digging in the past because geese were so numerous back then that they would have turned over the marsh simply by foraging.

Many North American Indigenous groups use fire as an important land management strategy (Cronon 1983; Doolittle 2000; Vale 2002; Turner, Davidson-Hunt, and O'Flaherty 2003; Berkes and Davidson-Hunt 2006; Natcher et al. 2007). However, the literature rarely mentions that the Crees employed it in this way. Toby Morantz (2002), for example, notes the historic use of fire to promote berry growth, which can also attract bears for hunting. Katherine Scott's (2008) detailed ethno-ecology of the role of fire in Wemindji indicates that fire was not used as a major management tool.

Several elders and tallymen with whom we spoke confirmed that fire played a relatively minor role in Cree landscape management. Lot Kakabat (pers. comm., July 7, 2007) recalled his father "experimenting" with fire to see how the grasses and berries that geese feed on would grow after being burned. This was before "his time," he said. As a young boy, elder tallyman Fred Stewart (pers. comm., June 28, 2007) had heard stories about people burning grasses a "long time ago" to promote new growth and to attract geese. He identified a location where this was done. Raymond Atsynia (pers. comm., July 5, 2007), another elder tallyman, had once experimented during the 1960s with burning grasses in the fall. The next spring, when the "snow melted [the] burned grasses washed away and geese were feeding there ... There are more things that [grew] faster," he said. His grandfather did not use fire to attract geese, however, because it was not necessary back then; "the feeding areas were very good for the geese" during his grandfather's time. A younger tallyman (pers. comm., July 18, 2007) recalled seeing his older brother burning grasses around a goose hunting area, but he had never heard of his grandfather burning grasses to attract geese. Very recent experiments with burning goose habitat are discussed later in the chapter.

Several elders spoke of using fire to clean up camps or to control grasses when they grew too long. The danger of losing control of fire came up frequently in conversation, and there was general agreement that the wisest course was to refrain from interfering too much. When asked whether fire was used to improve the berry crop, one elder woman said, "[we] just let nature do the work" (pers. comm., July 19, 2007). Elder Annie Shasha-weskum (pers. comm., June 4, 2007) explained that when "willows would grow over where the berries were ... there is not much [we] could do, so [we] would just go to another place."

LANDSCAPE MODIFICATION PRACTICES: DIKES AND TUUHIIKAAN

Dikes and tuuhiikaan are the two primary forms of landscape modification employed by Cree hunters. Colin Scott (1983) briefly mentions their construction and use in relation to goose hunting. Austin Reed (1991, 346) also reports on dikes: "The Cree are seeking more than open water for hunting; they are placing emphasis on creating high quality wetland habitat which will provide additional food resources for migratory waterfowl." He also mentions the employment of "shrub clearing projects to slow down the invasion of certain salt marshes and fens by willows." Margaret Forrest (2006, 35) explains the function of tuuhiikaan in terms of "leading geese to their preferred feeding grounds." Apart from these isolated statements, the literature offers no comprehensive accounts of the construction, form, age, or function of these modifications, with the exception of further details in Jesse Sayles and Monica Mulrennan (2010) and Jesse Sayles (2015).

Dikes

In the early spring, geese arrive along the coast of James Bay on their migration north. The coastal marshes are still covered in snow, and the birds are on the lookout for small ponds where they can feed. With this in mind, and in the context of ongoing land emergence, generations of Cree hunters have built and maintained dikes in the marshes to attract geese and facilitate hunting.

Dike Construction and Form

Our fieldwork confirmed a wide variety in dike form, size, and construction materials. We measured eight dikes, in four different coastal hunting territories, from a total of twenty-two dikes that we identified among the seven coastal territories.

Dikes were primarily made of mud or sod dug from the immediate area. Digging up sod blocks from the planned impoundment served a triple purpose, according to Lot Kakabat (pers. comm., June 17, 2007). The elevation of the impoundment was lowered; any undesirable vegetation (from the perspective of both hunter and goose) was removed; and building material for the dike was made available. Logs and stones, as well as wooden boards in one dike we observed, were used in construction.

Most dikes were u-shaped, approximately 15 to 30 centimetres high, and between 4 and 269 metres long. The 4-metre dike, however, demonstrated a different strategy. It was a linear obstruction across a creek that drained a large natural pond used for hunting. The relatively minor investment of time, energy, and materials involved in its construction has been rewarded by impounding a larger body of water than was there under natural conditions. The largest dike we visited was a 269-metre circular mud embankment. Lower than the others (6 to 9 centimetres), it created a 3,239-square-metre pond (Table 9.1).

Age of Dikes

As Table 9.1 demonstrates, the construction dates for the twenty-two dikes cover a wide time interval. Reflecting on the oldest known examples, informants from three hunting territories recalled dikes from the early to mid-1900s, and informants on two other territories knew of dikes established as far back as the late 1700s and early 1800s. Of the two remaining coastal territories, tallywoman Daisy Atsynia (pers. comm., June 20, 2007) reported that there were no dikes on her territory until the 1980s, after her husband, the tallyman, died. In the other case, only a temporary dike had been erected in recent decades. On territories with older dikes, informants confirmed that dike building had continued over the years, with the construction of new ones proposed on several territories. Thus, notwithstanding the absence of dikes on one territory, and the presence of only a temporary example on another, dike building has been, and continues to be, an important landscape management approach.

Dike Function

Dike placement took advantage of natural drainage as marshes thawed in the spring, usually by blocking a small creek to create or maintain a pool that attracted waterfowl. Depositing decoys in the pond provided an additional lure. As we observed in the field, and confirmed with our Cree

TABLE 9.1
Dike dimensions and dates of construction

Dike	Territory	Length (m)	Height (cm)	Date of construction
1	A	269	6–9	Generations ago (1800–1900s)
2	A	44	15–30	Generations ago
3	A	–	–	Will build in fall (2007)
4	A	–	–	No data
5	B	24	30	1988 or 1989
6	B	180	16–25	1981
7	B	–	–	1988 or 1989
8	B	–	–	Early to mid-1900s
9	C	91	20–30	Early 1960s, youngest of 3 in marsh
10	C	87	10–20	Early 1960s, 2nd oldest of 3 in marsh
11	C	–	–	Early 1960s, oldest of 3 in marsh
12	C	–	–	1940s
13	C	–	–	Early 1900s
14	C	–	–	No data
15	D	–	–	Late 1700s, early 1800s
16	D	–	–	2000
17	D	–	–	2000
18	E	24	15–30	2001 or 2002
19	E	4	15–30	Early to mid-1980s
20	F	–	–	Grandfather's time (1940s–50s)
21	F	–	–	1999 or 2000
22	G	–	–	Mid- to late 1900s; temporary

Source: Based on interviews and/or field surveys.

informants, the retention of water inhibits the establishment of a defined channel and well-drained upland heath, and the wetland created by the dike supports plant species that geese favour, such as chaffy sedge, mare's tail, and spikerush (Reed et al. 1996). In this way, dikes play a dual role at the pond scale by creating or maintaining a water body to attract geese and by promoting the growth of food plants that also attract and retain the birds.

Dikes create an appealing environment at the pond scale, but their function extends beyond this. They are built in areas that hunters perceive as desirable hunting locations. The surrounding landscape influences how geese approach the pond and where hunters should station themselves to get the best shot. Diked ponds that are located in a narrow coastal embayment, for example, are seen as advantageous because they permit hunters to entrap geese in a cross-fire. In this strategy, hunters position their blinds on either side of the pond, with the experienced hunters upwind and the less experienced ones downwind. Geese habitually land into the wind. The lead hunter has the patience to wait until the flock is close enough so that when he shoots, hopefully killing some birds, the remainder will backtrack across the pond where other hunters are waiting. If the lead hunter shoots too soon, the flock will simply fly off and will be out of range for the other hunters. Once the lead hunter fires, the birds will flee from the shots, heading for the downwind side of the pond. Now, the young hunters take aim, trying to hit them all. Birds that they miss will retreat toward the experienced hunters, who can then fire a second time. This strategy is designed to give everyone a chance to shoot, but it also maximizes the chance of killing all the birds. Geese learn to avoid hunting sites, and Crees believe that any surviving birds, along with the new flock they join, will stay away from the area in the future (Colin Scott 1996).

The proximity of trees and shrubs to the pond also affects hunting success. Brush clearing at the ponds and on the adjacent ridges encourages geese to approach the water more slowly; indeed, the diameter of the clearing is said to have a direct bearing on their flight speed. Removing shrubs to aid hunting is not a new practice. For example, Sinclair Mistacheesick (pers. comm., August 2, 2007), an elder tallyman, talked about using an axe to clear the willows and small trees around a favourite hunting pond, "before chainsaws" in the late 1950s. Brush clearing has occurred on all the coastal territories in recent years (Poly-Geo Inc. and Goyette 2003).

Some ponds are also built to function at a regional landscape scale by taking account of macro-level wind conditions and goose flight patterns. For example, during our field research, one tallyman (pers. comm., July 18, 2007, August 2–8, 2007) explained that he was in the process of deciding whether to maintain an old dike or construct a new one in a coastal bay that had been used as a hunting spot for generations of hunters. He wanted to ensure that there would be a good spot for the next generation to hunt. The two sites under consideration had different advantages for hunting; when the wind was blowing from the south, hunting at the site of the proposed dike was among the best in the area, whereas hunting at the old dike site was good when the wind blew from virtually any direction. A further consideration was the increasing unpredictability of goose flight patterns and weather conditions that favoured maintenance of the old site because it served hunters well under a greater range of conditions.

Several elders highlighted the connection between the role of dikes in maintaining good hunting sites and inter-generational continuity. When asked about the possibility of having to abandon a set of dikes in one bay due to ongoing land emergence and relocating to a new site in the future, Lot Kakabat (pers. comm., June 23, 2007) said, "The routine is going to keep going. People will keep hunting there, where people used to hunt; we might just upgrade it a bit I guess." Future generations will hunt where past generations did; meanwhile, hunters will continue to invest in this area to ensure its viability and maintain this continuity.

Essentially, dikes function on three scales. At the smallest scale, they create or maintain a pond that draws waterfowl by delaying wetland succession. The use of decoys enhances this attractiveness. At the scale of an individual coastal embayment, the dike maintains a pond as the core feature of a hunting strategy that is designed to limit disturbance to the geese and that responds to the differing experiences of hunters. At the inter-bay scale, the dike is tied to the location of other ponds and their relationship to wind direction and goose flight patterns. Decisions about dike construction and use are motivated by a desire to create, maintain, and enhance desirable aspects of the landscape at nested scales to establish predictable locations for goose hunting and to preserve the link with past generations of hunters.

Tuuhiikaan

Cree hunters know that geese like to fly relatively close to the ground and through natural gaps in the trees. To capitalize on these habits, they create tuuhiikaan, expansive corridors cut through the coastal forest.

Tuuhiikaan Construction and Form

We inventoried thirty-one tuuhiikaan via interviews and field surveys. Although this inventory captures the majority of tuuhiikaan on Wemindji hunting territories, including one proposed example and two that were destroyed by forest fires in the mid-1980s, it is unlikely to be exhaustive.

Tuuhiikaan tend to be oriented southwest to northeast or south to north, following the dominant flight paths of geese as they migrate north in the spring. Most are on headlands, with bays and/or marshes to the south and north. Alternatively, they can lie on a ridge on the north side of a bay, with the southern end abutting the coastal marsh and the northern end often opening toward a lake or marsh. These are only general parameters, however; the precise placement of a tuuhiikaan is informed by Crees' observations and understandings about the behaviour of geese.

Tuuhiikaan vary greatly in size (Table 9.2). To some extent, the size of tuuhiikaan differs from territory to territory, possibly reflecting accommodations to local spatial variations in goose behaviour or perhaps the personal preferences of tallymen and hunters. The shape or form of tuuhiikaan also vary. Ernie Hughboy (pers. comm., February 12, 2008), a senior hunter, constructed an hourglass shaped tuuhiikaan that was ninety metres in the middle and about two hundred metres at each end. On the basis of his observations of goose behaviour, he claimed that this shape was the most effective. The southern end was cut somewhat wider for increased visibility.

Large tuuhiikaan (e.g., >5 ha) have been cut since the practice began in the early to mid-twentieth century. In recent decades, no new small

TABLE 9.2
Summary statistics for tuuhiikaan dimensions

	Width at narrowest part (m)	Width at widest part (m)	Length (m)	Area (ha)
Minimum	30.00	45.00	80.00	0.34
Maximum	270.00	370.00	1,100.00	32.16
Average	120.60	149.80	399.80	5.17
Standard deviation	65.10	78.00	264.03	6.37

Note: Minimum, maximum, average, and standard deviation are reported for width (metres), length (metres), and area (hectares). Many tuuhiikaan tend to widen at one end or have an hour glass shape, so width was measured for both the narrowest and widest parts.
Source: Based on interviews and/or field surveys.

tuuhiikaan (e.g., <1 ha) have been constructed; the three smallest tuuhii-kaan were built between 1930 and 1970. This suggests a contemporary trend to not cut new small tuuhikaan; however, existing small tuuhikaan were still maintained in the late 1990s and early 2000s by removing new tree growth.

The absence of a trend over time toward the cutting of larger tuuhiikaan is interesting, given that chainsaws, which hunters began to acquire in the late 1970s and the 1980s (Scott and Feit 1992), greatly reduce the effort of creating a tuuhiikaan. This suggests that heavy labour did not hinder the Crees from making these large hunting investments. In fact, the largest tuuhiikaan was cut in 1962, "us[ing] an axe, no saw" (Raymond Atsynia, pers. comm., July 3, 2007), further weakening any correlation between size and the availability of labour-saving technology.

Ruling out labour or effort as a major determinant of size strengthens the argument that tuuhiikaan creation is influenced by other factors. Variations in tuuhiikaan size and form within and across individual terri-tories suggest that Cree hunters are primarily responding to interactions between local landscape conditions and patterns of goose behaviour. Personal preferences probably play a role as well, though likely as a sec-ondary influence.

Tuuhiikaan Age

When asked about the origin of tuuhiikaan, an elder (pers comm., August 7, 2006) stated that his family had been cutting one example since "time immemorial" and that it had been maintained twice in his lifetime, once in the late 1950s and again in 2003. Given that the earliest known tuuhii-kaan on the other territories were first established between the 1930s and 1950s, it seems unlikely that this particular example is significantly older. It is plausible, however, that the forest in this location naturally resembled and functioned like a tuuhiikan. Generations later, in response to tree growth associated with land emergence, Cree hunters could have removed the trees. Once this occurred, the corridor became a real tuuhiikan since, according to Lot Kakabat (pers. comm., June 17, 2007), "[a tuuhiikan] is where someone cut the trees down with a saw or axe. If you see the same thing but the trees were knocked down naturally it is not a tuuhiikan."

As with dike construction, tallywoman Daisy Atsynia (pers. comm., June 20, 2007) stated that no one cut tuuhiikaan on her hunting territory until her husband, the tallyman, passed away in the 1980s. "No, others did that," she said. "People wanted to act like they own the land. The only thing [my husband and my grandfather] would do is make a trail to where they hunted." The dates of the two tuuhiikaan on her territory, the early 1980s and 2003, as reported by other senior hunters, corroborate her account. Clearly, as Daisy's comment confirms, management perspectives and practices can differ from territory to territory. Nonetheless, cutting tuuhiikaan is widely practised across Wemindji coastal territories.

There was a surge of tuuhiikaan creation from the 1930s to the 1960s and again in the 1980s, with a tapering off during the 1990s (see Sayles 2008 and Sayles and Mulrennan 2010 for details). This pattern roughly correlates with the decline in goose availability. As mentioned earlier, lesser snow geese started to diminish in the 1950s and 1960s through to the 1980s, with Canada geese declining in the 1980s and 1990s. It is possible that a decline in the number of geese locally along the Wemindji coast prompted more tuuhiikaan cutting to increase the likelihood of intercepting geese. Although we cannot infer causality from our limited data on "cut" dates, this hypothesis is supported by reports from elders of an abundance of geese in the past. According to Sam Hughboy (pers. comm., June 20, 2007), for example, people did not need goose decoys in the days before the tuuhiikaan: "There were so many geese you just needed to call them." People knew where to hunt the birds "simply by looking at them," he explained.

In 1979, a remedial works program was established, supported by funding commitments associated with the James Bay and Northern Québec Agreement (SOTRAC 1980), a comprehensive land claim settlement resulting from the James Bay hydro-electric project. The program was intended to offset the negative impact of the hydro development on Cree natural resource use. Among other things, it provided financial support for labour and costs related to the construction of dikes and tuuhiikaan. The advent of this program may explain why some cut dates cluster in the 1980s. Perhaps its financial incentives encouraged more people to cut tuuhiikaan and build dikes. However, according to local informants, the program functioned more as a deterrent because its approval process for projects was so slow. For example, when senior hunter Leslie Kakabat wanted to recut multiple tuuhiikaan on his territory, he had to wait two years for approval by the board of the remedial works program. Though the tuuhiikaan urgently needed attention, he waited until he received board approval because he was concerned that commencing work might make him ineligible for the associated compensation payments, a significant income supplement for hunters (pers. comm., August 12, 2006, June 13, 2007). His plans were further delayed because the program dictated the order in which his various projects must be carried out to be eligible for payments. His tuuhiikaan had to wait until after he built the new snow-mobile trail that he had also submitted for approval (pers. comm., March 2009). A tallyman who planned to build a new dike expressed similar frustrations. At the time, he was waiting for final approval from the program before he could enlist family members to work on the project and receive much-needed financial supplements (pers. comm., August 8, 2007).

Tuuhiikaan Function

When asked why a tuuhiikan is located where it is, multiple informants responded "because people saw geese flying there." Tuuhiikaan are often established in locations that were used for hunting in the past. Their purpose is not to bring geese to a certain spot, and no one would install a tuuhiikan where the birds did not normally fly. Rather, the tuuhiikan is intended to take advantage of an area that geese already fly through and to increase the likelihood that they will pass over a given spot in that area.

"The tuuhiikan gives you a place to sit," explained Leslie Kakabat (pers. comm., June 23, 2007). His father, tallyman Lot Kakabat (pers. comm., June 23, 2007), explained that before the invention of the tuuhiikan, people would run out onto the ridge when they heard the geese coming and conceal themselves wherever they thought the birds would pass. Sometimes, the geese did fly overhead and they got a good shot. Sometimes, however, the birds chose a different route, and by the time this became obvious, it was often too late to change position. Cutting a tuuhiikan created a more predictable location on the ridge for geese to pass. Leslie Kakabat (pers. comm., June 13, 2007) credited the tuuhiikaan with an 80 percent success rate, meaning that 80 percent of geese passing through a particular area would opt to fly through it. Tuuhiikaan also aided hunters in observing and monitoring the flocks. Their southern ends were commonly flared to enhance the visibility of the approaching birds, which gave hunters enough advance warning to reposition themselves if the geese did not pass through the tuuhiikan.

DIKES AND TUUHIIKAAN IN THE CONTEXT OF LAND EMERGENCE

As the coast changes due to land emergence, dikes and tuuhiikaan reduce the uncertainty of geese harvesting. "Maintaining the pond makes it so the land does not grow," said elder Sam Hughboy (pers. comm., June 20, 2007). "Cutting trees keeps it the same. If you don't cut them, they will grow back," he added. "Nothing is destroyed when you cut the tuuhiikaan; [the] land keeps growing," said elder tallyman Raymond Atsynia (pers. comm., July 3, 2007). "The trees don't grow back, though, but the land still grows," he continued. "When you make a dike for a pond," explained Lot Kakabat (pers. comm., June 17, 2007), "when you cut off ground from the earth, the earth would turn into mud; it was like it [the land emergence] was all starting over again."

Dikes prevent desirable hunting areas from deteriorating into less desirable spots dominated by *Salix* spp. and *Festuca rubra*. In so doing, they provide a predictable location where geese can be successfully hunted from season to season. Similarly, tuuhiikaan allow an established hunting area to remain attractive for geese as the coastal forest grows. They prolong the useful lifespan of hunting areas and increase resource predictability (Figure 9.1). However, though dikes and tuuhiikaan are intended to retard change and prolong the utility of hunting sites, they are ultimately abandoned when the cost-benefit ratio becomes unfavourable, after which new ones are constructed. This occurs when larger-scale coastline shifts and associated habitat changes trigger alterations in goose flight paths, for example (Sayles and Mulrennan 2010).

A NEW ERA OF EXPERIMENTATION

Local hunters have begun to experiment with the restoration of coastal marshes to help manage geese. Two restoration projects are currently proposed on two hunting territories. They involve two stages: cutting back the thick willow growth that has invaded the area and then bringing in

FIGURE 9.1

Diagram of landscape change, without (top) and with (bottom) Cree modifications

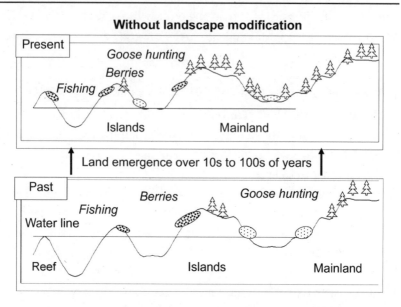

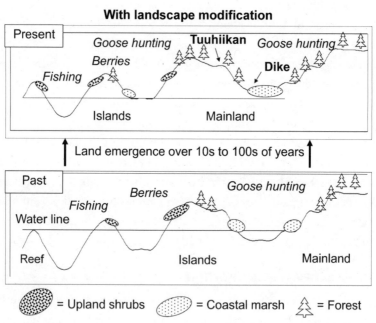

= Upland shrubs   = Coastal marsh   = Forest

Note: Modifications prolong the life of useful hunting areas, as well as that of associated landscape configurations, such as camps.

heavy machinery to dig up the dense willow roots so that the trees cannot regenerate. The Cree Trappers Association is in the process of determining whether any provincial environmental review processes are required with respect to the second stage.

The first of these two projects involves the creation of a *minchiichaau*, which translates as a "seasonally protected space" (Edward Georgekish, pers. comm., February 16, 2008). In this forty-hectare area, Cree hunters take great care not to disturb wildlife, except possibly for a period in the late spring if hunting has been poor. If the restoration project improves goose hunting on the territory, the Cree Trappers Association will then move to a second area, Aanaaschuukaach, which consists of seventy-five hectares, and from there to other locations. The place name Aanaaschuu-kaach, which means "muddy point," is reflective of the changing nature of the coast: no longer muddy, the site is now characterized by high marsh plants. According to Edward Georgekish (pers. comm., July 30, 2007), director of Wemindji's Cree Trappers Association at the time, "As long as we make the effort to restore the areas, then we will see what happens ... Maybe the geese will fly along the coast again like they did in the old days. We'll see what happens; won't know till we do it." Given this reasoning, the projects are being treated as experimental.

Ploughing up the marshes is widely discussed among Cree hunters and seen as holding potential for maintaining and possibly increasing geese numbers along the coast. Interestingly, even those who dismissed the historic value of using shovels to dig up marshes favour the proposed ploughing as a way to improve the hunt. There is widespread support for the use of heavy machinery, or at least a plough pulled by an all-terrain vehicle, as shovels are considered inappropriate for the scale of the project.

Other recent experimental measures along the Wemindji coast include laying corn for geese and prescribed burning. In recent years, some hunters have used corn to attract geese. The intention is to draw large numbers to an area and to refrain initially from shooting them, so that they will see it as a desirable feeding ground. If a large flock is feeding, hunters will drive it away rather than shoot. Later, over the course of the day, the birds will return in smaller and more manageable numbers that can all be killed, preventing survivors in accordance with the practice discussed above (i.e., preventing survivors from associating the site with hunting, subsequently avoiding the site, and possible teaching other geese to avoid it). Although many informants claimed that the use of corn was consistent with the fundamentals of Cree hunting, others disagreed and felt that it should not be used, because it was not part of traditional hunting.

Recently, people have also started to burn grasses to attract geese. Burning dead grass promotes the growth of new shoots, which appeal to the birds. For example, in late spring 2007 Jesse Sayles was shown a large area where the dune grass (*Elymus* spp.) had been burned for this purpose. Two hunters also reported to Monica Mulrennan that they had experimented with burning goose habitat on an offshore island in recent years, in part because they were inspired by a Discovery Channel documentary about Plains Indian use of fire (pers. comm., July 10, 2008). Their fire spread so rapidly that they decided not to repeat the experiment. As in discussions about the historic use of fire, other hunters feared that it might get out of control and had no interest in using it to lure geese. These differing opinions provide another illustration of the diversity of management perspectives in Wemindji.

CONCLUSION

Generations of Cree hunters have invested considerable time and effort in modifying their landscape to improve resource-harvesting opportunities, as evidenced by stone weirs and mud dikes that date back to the distant past. Fire seems to have played a minor role, and only a few Wemindji Crees remember the historic digging-up of marshes with shovels. In more recent decades, landscape modification has been dominated by dikes and tuuhiikaan. The creation and maintenance of camps and trails, discussed elsewhere (Sayles 2008), provide further evidence of a landscape known, used, invested in, and altered. The result includes a legacy of, and an ongoing commitment to, landscape modifications that attest to the agency and innovation of Cree hunters.

These modifications illustrate the depth of Crees' understandings of the system in which they live. This knowledge not only enables hunters to harvest resources, but also allows them to create and maintain certain key places. For example, the tuuhiikaan requires astute observation of geese flight patterns and a sophisticated appreciation of how the birds respond to topography. Dikes are not only guided by an understanding of goose behaviour, but are also intricately choreographed, based on multiple nested scales of ecological interactions.

In response to ongoing changes, Wemindji hunters are experimenting with new approaches. Not everyone concurs on their value, however, which reminds us that Cree communities are not homogeneous (Agrawal and Gibson 1999). Furthermore, significant differences regarding management can sometimes arise between and within territories. For example, tuuhii-kaan were not used on one territory until the tallyman died in the 1980s. In another territory, multiple opinions were expressed regarding whether a particular dike should be maintained. In light of this diversity of opinions, processes of consultation and negotiation orchestrated by the tallyman commonly result in compromise and consensus. Although many hunters and tallymen were apprehensive about some new management approaches, there was a general openness to experimentation. Diverse views about management probably help to support a balance between resisting landscape change versus adapting to and embracing it, between continuity with the past and opportunities for the future, and between tradition and innovation.

The restoration projects represent an interesting development. Employing machinery to dig up marshes is comparable to the earlier method of using shovels. The willingness of hunters to use heavy machinery and work within current administrative frameworks, such as applying for environmental permits, is also noteworthy. In this sense, the future is being informed by past practices but facilitated and sometimes constrained by present technologies and socio-political realities. The use of corn is an example of a new practice made compatible with those of the past. Thus, what might initially be perceived as novel, from sources as distant as the Discovery Channel, is rendered meaningful and familiar through an often seamless integration into past and existing practices. This particular phenomenon is relevant to our understanding of adaptive learning and consistency in resilient systems (Holling 2001; Folke 2006; Sayles and Mulrennan 2010).

Our analysis reveals that the primary goal of landscape modification by Cree hunters is to increase the predictability of resources. Building deflectors to funnel fish into natural areas aims at creating predictable locations for harvesting. Likewise, burning grasses creates predictable sites for goose hunting. The construction of dikes retains or creates ponds and wetlands that would otherwise vanish, maintaining desirable hunting locations for seasons, years, or generations. Tuuhiikaan perform the same function, as trees develop on the maturing landscape. Recent burning practices, laying corn, and restoration efforts also hinge on enhancing the attractiveness of specific areas for geese. These practices need to be fully integrated into management approaches for the biodiversity reserve. Increasing the predictability of resource harvesting ensures that an important food source is secured, that an important cultural activity is facilitated, and that life on the land is maintained – outcomes that are fully consistent with the goals of the reserve.

We suggest that the landscape modification practices of Cree hunters illustrate a dynamic between investing in place and remaining flexible, willing to move, and to experiment as the environment changes. Crees are clearly dedicated to preserving advantageous hunting locations, which is the key reason for constructing dikes and tuuhiikaan. There is security in the known – security about procuring important resources and participating in important cultural activities from year to year and generation to generation – and insecurity in the unknown. Of course, Crees clearly recognize where and when resistance to change returns little benefit. Certain structures and sites are thus abandoned in favour of new ones, or a new practice is introduced, sometimes leading to different landscape modifications. There is a complementary balancing act between being reactive to change and actively shaping the coast. This balancing act is an essential aspect of life for the Crees and their response to a rapidly changing environment.

NOTES

1   *Chiniskumitin* (thank you) to Fred Asquabaneskum, Leonard Asquabaneskum, Winnie Asquabaneskum, Daisy Atsynia, Sr., Lillian Atsynia, Raymond Atsynia, Andrew Atsynia, Nancy Danyluk, Edward Georgekish, Sam Georgekish, Billy Gilpin, Ernie Hughboy, Sam Hughboy, Leslie Kakabat, Lot Kakabat, George Kudlu, Alan Matches, Clayton Matches, Beverly Mayappo, Irene Mistacheesick, Sinclair Mistacheesick, William Mistacheesick, Annie Shashaweskum, Bill Stewart, Fred Stewart, George Stewart, Henry Stewart, Danny Tomatuk, Morris Tomatuk, and Sarah Tomatuk. *Chiniskumitin skuutamaahaakaatuuwits iyiyuu iihtuun.*

2 "Pers. comm." refers to interviews conducted by Jesse Sayles, unless otherwise indicated. For some pers. comms., the speaker's name is not given because they wished to remain anonymous.

WORKS CITED

Abraham, Kenneth F., Robert L. Jefferies, and Ray T. Alisaukas. 2005. "The Dynamics of Landscape Change and Snow Geese in Mid-continent North America." *Global Change Biology* 11: 841–55.

Agrawal, Arun, and Clark C. Gibson. 1999. "Enchantment and Disenchantment: The Role of Communities in Natural Resource Conservation." *World Development* 27 (4): 629–49.

Berkes, Fikret, and Iain Davidson-Hunt. 2006. "Biodiversity, Traditional Management Systems, and Cultural Landscapes: Examples from the Boreal Forest of Canada." *International Social Science Journal* 58 (1): 35–47.

Cronon, William. 1983. *Changes in the Land; Indians, Colonists, and the Ecology of New England.* New York: Hill and Wang.

Doolittle, William. 2000. *Cultivated Landscapes of Native North America.* New York: Oxford University Press.

Folke, Carl. 2006. "Resilience: The Emergence of a Perspective for Social-Ecological Systems." *Global Climate Change* 16: 253–67.

Forrest, Margaret. 2006. "Stewardship as Partnership: A Comparative Study of Positive Human-Environment Relationships in East Cree and Suburban Montreal Communities." Master's thesis, McGill University.

Handa, I. Tanya, R. Harmsen, and Robert L. Jefferies. 2002. "Patterns of Vegetation Change and the Recovery Potential of Degraded Areas in Coastal Marsh Systems of the Hudson Bay Lowlands." *Journal of Ecology* 90: 86–99.

Hik, David S., Robert L. Jefferies, and Anthony R.E. Sinclair. 1992. "Foraging by Geese, Isostatic Uplift and Asymmetry in the Development of Salt-Marsh Plant Communities." *Journal of Ecology* 80 (3): 395–406.

Holling, Crawford. 2001. "Understanding the Complexity of Economic, Ecological and Social Systems." *Ecosystems* 4 (5): 390–405.

Morantz, Toby. 1983. *An Ethnohistoric Study of Eastern James Bay Cree Social Organization, 1700–1850.* National Museum of Man Mercury Series. Ottawa: National Museum of Canada.

–. 2002. *The White Man's Gonna Getcha: The Colonial Challenge to the Crees in Québec.* Montreal and Kingston: McGill-Queen's University Press.

Natcher, David C., Monika Calef, Orville Huntington, Sarah Trainor, Henry P. Huntington, La'ona DeWilde, Scott Rupp, and F. Stuart Chapin, III. 2007. "Factors Contributing to the Cultural and Spatial Variability of Landscape Burning by Native Peoples of Interior Alaska." *Ecology and Society* 12 (1): art. 7. http://www. ecologyandsociety.org/vol12/iss1/art7.

Poly-Geo Inc., and D. Goyette. 2003. "Technical Descriptions and Cost Evaluation of Wemindji Projects Submitted to the Apatisiiwin Board of Directors." Progress Report, May 28, Montreal.

Reed, Austin J. 1991. "Subsistence Harvesting of Waterfowl in Northern Québec: Goose Hunting and the James Bay Cree." In *Transactions of the 56th North American Wildlife and Natural Resource Conference,* ed. R.E. McCabe, 344–49. Washington, DC: Wildlife Management Institute.

Reed, Austin J., Réjean Benoit, Michel Julien, and Richard Lalumière. 1996. *Goose Use of the Coastal Habitats of Northeastern James Bay.* Ottawa: Canadian Wildlife Service, Environment Canada.

Sayles, Jesse. 2008. "*Tapaiitam:* Human Modifications of the Coast as Adaptations to Environmental Change, Wemindji, Eastern James Bay." Master's thesis, Concordia University.

–. 2015. "No Wilderness to Plunder: Process Thinking Reveals Cree Land-Use Via the Goose-Scape." *Canadian Geographer* 59 (3): 297–303.

Sayles, Jesse, and Monica E. Mulrennan. 2010. "Securing a Future: Cree Hunters' Resistance and Flexibility to Environmental Changes, Wemindji, James Bay." *Ecology and Society* 15 (4): art. 22. http://www.ecologyandsociety.org/vol15/iss4/art22/.

Scott, Colin H. 1983. "The Semiotics of Material Life among Wemindji Cree Hunters." PhD thesis, McGill University.

–. 1996. "Science for the West, Myth for the Rest? The Case of James Bay Cree Knowledge Construction." In *Naked Science: Anthropology Inquiry into Boundaries, Power, and Knowledge,* ed. Laura Nader, 69–86. London: Routledge.

Scott, Colin, and Harvey Feit. 1992. *Income Security for Cree Hunters: Ecological, Social and Economic Effects.* Montreal: McGill Programme in the Anthropology of Development, McGill University.

Scott, Katherine. 2008. "Fire, Plants and People: Exploring Environmental Relations through Local Knowledge of Postfire Ecology at Wemindji, Québec." Master's thesis, McGill University.

SOTRAC (Société des travaux de correction du Complexe La Grande). 1980. "Annual Report 1980." Montreal.

Turner, Nancy J., Iain J. Davidson-Hunt, and Michel O'Flaherty. 2003. "Living on the Edge: Ecological and Cultural Edges as Sources of Diversity for Social-Ecological Resilience." *Human Ecology* 31 (3): 439–61.

Vale, Thomas R. 2002. "The Pre-European Landscape of the United States: Pristine or Humanized?" In *Fire, Native Peoples, and the Natural Landscape,* ed. Thomas Vale, 1–39. Washington, DC: Island Press.

von Mörs, Iris, and Yves Bégin. 1993. "Shoreline Shrub Population Extension in Response to Recent Isostatic Rebound in Eastern Hudson Bay, Québec, Canada." *Arctic and Alpine Research* 25 (1): 15–23.

10

Aa wiichaautuihkw
Cultural Connections and Continuities
along the Wemindji Coast

VÉRONIQUE BUSSIÈRES, MONICA E. MULRENNAN,
and DOROTHY STEWART

This is a story that begins "[m]any moons ago, when people
were still using bows and arrows, before white men brought
guns and powder."

— Irene Mistacheesick (pers. comm., September 23, 2008).[1]

We are told that people have lived along the shores of James Bay "since
time immemorial." Likewise, local legends extend back to a "time when
animals were still talking and living like people" (Fred Blackned, pers.
comm., September 25, 2008). Local stories and place names, as well as
recent archaeological findings in the area (see Chapter 5, this volume),
attest to this continuity and to the connections that Wemindji Eeyouch
maintain with their territory.[2]

The present chapter tells a story about the relationship between Wemindji Eeyouch and the coastal and marine portions of their traditional
territory. It is about connections to places that have been inhabited and
culturally constructed over many generations. These connections are
deeply rooted in tradition, but they also testify to the dynamic nature of
this environment and to the adaptive capacity of the generations that
have lived there. Our story draws mainly on accounts and perspectives
of Cree people that have been shared through the research partnership
established by the Wemindji Protected Areas Project, supplemented by

relevant literature. To describe our work, Dorothy Stewart uses the word *aa wiichaautuihkw,* which means "travelling together." We have undertaken many journeys together, the most important of these being canoe trips down Paakumshumwaashtikw (Old Factory River), a route used by Cree hunters and their families for generations and now the motivation for the establishment of the Paakumshumwaau-Maatuskaau Biodiversity Reserve and the proposed Tawich (Marine) Conservation Area.[3]

In this chapter, we paddle and portage our canoes along this special river, down to Old Factory Bay (also known as Paakumshumwaashtikw). From there, we travel up the coast to the village of Wemindji, some forty kilometres north. Our story covers a long period. In fact, it goes back to "time immemorial," to borrow an expression often used by community members. It plays out at different human levels and geographical scales: at a regional level, coastal Crees feel a strong connection to the sea and coast in general; at the community level, certain rivers, streams, and islands hold special significance; at the extended family level, specific places on the family hunting territory, such as the birth or burial place of a relative, will be of particular importance. Our narrative tries to connect with collective and individual stories of the Wemindji Eeyouch as they relate to the coast.

Like the "scapes" described by others, the coast- and seascapes that are central here have been culturally constructed over many generations. Harvey Feit (2004) explains that relationships between Cree hunters and their territories develop from personal memories through continued occupation of the land, as well as from previous generations through stories, place names, and legends. Keith Basso (1996) further suggests that through vestiges of past inhabitation, stories, and place names, landscapes serve as reminders of the past. Memories, thoughts, and experiences are held in such places, transforming them from neutral spaces to significant places (Casey 1996; Cruikshank 2001). In her work on landscape, history, and change in the Pacific Northwest, Julie Cruikshank (2000, 2005) shows the power of landscape in connecting people to the past. Such links are dynamic, maintained by keeping knowledge and history alive through time spent nurturing relationships on the land (Feit 2004; Basso 1996). And when outsiders are invited, they too become part of this network, connecting Cree hunters to other places and times (Feit 2004). In the context of developing a protected area, we thus suggest that the coast and sea are archives of past and present human-environment relationships.[4]

For many Indigenous peoples, their relationship to land, sea, and resources is fundamental to their health, well-being, and identity (Adelson 2005; Richmond 2015). This connection is encapsulated in Kathleen Wilson's (2003, 90) concept of "therapeutic landscapes," whereby customary activities such as hunting, fishing, and gathering, as well as certain ceremonial and symbolic practices, "allow individuals to pursue simultaneously physical and spiritual connections to the land that are important for emotional and mental health." Indigenous scholars Taiaiake Alfred and Jeff Corntassel (2005) also underscore the importance for the identity of Indigenous peoples of maintaining their relationships to the land. Similarly, Ahkil Gupta and James Ferguson (1997) assert that identity is strongly linked to the process of place making and to the development of relationships that make place "local." Interestingly, and we believe this applies to Wemindji Eeyouch, Wilson (2003) suggests that the land (and sea in our context) does not merely influence or shape identity – it is an integral part of it. Because of the importance of such relationships, it is not surprising that many Indigenous groups across the world oppose dominant narratives of their territories as unused wilderness (see Chapter 1, this volume) and the associated silencing of their counter-narratives of occupation and use of, and attachment to, their lands, coasts, and seas through past and present use.

SOME CONSIDERATIONS BEFORE EMBARKING ON OUR JOURNEY

Reflecting upon her own experience with Indigenous peoples in northern Ontario, Kathleen Wilson (2003) captures a pitfall of participatory research in her remarks about the danger of "speaking for others." She rightly questions the appropriateness of such an approach. Thus, we need to clarify our position with regard to the Wemindji community and our role as narrators here. In his discussion of relationship building for the Crees, Feit (2004, 94) explains that, for various reasons, Crees have invited outsiders to "build understanding and connections" with their land. These outsiders then become part of a Cree person's "life project," part of the intertwined web of relationships that Crees foster with their environment, human and non-human, animate and inanimate.

Dorothy Stewart, a co-author of this chapter, was born beside Paakum-shumwaau (Old Factory Lake) and is the daughter of this hunting territory's former tallyman. She has taken particular interest in working with and learning from elders. She thus served as a guide and teacher to the non-Cree participants on this journey. Monica Mulrennan has worked as a researcher with the Wemindji community since the mid-1990s. She is also Véronique Bussières's graduate thesis supervisor. She is both learning and mentoring. Véronique has been a regular visitor to Wemindji since 2003, initially as a master's student and now as a doctoral candidate. Monica and Véronique do not claim to fully understand the relationship between the Wemindji Crees and the coast. Nor does Dorothy feel comfortable in speaking for her entire community, although she comes from Wemindji. Nevertheless, the system of relationships that govern Eeyouch interactions with each other and outsiders provides us with some authority to tell this story. As part of this collaborative research project, Monica and Véronique have been invited to learn about this place. Dorothy is one of the many people in Wemindji who have made us part of their web of relationships. We narrate this story together, discovering and interpreting Cree connections to the coast and sea, from the vantage point of our particular positions within this intertwined network.

The journey presented here is actually a composite of several trips, with different guides and travelling partners. Our account is enriched by conversations that took place in the village of Wemindji, in Montreal, over the phone, in emails, and on Facebook. It also includes recollections and stories from community members.[5] It represents a metaphorical journey of the Wemindji Eeyouch over time and across a portion of their territory, as well as our own journey as research partners across Paakumshumwaau, down Paakumshumwaashtikw (the Old Factory River) and into Old Factory Bay. This chapter is organized spatially, as it follows our journey from the lake to Wemindji. It also makes several jumps in time, taking us back to the days when animals spoke like humans, to various points in the fur trading era, and to recent decades. Because the land is growing in Eeyou Istchee due to coastal emergence (see Chapter 5, this volume), earlier events are imprinted on land found farthest inland, whereas more recent ones take place on recently emerged islands.

CROSSING PAAKUMSHUMWAAU

We begin our tale some eighty kilometres east of James Bay, on the banks of Paakumshumwaau (see Figure 10.1). On a trail amid black spruce, caribou moss, and blueberry bushes, black flies buzzing in our ears, we portage our canoes from the modern highway. We encounter the cabin of the Matches family; the eastern portion of the lake is part of their hunting territory. Farther down the path, we wade a hundred metres through a muddy bog before reaching the clear waters of the lake, where we launch our canoes to truly begin our journey to the brackish inland sea of James Bay. Our first camp is set up at Uupimiskaahsh, a sandy point on the northern shore of the lake. Early next morning, Fred Stewart calls "*Winishkaau!*" (Wake up!), so we break camp and continue crossing this extensive lake, home to prized fish species such as walleye, cisco, and sturgeon. Up on a ridge, we see the dark silhouette of a bear, probably feasting on the abundant berries; both are highly valued by the Crees.[6]

At a wider part of the lake, we see Aapimisaachiiush on our left, around a bend, where Fred's cabin is located. He is Dorothy's brother and also the tallyman of this trapline. Most of our journey takes place in Stewart territory. Straight ahead lies the cabin of their brother Henry Stewart, on a sandy outcropping that forms a bottleneck in the lake. He is gradually taking over the role of tallyman, as Fred's health is fading. He has a full-time position with the Department of Public Works in Wemindji but spends much of his free time here, hunting, trapping, and fishing. His steps and ours now follow much more ancient ones. Indeed, clues to the earliest inhabitants of the Wemindji coast can be found along the banks of this lake because, as people from Wemindji will tell you, "the land is growing" (see Chapter 5, this volume, for a description of isostatic rebound and early habitation). As a result, the remnants of human activities that occurred on the shores of James Bay in ancient times are now found as much as a hundred kilometres inland, and all of this land we are travelling through was coastline at some stage in the past.

PADDLING DOWN PAAKUMSHUMWAASHTIKW

Leaving the lake, we enter the river channel, where our guides remind us of the river's significance to the Eeyouch. Rodney Mark (pers. comm., April

16, 2004) explains, "Paakumshumwaashtikw is a historical route. Many families used it. My grandfather used to always tell me stories about paddling down and portaging along Paakumshumwaashtikw. Yes, most of the stories were about paddling and portaging from the lake to the bay. This river is really a historical route that was used by more than one family."[7]

Elder Clifford Georgekish (pers. comm., August 6, 2003) stresses the importance of the river as "a highway from Paakumshumwaashtikw and back into the inland hunting territories." While we paddle, place names, anecdotes, and stories told by our guides offer glimpses of their strong relationship with the area and its past. For example, Fred, Henry, and Dorothy Stewart highlight the history and significance of the portages used to get around rapids as well as the various campsites used when canoeing down the river, most of which have specific place names and are associated with stories. According to Fred Stewart, Kaamuhkuutaakin, where a carver drowned and his body was never found, serves as a lesson to paddlers. Because of this story, "we don't go over these rapids," but instead use the portage, he says. Farther downriver is a place called Kaawaa-yaapiskaach, where rocks have been positioned to form pools and entrap fish. Again, Fred explains that, in the past, "people knew they would find fresh fish there" (pers. comm., July 18, 2010). According to Dorothy, this spot was so productive that there was no need for a fishing rod. She explains the importance of such places; people who live on the land spend a great deal of energy searching for food, and sites where it can reliably be found are sometimes a matter of life or death.

After several more days of paddling, punctuated by nights of camping and storytelling, we reach the mouth of the river and enter the bay. Although this area has emerged more recently from James Bay than the areas further inland, it too is connected to the distant past. A good example, Chiipiitukw (Figure 10.1), is located on a point just north of the river mouth. The name means "Ghost dwelling" or "Place where two families once starved"; this is possibly the oldest known campsite in the bay area (Bussières 2005).

Paddling toward James Bay, we are surrounded by islands. To the south is an island known as Kampaniiu minishtikw. To the north lies Ataawaa-siiu minishtikw, where the red roof of an old warehouse is visible. And, west of this one, farther out toward the bay, is Upishtikwaayaaukimikw.

FIGURE 10.1
Paakumshumwaashtikw and Old Factory Bay

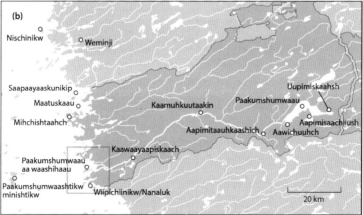

Source: (a) Adapted from Figure 11.1, this volume. Place names collected by Colin Scott and Monica Mulrennan in 1999–2000. (b) Adapted from Véronique Bussières (2005).

Their names, as we shall see later, are only a few of the many clues to the historical significance of this area.

REACHING THE ESTUARY

At the estuary, where we see the open waters of "tawich" in the distance, we are reminded of the story of Wolverine and Giant Skunk, which explains how the waters of James Bay became salty. According to Fred Blackned (pers. comm., September 26, 2008), Giant Skunk was a great threat to all the land animals. Wolverine, as the head of the animals, warned them not to cross the tracks of Giant Skunk. Unfortunately, they did not heed his warning. When they ambushed Giant Skunk, Wolverine was sprayed in the face by Skunk. The other animals told Wolverine not to wash his face in the lakes or rivers but to go out to the bay. Blinded by the spray, Wolverine talked to the various tree species he encountered; black spruce dominated inland and was replaced by white spruce as he neared the coast. He eventually reached the shore and washed his face in the bay, whereupon its waters became salty. Blackned speculates that the animal characters may have been humans, "because humans may have had animal names back then." He also suggests that Wemindji Eeyouch may be descended from these characters (for other accounts of this story, see Chapters 6 and 12, this volume).

An event occurred at the bay that would significantly change this place and its people – the first encounter between Eeyouch and Europeans. There are several versions of this story. One says that Chahkaapash, a mythical Cree hero, first saw a boat on James Bay, was invited on board, and exchanged food with its crew (Scott 1992). In other versions, such as those told by Elders Geordie and Jacob Georgekish, and reported by Colin Scott (1992), an Eeyouch man was warned by a *mistaapaau* (spirit helper) that white men would come on a large boat. Both elders agreed that this took place at Old Factory Bay. They said that the Crees were given clothes (and a rifle, according to Jacob Georgekish) in exchange for their fur clothing, the start of a long relationship of contact between Crees and Europeans/Canadians. These first white people are said to have established their settlement on the island called Upishtikwaayaaukimikw, meaning Frenchman's Island (Figure 10.1).

Following this first contact, other white men came to the area. Eventually, representatives from the Hudson's Bay Company (HBC) arrived, and from the mid-1600s until the mid-1900s, the east coast of James Bay was incorporated into the wider fur trading activities of the James and Hudson Bay basin (Francis and Morantz 1983). Many Wemindji Eeyouch divide the fur trade era into two distinctive periods: the first, spanning 1600 to 1934, is associated with early HBC trade expeditions in the area and the presence of a French trader at Upishtikwaayaaukimikw; the second began in 1934, with the establishment of the trading post at Old Factory Bay, and ended when the post closed in 1959.

Early Fur Trading: 1600s-1900s

In 1668, the HBC established its first trading post on the James Bay coast, at Charles Fort, later called Rupert House (today it is the Cree community of Waskaganish, about 120 kilometres south of Wemindji; Denton 2001). Elder Billy Gilpin (pers. comm., May 4, 2004) describes the fur trading activities in the bay:

> The Frenchman was here around 1670 and was the first trader. The Hudson's Bay Company then came here for fur trading. The company was already trading south of here, as we were told ... They say [that the Frenchman] discovered the Crees here in the area. The one [island] they call Kaachinwaanikaach, it was there that the ship was frozen solid in the ice, all winter long ... He was the first to be trading with the Eeyouch in the area. This is what I heard talked about, before I was born.

It's not clear whether this account refers to the expedition that Geordie and Jacob Georgekish mention above, but it suggests that fur trading occurred in the area for more than three hundred years.

During the early fur trading period, the bay was a busy place. However, as Toby Morantz (2002) highlights, accounts of life at the trading post reflect a Eurocentric perspective, as they rarely mention what was going on in the lives of local Crees at the coast or elsewhere. Cree families were established at other locations along the coast, and they engaged in a variety of land-based activities that were unrelated to the post. For example, several old coastal campsites, some dating from the 1700s and 1800s, and a few as far back as the 1600s, have been identified by elders and tallymen around the Maatuskaau and Saapaayaaskunikip Rivers, both of which drain into Moar Bay. Fred Asquabaneskum (pers. comm., August 13, 2003) pointed out one of these campsites just south of Maatuskaau, approximately seven kilometres inland from the present river mouth, which he identified as dating from the 1670s. This could be Aaskwaapisuaanuuts, which, according to archaeologist David Denton (2001), was a major occupation site between 1740 and 1785. Sinclair Mistacheesick identified the location of another camp, on the south shore of Moar Bay, again near Maatuskaau. Translating for her husband, Irene (pers. comm., September 23, 2008) explains, "On the south side of Moar Bay, his uncle mentioned that there was an old campsite, with a longhouse, shaped like a *shaptuan* ... Two to three families were staying in that area, way back when his grandfather was still young."[8]

Later Fur Trading: 1934–59

Fur trading was at its peak in the bay during the mid-1930s. Older community members retain vivid and often fond memories of this time, when the bay was a busy centre of activity. Trading posts were established in 1934 by an independent trader and in 1936 by the HBC. A Catholic mission house and an Anglican church were also built. Vestiges dating from this period, in the form of old buildings, burials, and abandoned campsites, can be found throughout the estuary (Bussières 2005). One example is Waapischinikaau or Rabbit Island. Dorothy Stewart explained that it got its name because when it snowed, it stood out from its surroundings and looked like a rabbit. This place was of particular significance to her family, as it was "almost a permanent dwelling" when they were young. Her family spent a lot of time there. She remembered going to Ataawaasiiu minishtikw for Christmas one year. The sleigh was sticking on the ice, so they had to get out and push. Although the campsite hasn't been used for several years, its numerous *miichiwaahp* (teepee) rings attest to its importance in the past. Several other place names connect to this later fur trading period. For example, as mentioned earlier, a trading post was located on Ataawaasiiu minishtikw, which means "Trader's Island." The HBC established its post on Kampaniiu minishtikw (Company Island). And the name of Paaschikin minishtikw, or Gun Island, refers to the practice of having to deposit one's gun there before visiting the trading posts.

During this period, most families spent the winter months on their hunting territories, either on the coast or inland, and gathered at Old Factory Bay during the summer. From the women's stories, it is easy to imagine large boats arriving and leaving, hunters and trappers coming to exchange their furs for supplies, the joyful reunions of families and friends, and the frantic activity prior to their return inland or up the coast. As Irene Mistacheesick (pers. comm., September 23, 2008) explains, they went to Old Factory Bay for supplies, to sell pelts, to visit the doctor and dentist, and to celebrate weddings and other events with family and friends. Also, families based on the coast exchanged food with those whose hunting territories were inland. Dorothy Stewart's sister, Clara Visitor (pers. comm., May 4, 2004), remembers the excitement of spending time there when she was a child: "People would go to the mission to get medication and other things. There were children running everywhere. We would go play on the big boulders on the beach. And then, in the fall, we would go back inland."

Daisy Atsynia (pers. comm., September 23, 2008) moved from Eastmain in the late 1940s to her husband's camp, located north of Wemindji. She remembers the summers they spent at Old Factory Bay. Back then, they remained on the land from September to June, hunting and trapping. In the summer, she and her husband would go to the bay to sell furs and also for fishing, because "fishing was better there."

For many, especially the women, this was a time for rejoicing. According to Daisy Atsynia, Nancy Danyluk, and other women, everyone looked forward to it – people gathered, talked to kin, and exchanged stories after the long months of relative isolation on their hunting territories. Dorothy remembered dances that lasted until dawn. Two of her sisters, Sarah Stewart and Clara Visitor, who were young girls then, laughed as they remembered their mother going to these dances after they had been put to bed. Sometimes, she did not come back until the morning. There were lots of chores and responsibilities to attend to, too. People were busy fishing, both for immediate consumption and as a reserve since fish and fish eggs were often dried and kept for later.

Inuit Neighbours from Nearby Cape Hope Islands

Standing on Ataawaasiiu minishtikw, next to the old warehouse and what remained of the trader's house, Clara Visitor (pers. comm., April 30, 2004) recalls, "Both Crees and Inuit came here to bring their furs [on Ataawaa-siiu minishtikw]. There were a lot of people. Our parents used to set up our miichiwaahp over here. Further at the back, over there, it would be the Inuit."

This reminds us that local Eeyouch history is intertwined with another story, that of several Inuit families who shared a portion of their territory until recent decades. Inuit men piloted the big HBC boats from Hudson Bay to the shipping port on Charlton Island in James Bay (Minnie Freeman 1978). The HBC also hired them to perform certain jobs, such as piling snow against buildings in the winter. The sealskin boots made by Inuit women were much prized (Morantz 2002). A major Inuit settlement was established on Charlton Island, in the Fort Rupert area. When the railway reached Moosonee in 1933, the island lost its status as the main shipping port for James Bay, and the community moved to Cape Hope Island (Wiipichiinikw/Nanaluk), under the leadership of George Weetaltuk. By 1935, Wiipichiinikw (the Cree name for this island, Nanaluk is its Inuit name), a few kilometres southwest of Old Factory Bay, had become a permanent settlement for eight Inuit families (Milton Freeman 1983). George Kudlu, who married Weetaltuk's granddaughter Louisa, today lives in Wemindji, though he maintains a summer camp on the island, which is steeped in family history. Louisa passed away a few years ago. On the island, the ruins of several houses and buildings, including a boat factory, are barely visible these days. However, the stories told by George offer a vibrant glimpse of what life was like there over fifty years earlier. Walking through the site, he provides details of the boat factory, which flourished during the 1930s. Two of its boats were sold for $250 each to white people, a large sum at the time. Canoe frames are still visible beside the steamer used to bend them. The wreck of the *Venture*, a boat that was built there, stands proudly next to the Kudlu cabin. Cape Hope's Inuit community, the most southern Inuit settlement in Canada, was relocated by the federal government in 1960 (Milton Freeman 1983), but the Kudlu family returned to Wemindji over twenty years ago.

The legacy of these Inuit families is imprinted on the cultural land- and seascape. Four inuksuk mark the cardinal points around a peninsula just east of the Kudlu camp. Both Inuit and Cree burial sites are found on Upishtikwaayaaukimikw, and the intertwining of Cree and Inuit lives is a common feature of stories recalled from places such as Ataawaasiiu minishtikw, in Clara Visitor's quote above. Clara's brother, Henry Stewart, remembers that his father, Sinclair, travelled with Inuit hunters to bring the mail up from Moose Factory. Indeed, Clara inherited a pair of sealskin boots from her father, who had received them from an Inuk friend.

WEMINDJI: A NEW TOWN

The HBC's Old Factory post was not a permanent settlement. Most Crees visited it only during the summer and returned to their traplines for the rest of the year. In 1959, various reasons motivated the relocation of the HBC store, mission house, and other facilities some forty kilometres north to what would become Wemindji village. Several community members explained that boat access became increasingly difficult in Old Factory Bay due to coastal uplift. Toby Morantz (2002) mentions increasingly unsanitary conditions, combined with a shortage of wood and drinking water, as well as spread of diseases, as important reasons for the move. The federal Department of Northern Affairs gave community members the option of joining Eastmain House, forty-five kilometres south of Old Factory Bay (Morantz 2002). Some accepted this option (which explains why many Wemindji residents have relatives in Eastmain), but approximately three hundred chose to move farther north, to the shore of the Maquatua River. This sheltered site enjoys a particularly deep harbour and no shortage of drinking water, wood, and space.

The establishment of Wemindji village was accompanied by rapid demographic, economic, and social changes. Many people became involved in the cash economy and adopted a lifestyle that limited their time in the bush. Despite this, the coastal area, and especially Old Factory Bay, retains a particular significance to Wemindji residents as the place where many of them grew up and where customary and contemporary cultural activities are still practised on a regular basis.

TRAVELLING TAWICH TODAY

Attending the Summer Gathering

We return to our canoes and resume paddling through the estuary, a challenging task given the shallowness of these waters, a tangible reminder that this coastal land continues "to grow." In the distance, farther out in the bay, we see our destination, a large white tent on an oval-shaped island covered with spruce forest. Our arrival coincides with the Old Factory Gathering, an annual event commemorating the importance of the early settlement at Old Factory for Wemindji Eeyouch. The gathering was initiated in the 1980s and has been held on most summers since then. For this occasion, community members assemble on Akwaanaasuukimikw, otherwise referred to as the Gathering Island. The tent has been erected, along with several miichiwaahp and temporary cabins. Festivities last a week or two and include communal cooking and feasting, as well as games and dancing. Several people take the opportunity to visit the nearby burial sites of family members and other historical sites. This period of rejoicing and remembering supports social exchanges that are rarely possible back in Wemindji. For this reason, it holds a special socio-cultural significance for the community. Nancy Danyluk (pers. comm., August 13, 2003) describes it as a "going back, homecoming." She explains,

> When you live in Wemindji, you have your own house. When we go back to Old Factory, it's like going back home, a long time ago, because in those days when you lived in Old Factory, you just lived in tents or tee-pees. People are visiting each other more, you talk to more people than you would in town, all the kids are together without getting into trouble. That's how it used to be when we lived in Old Factory. Some people you didn't talk to all year, you would be talking to them at the gathering, and it's kind of nice.

Attendance at the gathering has fluctuated over the years due to the challenges of organizing such an event, as well as unanticipated developments.

There is, however, a deeply felt commitment, especially among the women, to continue this tradition. This was particularly evident in efforts made to celebrate the gathering in 2009, planned in conjunction with the commemoration of the fiftieth anniversary of the move to Wemindji. In 2017, for the thirtieth anniversary of the gathering, celebrations lasted for two weeks, and more than two hundred people attended. Several community members, who had not been to Old Factory Bay for many years, expressed their joy in being there and reconnecting with the place, as well as their intention to come more often in the future. Research team members have attended these celebrations over the years, including those for the fiftieth anniversary of Wemindji and the thirtieth anniversary of the gathering, in response to generous invitations from the community.

Youth Canoe Expedition

One afternoon, people assemble on the shore, apparently waiting for something. We notice a group of canoes coming from the river mouth. People tell us that "the paddlers are finally arriving!" Every year since 1995, at midsummer, a canoe expedition down the river commemorates its importance as a historical travel and trade route. This journey was revived to show young people how it was done. A group of guides accompany them on the river for over a week, stopping at campsites and portages, from Paakumshumwaau to the Gathering Island in James Bay. The initiative brings together youth and older community members, including elders, who are knowledgeable about the land. It provides an opportunity for knowledge and values to be transmitted to the youth and for them to experience being on the land and using their language. Chief Rodney Mark (pers. comm., April 16, 2004) underscores the value of this event for fostering the connections between Cree youth and Eeyou Istchee:

> For the youth to be interested in the land, we have to work on [Cree] values. We have had and have programs to get the youth interested in the land. However, they tend to be like preaching. We tell them the way of life is this and that. However, for the youth, it has to be fun. They have to simply be out on the land and enjoy it to start feeling a connection with it again. This is what we hope to achieve with the canoe trip down Paakumshumwaashtikw.

Participation in the canoe expedition has increased in recent years, with three canoes making the first trip and thirteen in 2017. Many community members, encompassing all generations, acknowledge the sociocultural importance of this event and hope that it will continue. The annual gathering, the canoe expedition, and periodic visits at other times reaffirm attachments to this place and its past. These events and activities have also been valued opportunities for team researchers to spend time with community members.

What the Coast Provides

As we prepare to leave Old Factory Bay, an older man and his son approach in their freighter canoe, with whitefish and sea-run brook trout they have netted. A few women have gathered in a miichiwaahp to clean and process the fish, some of which will be smoked over the fire. The rest will be hung on a wooden rack to dry in the sun. One woman invites us to take some dried fish, explaining that her family keeps only what they need. The remainder will be sent to the village to be distributed among community members, as is customary.[9] We thank her for the fish and set off by freighter canoe for the final leg of our journey. Now we head north, navigating through the islands and shoals that are scattered in the waters of the many embayments that define this part of the coast.

Pointing to the shoreline, Henry Stewart (pers. comm., August 12, 2003) tell us, "We've got goose blinds in here, you might not have seen, all along this bay, and the same thing in here. And they go out to these islands to go hunting too. Yeah, to all these islands out here, and every little bay there is." We soon realize that places all along this coast are of significance for various hunting, fishing, and/or gathering activities (see Chapters 8 and 9, this volume). Translating for her brother Fred, Dorothy points to Blackstone Bay, the next bay north of Old Factory Bay, as one such area. She explains that it is a very good place to hunt waterfowl in the fall, including various duck species, snow geese, and longnecks (a subspecies of Canada geese). Some people go there by canoe or even walk all the way from Old Factory Bay. This is because it is regarded as a reliable spot for hunting, where people know they will get food. There are several such places along the coast between here and Wemindji. Fred says that Moar

Bay, just north of Blackstone Bay, is also an important fall hunting area. Farther north again, the eggs of nesting marine birds were gathered on the archipelago at the mouth of the Maquatua River, as a supplementary food source. North of the river, Goose Island has been a prime waterfowl hunting spot for generations, and we've heard several local claims that the fish caught there taste better than those from anywhere else.

All these places give us a sense of the seasonal cycle of resource harvesting that forms the basis of Cree customary life (see Chapter 7, this volume). Mid-October to March is the season for moose hunting and beaver trapping. Although this generally takes place in the inland portion of the territory, Fred Stewart notes that on his trapline, it is practised more intensively in areas closer to the coast, thus near Old Factory Bay. Mink, otter, fox, snowshoe hare, and "whitebirds," or ptarmigan, a Cree favourite according to Sam Georgekish, are also hunted. As spring arrives, preparations begin for the goose hunt, the most intensive harvesting period along the coast, at least until recently (see Chapter 8, this volume). During this time, Wemindji village is vacated as families spend two to several weeks in the bush, which is facilitated by a school "goose break" and flexibility in the work commitments of many adults. Places such as Old Factory Bay and Blackstone Bay become busy centres of activity as skidoos (and now helicopters due to thin ice) travel up and down the coast, bringing people and supplies to camps. Cabins that had been closed for months are swept and aired, firewood is cut and stacked, and spruce boughs are collected for the floors of the miichiwaahp already under construction. In each family territory, the tallyman carefully orchestrates the hunt to optimize the catch while minimizing long-term disruption to the birds, which use the coast as a staging area on their northward migration (see Chapter 9, this volume). After the geese have left, people who work in town go back to Wemindji, whereas full-time hunters move to the outer islands for the loon and duck hunt. Next comes the summer fishing season, when most other harvesting activities are abandoned, and families move to fish camps on the coast or on nearshore islands, such as Sheppard Island or the Gathering Island. The fall goose hunt lasts from September to mid-October but is less intensive than the spring one because, according to Fred Stewart, "there are more geese in the spring than in the fall, and they are fatter in the spring, too" (Bussières 2005, 80). People then begin to prepare their gear for the moose hunting and trapping season, and the seasonal cycle starts again.

Several people mentioned the importance of bush food for their well-being, especially fish from the coast or estuarine areas. As a coastal tallyman explains, "people in town want to eat fish, and they say fish from Maatuskaau tastes good. Around Wemindji, fish doesn't taste as good" (pers. comm., September 23, 2008). The diversity and abundance of berries – including blueberries *(iyimin)*, cloudberries *(shikuutaau)*, strawberries *(utaachiimin)*, and cranberries *(wiisichimin)* – encourage many families to visit the islands. The coast also provides essential resources, such as driftwood for fuel, and drinking water is collected from rain puddles on rocky islands. As with all other resources, great care is taken to ensure their long-term availability and quality; for example, people are careful not to contaminate the drinking water, and local stories and taboos remind them of the perils of carelessness (Mulrennan and Bussières 2018).

Attachments to the coast extend beyond its importance for sustenance. Several people tell us that it is important to their personal well-being. Irene Mistacheesick (pers. comm., September 23, 2008) says that "in the summer, it is more enjoyable on the islands; there is a better view!" She adds that it is healthier to be by the sea, where there is salt water, and cites as proof the good condition of plants and the abundance of berries on the coast and the islands. The preference of many for being on the coast and bay is reflected in the regularity with which people visit these areas, making them significant to the well-being of individuals, and by extension to the community as a whole.

Through generations of resource harvesting and management, Wemindji Eeyouch have left an indelible mark on this coastal landscape. Countless current and previous seasonal campsites are scattered along the coast and on offshore islands. They are associated with various harvesting activities; those for the goose hunt are generally on the coastal mainland, whereas many fishing camps, particularly those also used for loon and duck hunting, are on islands (Bussières 2005; Sayles 2008). Other modifications to the coastal landscape include goose flyways cut into the white spruce forest, the construction of dikes, the digging up of marshes to improve the foraging habitats of geese, and numerous hunting blinds along the bay and on the islands (see Chapter 9, this volume).

Navigation Hazards and Wariness of the Sea

Almost halfway between Old Factory Bay and the village of Wemindji, we reach Mihchishtaahch (Sheppard Island, where people commonly stop and rest during their travels up and down the coast. Fred Stewart (pers. comm., July 18, 2010) explains, "they knew people were there to meet and feed them. When it was windy and the weather was bad, it was a place where people would stop." This practice continues today; many people, ourselves included, have stopped here while waiting for rough weather on the bay to subside. Winnie Asquabaneskum and her family welcomed us with hot tea, allowing us to warm up and rest before continuing to Wemindji.

The dangers of travelling on James Bay are ever present for the Crees. According to tallywoman Daisy Atsynia and tallyman Sinclair Mistacheesick, the inshore islands are used regularly, whereas the outer islands are seldom visited; fear of bad weather and rough waters are the principal reasons. Certain place names, stories, and legends reflect a degree of wariness and fear of the sea. Ishkwaashiiuminishtikush refers to a place where a "woman got stuck picking berries when tide came in." Daisy Atsynia (pers. comm., September 23, 2008) says that she told her grandson not to go to Niishuchaash (Twin Islands) because "there is something not good over there, under water ... a creature. It is bad luck to go there." Many community members acknowledge their reluctance to venture onto the bay when the weather is windy and the waves are high. Misadventures on the bay can also occur in fair weather, as Sinclair Mistacheesick reminds us in a story from his childhood. Translating for him, Beverly Mayappo (pers. comm., September 23, 2008) says, "Once, coming from Solomon's Temple islands [by boat], the motor broke down. They had to float on waves to the coast. He was eleven years old. Even today, he remembers being scared and crying."

Another incident from about four or five decades ago involved a shipwreck near Mihchishtaahch , which remains prominent in the memory of many people. Dorothy vividly recalls seeing various floating objects after the wreck. Mary Hughboy, who was present at the time, remembers helping the crew and using some of the supplies that were found floating nearby, which lasted her family for about a year (Cree Nation of Wemindji 2009). Such misadventures are also familiar themes in local legends about the sea and are said to include valuable lessons for listeners. For example, one story of Chahkaapash, as told by Fred Blackned (pers. comm., September 26, 2008), relates to learning how to fish:

> One day, Chahkaapash shot an arrow over James Bay. It fell right where the seabed drops, where the big fish swim. So, he jumped in the water and swam to get it. As he was looking for his arrow, he got swallowed by a big fish. His sister knew what happened. To get Chahkaapash out, she made a fishing rod from the pole of a teepee. Chahkaapash, who could talk to the fish from inside his belly, told him to swim to the hook and swallow it. His sister could then get the fish out. It had a big belly. She cut it open, and Chahkaapash jumped out. His sister was wise; she knew and told him he should not have been swimming out there.

Recent observations of weather and climate changes by community members suggest that the hazards of travelling on water and ice have increased and will probably worsen in the coming decades. As Andra Syvänen (2011) documents, abrupt and unpredictable changes in the seasons, as well as in ice cover and quality, are having dramatic impacts on travel during the spring and winter. Due to the ruggedness of the terrain, ice and water have been the preferred travelling routes for Crees. However, this may soon change, as indicated by stories of people who were stranded on islands due to poor ice conditions and of others who limit their winter travel on the bay for fear of accidents (Syvänen 2011). Summer conditions are also affected, as the reduced precipitation during winter and spring results in lower water levels, which make travel by canoe dangerous. Thus, climate change is inspiring new stories and generating anecdotes that reinforce existing cautions about the sea.

REACHING THE WEMINDJI SHORE

One chapter in the story of Wemindji Eeyouch concluded with the relocation of the community to its current site. Despite subsequent changes, most

Wemindji residents retain strong ties to the bush, including the coast and bay area. These ties are embedded in rich localized histories, which, as Feit (2004, 95) reminds us, are manifested as places that are profoundly personal and local while also having "far reaching connections, recognitions and histories." Place names, stories, and legends bear witness to the past by recalling details of past events, people, and associations. Personal memories, combined with ongoing engagement in hunting and other activities on the land, ensure that connections to the past are maintained as vibrant and meaningful lived experiences.

As an important node in a much more extensive socio-ecological system, Old Factory Bay illustrates the richness and reach of these connections to other places, times, and people. An archive to past encounters with others, it also occupies a central position these days as the site of important cultural events and activities. These connections and commitments motivate and inform a will to protect the area for present and future generations. In doing so, they are generating new encounters with outsiders such as Véronique and Monica, along with other team members, who have been invited to enter Cree places and to form respectful relationships within Cree life projects that have become essential to Cree "living here and now" (Feit 2004).

At Wemindji, we drop anchor and throw a rope to a young Cree boy, who pulls us ashore. The connections of Wemindji Eeyouch to the James Bay coast unfold as a narrative. However, our story does not end here, since our relationship is ongoing and committed. The creation of the Paakumshumwaau-Maatuskaau Biodiversity Reserve, and hopefully before too long the Tawich (Marine) Conservation Area, represents beginnings of a new chapter that will build on the earlier ones outlined here. The significance of this new chapter is made clear in a statement from Edward Georgekish (pers. comm., April 23, 2004):

> It's important to protect certain areas. And, that's what we're saying with this project, to protect certain areas that will be of special significance to the people, our heritage, our way of life, to show that we've been here, that we have been on the land since time immemorial, and I think, as people, we have to show that to society as a whole.

NOTES

1 According to the Tri-Council Policy Statement on Ethical Conduct for Research Involving Humans (Canadian Institutes of Health Research et al. 2010, 58), participants may waive anonymity if, for example, "they wish to be identified for their contributions to the research." In accordance with the policy, informants whose names appear in this chapter have agreed to be identified. The interview of Irene Mistacheesick was conducted by Véronique Bussières and Monica E. Mulrennan on September 23, 2008.

2 Eeyouch, in Cree, means "people," the Cree people of James Bay; Eeyou Istchee is their territory.

3 Tawich, in Cree, refers to the sea or marine area adjacent to the coast. Fred Stewart describes it as "out in the bay."

4 The authors would like to thank Katherine Scott for her contributions to this essay, particularly to this paragraph.

5 Two types of stories recognized in Cree culture reflect differences between Western Cree conceptions of time and space: *tipaachimuu* and *atiukan* (Preston 2002). The former relates to an anecdote or event that occurred at a particular time and place known to living people. The latter is closer to the Western idea of myth or legend, with a more timeless and vaguely defined spatial dimension. These are described as very old narratives "whose personalities and places are not known to people living now" (Preston 2002, 158). In this chapter, we present both; tipaachimuu fit readily within our chronological framework, whereas atiukan are more difficult to integrate. Instead, we incorporate them into our storyline on a thematic basis. As a result, it jumps back and forth between specific points in time and space and a more timeless, poorly defined spatial dimension, while progressing from the distant past toward the present.

6 A portion of the lake was initially excluded from the registration of the biodiversity reserve due to the presence of mining claims. In its negotiations, Wemindji insisted on the high significance of this area and on the importance of protecting an entire watershed; the area was subsequently included in the projected reserve. For more details, see Chapters 3 and 11, this volume.

7 Interview conducted by Véronique Bussières.

8 A shaptuan is an elongated *miichiwaahp* (teepee), a longhouse with an entrance at each end.

9 Through a subsidized program, a number of families are remunerated for fishing intensively during a certain number of weeks in the summer. The fish are then redistributed throughout the community (Dewan 2016).

WORKS CITED

Adelson, Naomi. 2005. "The Embodiment of Inequity: Health Disparities in Aboriginal Canada." *Canadian Journal of Public Health/Revue Canadienne de Santé Publique* 96 (suppl. 2): S45–S61.

Alfred, Taiaiake, and Jeff Corntassel. 2005. "Being Indigenous: Resurgences against Contemporary Colonialism." *Government and Opposition* 40 (4): 597–614.

Basso, Keith H. 1996. *Wisdom Sits in Places: Landscape and Language among the Western Apache.* Santa Fe: University of New Mexico Press.

Bussières, Véronique. 2005. "Towards a Culturally-Appropriate Locally-Managed Protected Area for the James Bay Cree Community of Wemindji, Northern Québec." Master's thesis, Concordia University.

Canadian Institutes of Health Research, Natural Sciences and Engineering Research Council of Canada, and Social Sciences and Humanities Research Council of Canada. 2010. "Tri-Council Policy Statement: Ethical Conduct for Research Involving Humans." http://www.pre.ethics.gc.ca/pdf/eng/tcps2/TCPS_2_FINAL_Web.pdf.

Casey, Edward S. 1996. "How to Get from Space to Place in a Fairly Short Stretch of Time: Phenomenological Prolegomena." In *Senses of Place,* ed. S. Feld and K.H. Basso, 13–46. Santa Fe, NM: School of American Research Press.

Cree Nation of Wemindji. 2009. *Wemindji Turns 50: A Community Where Tradition Lives On.* Milton, ON: Farrington Media.

Cruikshank, Julie. 2000. *Social Life of Stories: Narrative and Knowledge in the Yukon Territory.* Vancouver: UBC Press.

–. 2001. "Glaciers and Climate Change: Perspectives from Oral Tradition." *Arctic* 54 (4): 377–93.

–. 2005. *Do Glaciers Listen? Local Knowledge, Colonial Encounters, and Social Imagination.* Vancouver: UBC Press.

Denton, David. 2001. *A Visit in Time: Ancient Places, Archaeology and Stories from the Elders of Wemindji.* Nemaska, QC: Cree Regional Authority.

Dewan, Kanwal. 2016. "Towards an Improved Understanding of Community-Based Monitoring: A Case Study of the Wemindji Community Fisheries Program." Master's thesis, Concordia University.

Feit, Harvey A. 2004. "James Bay Crees' Life Projects and Politics: Histories of Place, Animal Partners and Enduring Relationships." In *In the Way of Development: Indigenous Peoples, Life Projects and Globalization,* ed. Mario Blaser, Harvey A. Feit, and Glenn McRae, 92–110. New York: Zed Books and the Canadian International Development Research Centre.

Francis, Daniel, and Toby E. Morantz. 1983. *Partners in Fur: A History of the Fur Trade in Eastern James Bay, 1600–1870.* Montreal and Kingston: McGill-Queens's University Press.

Freeman, Milton M.R. 1983. "George Weetaltuk (ca. 1862–1956)." *Arctic* 36 (2): 214–15.

Freeman, Minnie Aodla. 1978. *Life among the Qallunaat.* Edmonton: Hurtig.

Gupta, Akhil, and James Ferguson, eds. 1997. *Culture, Power, Place: Explorations in Critical Anthropology.* Durham: Duke University Press.

Morantz, Toby. 2002. *The White Man's Gonna Getcha: The Colonial Challenge to the*

Crees in Quebec. Montreal and Kingston: McGill-Queen's University Press.

Mulrennan, Monica E., and Véronique Bussières. 2018. "Social-Ecological Resilience in Indigenous Coastal Edge Contexts." *Ecology and Society* 23 (3): art. 18. https://doi.org/10.5751/ES-10341-230318.

Preston, Richard J. 2002. *Cree Narrative: Expressing the Personal Meaning of Events.* Montreal and Kingston: McGill-Queen's University Press.

Richmond, Chantelle. 2015. "The Relatedness of People, Land, and Health." In *Determinants of Indigenous Peoples' Health,* ed. Margo Greenwood, Sarah De Leeuw, Nicole Marie Lindsay, and Charlotte Reading, 3–17. Toronto: Canadian Scholars' Press.

Sayles, Jesse. 2008. "*Tapaiitam:* Human Modifications of the Coast as Adaptations to Environmental Change, Wemindji, Eastern James Bay." Master's thesis, Concordia University.

Scott, Colin H. 1992. "La rencontre avec les Blancs d'après les récits historiques et mythiques des Cris de la Baie James." *Recherches Amérindiennes au Québec* 22 (2–3): 47–62.

Syvänen, Andra L. 2011. "Wemindji Cree Observations and Interpretations of Climate Change: Documenting Vulnerability and Adaptability in the Sub-Arctic." Master's thesis, Concordia University.

Wilson, Kathleen. 2003. "Therapeutic Landscapes and First Nations Peoples: An Exploration of Culture, Health and Place." *Health and Place* 9 (2): 83–93.

PART 3

How to Protect

11

Wemindji Cree Relations with the Government of Quebec in Creating the Paakumshumwaau-Maatuskaau Biodiversity Reserve

JULIE SIMONE HÉBERT, FRANÇOIS BRASSARD, UGO LAPOINTE, and COLIN H. SCOTT

This chapter analyzes the experience and interactions of the Wemindji Cree community, university researchers, and Quebec government agencies in setting up a protected area. This process began in 2004, and though some of the context for protected area development has changed since then, much of that experience remains relevant and instructive.[1] Julie Hébert and François Brassard were project partners from the Québec provincial government, Ugo Lapointe worked for a non-governmental organization, and Colin Scott was an academic researcher. The protected area is made up of the Paakumshumwaashtikw (Rivière du Vieux Comptoir/Old Factory River) and Maatuskaau (Rivière du Peuplier/Poplar River) watersheds, comprising 4,392.5 square kilometres of Wemindji's traditional territory (WDPA 2017) (Figure 11.1). Our analysis highlights the obstacles overcome and the creative opportunities realized in achieving results, at both political and administrative levels.

QUEBEC GOVERNMENT AND COMMUNITY ROLES IN CREATING PROTECTED AREAS

The Quebec government understands its role to include the promotion and oversight of the exploitation of natural resources, with a view to ensuring the sustainable development of provincial territory. This perspective implies that the government is also responsible for the maintenance of biodiverse environments, notably through the creation of protected areas, where industrial activities are prohibited. In Quebec, the Ministère de

FIGURE 11.1

Proposed Paakumshumwaau-Maatuskaau Biodiversity Reserve

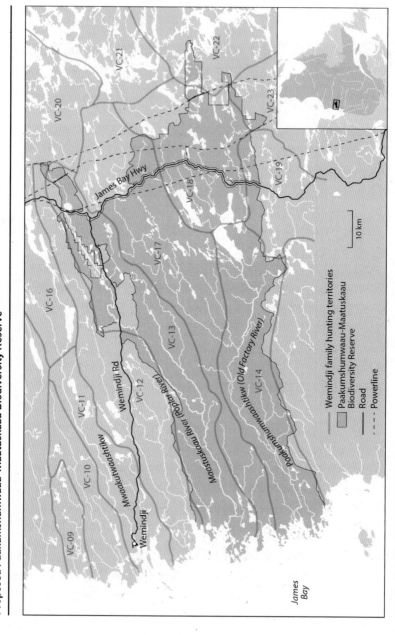

Source: Adapted from Christopher Wellen and Colin Scott (2006).

l'Énergie et des Ressources naturelles, which was the Ministère des Ressources naturelles et de la Faune (MRNF) in the years leading up to the creation of the Paakumshumwaau-Maatuskaau Biodiversity Reserve, assumes responsibility for the exploitation and management of natural resources. The Ministère de l'Environnement et de la Lutte contre les Changements Climatiques and the Ministère des Forêts, de la Faune et des Parcs are currently responsible for environmental protection and biodiversity conservation, to which end the creation and management of protected areas is a fundamental strategy. When the Paakumshumwaau-Maatuskaau Biodiversity Reserve was proposed, the Ministère du Développement durable, de l'Environnement et des Parcs (MDDEP) held this responsibility. In pursuing a strategy of protected area creation, the Quebec government must act in consultation with Indigenous and non-Indigenous communities. In the case of the Cree Nation of Eeyou Istchee, relations with the governments of Quebec and Canada are directed by nation-to-nation agreements, as well as by constitutional obligations in the field of Indigenous law.

In Quebec, identifying and prioritizing areas to be protected depends on several factors that include economic, social, and environmental aspects. Over the past decade or more, the participation and collaboration of Indigenous communities have greatly influenced decision making in this field, the Wemindji Protected Areas Project being a good example. The interests and needs of the communities, as well as the organizational, legal, and knowledge-based resources available to them, are thus key variables that government stakeholders should consider.

Indigenous communities are deeply implicated in matters of environmental management on their traditional lands and according to their own local practices (Berkes, Colding, and Folke 2000; Whiteman and Cooper 2000; see also Chapters 8 and 9, this volume). In Quebec, as elsewhere, their aspirations and needs have not always been well integrated with state-level decision making. Over the past three decades, however, several factors have generated significant improvements in relations between government and Indigenous communities.

In 1992, the Convention on Biological Diversity stressed the importance of the traditional knowledge of Indigenous peoples and recognized their role in the protection of the territory, as essential stakeholders whose rights and interests must be taken into account by state policy. Transnationally, Indigenous knowledge of natural environments has become a burgeoning field of research, involving a range of academic disciplines (see Chapter 4, this volume). The collaboration between researchers from McGill and Concordia Universities, the University of Manitoba, and members of the Wemindji Cree Nation is a case in point, where the intersection of local and scientific perspectives has targeted problems of environmental protection in terms adapted to community priorities and realities, both cultural and ecological.

It is important to note the considerable development of Aboriginal law during the last twenty-five years; this has provided new avenues for the participation of Indigenous communities in state-sponsored efforts directed toward the "conservation of nature." The Government of Quebec recognizes that because the rights claimed by Indigenous communities are affected by development projects requiring state authorization and by planning activities related to public lands (Groupe interministériel de soutien sur la consultation des Autochtones 2008), the creation of protected areas necessitates that the relevant government agencies systematically consult the communities concerned. Nowadays, in establishing protected areas, all communities participate from the outset – namely, from the first stage of selecting areas of interest for potential protection.

Protected area projects located on territory covered by the James Bay and Northern Québec Agreement (JBNQA) of 1975 are subject to an environmental and social impact assessment procedure that is defined by the agreement. Moreover, legal and constitutional obligations to consult Indigenous communities in general, including the many First Nations that are not subject to the JBNQA, and consequent governmental investment in new participation processes have shaped procedures both within and beyond the JBNQA territory. Generally speaking, over the last twenty years, community needs and interests related to the conservation of natural environments have achieved greater voice in the planning process. In some cases (such as Wemindji), communities themselves have initiated the creation of protected areas in Quebec (see Chapters 1 and 2, this volume).

Collaboration with the state in the establishment of protected areas makes certain demands on the resources of the community involved. Local knowledge and values, institutional structures, and perspectives on environmental protection and economic development can be decisive factors in planning and implementing protected areas. For communities to bring their organizational and cultural resources to bear on decisions and designs related to the conservation of nature, several challenges must be taken up, including those involving environmental expertise and leadership from within the communities and acquiring specific knowledge about the machinery of government.

A first challenge in advancing and negotiating community interests in contexts of inter-governmental (nation-to-nation) relations is for a community to communicate externally relevant aspects of its members' ecological knowledge. Leadership in the community must therefore develop approaches for putting Indigenous knowledge into dialogue with others. In doing so, leadership rallies and orchestrates the participation of community members around environmental issues that are of concern to them and simultaneously makes possible links to the state by presenting and defending the expectations of community members who use the territory.

The second challenge facing Indigenous communities in Quebec is to develop understanding of the federal and provincial government systems with which they must negotiate and collaborate in the management of their territory. The Crees have, of course, had several decades of experience in dealing with the Government of Quebec in matters of land and natural resources. Biodiversity conservation strategies, however, represent a relatively new policy arena that entails its own political dynamics and complexities because of its interface with a range of social interest groups. Indigenous communities have had to adapt to these dynamics under pressure because of the speed with which the Quebec government has invested in this arena, under pressure of its own to "catch up" in relation to international standards in the protected area field.

The decentralization of provincial processes and the presence of government representatives in this field are important factors that improve communication and allow communities to better understand the aims and responsibilities of the various government agencies. Nevertheless, several communities have had difficult relations with the provincial government.

This has slowed the process of creating protected areas in Quebec. In the case of the Paakumshumwaau-Maatuskaau project, Wemindji called on university partners as intermediaries in discussions with the Quebec government. University researchers complemented community understandings of provincial policies and structures, and identified their potential links to local realities.

The example of Wemindji affirms the possibility of productive integration between local cultural visions of conservation and the scientific processes and perspectives that contribute to the creation of protected areas (as reflected in the contributions to this volume). Nonetheless, this success cannot be generalized to all government initiatives in this field. Indeed, the space for such integration depends first and foremost on the character of relations between the government and each Indigenous community involved. These relations make it possible to establish a climate of trust, confidence, and respect that is essential for responding to community needs, especially on issues related to territory and resources. Relations with Indigenous communities are not always stable and positive. Poor relations can make the process difficult, particularly for communities whose territorial negotiations are unresolved or who do not possess developed leadership in the field of environmental politics.

INITIAL CONSULTATION PROCESS: NECESSARY CHOICES

The integration of Indigenous ecological knowledge is increasingly a condition for successful conservation measures. The creation of protected areas in Quebec is no exception here, and the Indigenous communities of affected territories have participated in this process in an engaged and occasionally avant-garde manner.

The objective of the initial consultation process is to define the area to be protected and to receive proposals from Indigenous community members who have ties to it. This step has become an established rule in initiating new protected areas in Quebec. Representatives from the ministry, in initiating an analysis of the ecological representativeness of a territory, first visit the relevant community or communities.[2] Consultation procedures are adapted to the affected community. In the case of the James Bay Crees, a presentation was first given by MDDEP representatives to community council members as well as to tallymen and other hunters with interests in the proposed protected area. This allowed everyone to have a good understanding of the agenda. Questions, comments, and needs specific to the community were discussed and documented at this stage. Next, each tallyman received a detailed map of his trapline (family hunting territory), on which he was asked to identify areas that could merit protection. These data were then compiled and geomatically superimposed on the results of the analysis of ecological representativeness. It is important to understand that a consultation is not a commitment that will always result in a protected area. Only a portion of the proposals coming from Indigenous communities will be adopted, which must be made explicit at the outset.

Clearly, the creation of a network of protected areas alone cannot meet all the conservation expectations of community members. Thus, it is important to establish a sustainable development strategy, which can protect the environment in the territory as a whole and respond to broader community expectations. Such a strategy has become central in negotiations between the Government of Quebec and the Grand Council of the Crees and its constituent communities in the elaboration and implementation of Plan Nord (Grand Council of the Crees 2011; Cree Nation Government 2015; CQFB 2017).

In most cases, tallymen understand the purpose of the exercise and can supply precise and pertinent reasons for protecting parts of their territories. Since Cree hunting, fishing, and trapping activities and associated land and resource management practices are considered compatible with protected areas, they are an attractive opportunity for fostering Cree autonomy in this domain. It is significant that the ban on exploiting mineral, hydro-electric, and forestry resources in the types of protected areas that Quebec is now creating will limit some industrial development that could otherwise have been of economic benefit to the Crees. The local and regional Cree economies rely in large measure on various industrial resource sectors for jobs, entrepreneurial activity, and revenue sharing. Economic and social development has become an important consideration in decision making about the future of families, communities, and the Cree Nation, including the option to forego development in particular areas in the interest of environmental protection (see the Introduction, this volume).

INTER-GOVERNMENTAL RELATIONS

The political representatives of the ten James Bay communities of the Grand Council of the Crees that lie within Quebec's boundaries are vital interlocutors for the implementation of Quebec's action plan on protected areas.[3] Although the consultation system is based on a community-by-community approach, the Grand Council is informed of the progress and of the main results of the consultations. It ensures that all the communities participate in the consultation sessions and may, when it deems it necessary, work on some files at the political level. Indeed, when a proposal from a community is supported by a majority but encounters roadblocks, the Grand Council can come to the assistance of the community.

The Grand Council is also consulted and informed by its representatives on the various committees, such as the James Bay Advisory Committee on the Environment, the Hunting, Fishing, and Trapping Coordinating Committee, and the Comité cri pour les aires protégées de la Baie-James (Cree Committee for James Bay Protected Areas, set up by MDDEP in 2007).

Turning to relations with the Municipalité de la Baie-James, the regional municipal government when Wemindji was advancing the Paakumshumwaau-Maatuskaau proposal,[4] we need to recall certain aspects of the history of the James Bay region to understand the views of its non-Indigenous residents regarding conservation. Although the region was traversed by English and a few French explorers and traders in the seventeenth and eighteenth centuries, it did not really attract the attention of governments and the public in Quebec until the early twentieth century. Indeed, the provincial government had no jurisdiction in the area prior to the passage of the Quebec Boundaries Extension Act of 1898, under which the territory north of the St. Lawrence drainage to the Eastmain River became part of the province. The Quebec Boundaries Extension Act of 1912 added the rest of northern Quebec, extending northward from the Eastmain River to the Hudson Strait.

In 1914, the Quebec government began research on the James Bay basin to evaluate its resources (see Chapter 4, this volume). This research was rewarding. The government found that the region possessed impressive natural resources, notably in energy, mineral, and forestry potential. Later, from 1954 to 1965, when Cree settlements were being constructed, five other towns were established on the southern periphery of the James Bay region. Originally, they were of the single-industry type, geared to industrial exploitation of resources. Since then, the mining, forestry, and hydroelectric resource sectors have attracted many people to the region and, in so doing, have increased the Jamesian population.[5] Today, the spatial organization of regional infrastructure – whether one considers the transportation system or the location of service centres – is explained by this connection between the settling of Jamesians and the exploitation of natural resources.

Jamesians are in a good position to grasp the importance of the region's various ecosystems. They understand the need to conserve elements of the natural environment, which provide them with jobs and from which they also benefit personally through the hunting, fishing, trapping, and vacationing activities practised by many residents. However, their interest in conservation is closely tied to their interest in the exploitation of resources, which ensures the region's economic survival. The concept of "resource region" has helped to develop their pride and identity as a cadre of pioneers and adventurers. However, they never depended on hunting and fishing for their survival. Thus, their connection with the environment differs from that of Indigenous peoples. In addition, when the Paakumshumwaau-Maatuskaau proposal was being developed, it was difficult to present protected areas as an economic diversification tool for the region since the work of MDDEP in northern Quebec was relatively recent. Were protected areas already proven to represent economic "value added" for the region, Jamesians would certainly be more receptive to their creation.

The Municipalité de la Baie-James represented the socio-economic interests of Jamesians. It was the main interlocutor with MDDEP concerning the Quebec action plan on protected areas and for consultations with non-Indigenous people in regional municipal boundaries. Its collaboration was necessary for MDDEP. From the inception of the action plan, the Municipalité de la Baie-James was informed of the progress of the work and was consulted on the creation of new protected areas.

The relationship between the Crees and the Jamesians is that of two groups who share the same territory but whose interests are in some ways divergent. No protected area project has been presented and developed through collaboration of the two groups. However, few projects have led to disagreements, even though the surface area of some projects has had to be modified to exclude zones implicated in projects that are incompatible with conservation objectives (campgrounds, roads, mining, hydro-electric, and forestry projects). In the case of the Paakumshumwaau-Maatuskaau project, the Municipalité de la Baie-James clearly expressed its reluctance to support protected area status on the eastern end of the proposed reserve, specifically along the Bay James Highway, arguing that municipal and tourism facilities and infrastructure were already in place or were to come. This explains the exclusion of a strip along both sides of the road, resulting from a compromise between the two parties.

THE TRAJECTORY OF NATURE CONSERVATION IN JAMES BAY

In 2000, the Quebec government set its sights on a first provincial strategy for protected areas. Its primary aim was to give Quebec a network of protected areas that was representative of the biodiversity of its 1.6 million square kilometres. More than half of this is covered by the JBNQA.

A first target – that 8 percent of the province would be devoted to protected areas – was set for 2005 and reached in 2009. In 2011, this was revised to 12 percent by 2015; as of March 31, 2017, 155,885 square kilometres were protected areas, amounting to 9.35 percent of the province's total surface area (MDDELCC 2017).

The delay in achieving these targets is explained in large part by the challenge of reserving zones for nature conservation in a province that is already heavily committed to the exploitation of natural resources. It reflects tensions in the two-fold mission of the Quebec state: that of developing natural resources and that, much more recently, of protecting natural heritage.

The two watersheds chosen by the Wemindji community to create a biodiversity reserve possess ecological and cultural characteristics that are compatible with the protected area concept. Since neither watershed has been affected by hydro-electric developments, both have a relatively high level of development "neutrality" in a regional context that is characterized by the intensive exploitation of hydrological capacity. Moreover, the ecosystems of the littoral area embraced by the Wemindji proposal were not yet represented in the network of protected areas of the Natural Region of the Grande Rivière Foothills. The proposed area was large enough to create a representative sample of these ecosystems. In addition, archaeological digs have revealed traces of continuous use by humans over the past five thousand years (see Chapter 5, this volume). Finally, Wemindji residents use this territory to carry out activities that are integral to their stewardship responsibilities and inter-generational knowledge transmission. Consequently, from an ecological and cultural standpoint, the essential conditions for creating a biodiversity reserve were in place. However, this territory also presented significant potential for the exploitation of natural resources.

In the early 2000s, indicators of diamond-bearing minerals (2002) and a major gold deposit (2004) were discovered in southern and eastern parts of Wemindji territory. The latter discovery stimulated mineral exploration in this entire portion of the James Bay region. The provincial government granted hundreds of mining exploration permits, including some on cultural sites of great value to Wemindji as well as for archaeological and historical heritage in general. The protection of these heritage sites was a central objective of the biodiversity reserve project.

The mining industry had become a major "stakeholder," and the uncertainty and speculation surrounding mineral resources seemed to remove all hope for the creation of the protected area.

EXPERIENCE OF THE WEMINDJI COMMUNITY

The Cree community of Wemindji contacted MDDEP in 2004, expressing its wish to create a biodiversity reserve on its traditional territory. The planned reserve, which was decreed on May 14, 2008, covers 4,392.5 square kilometres (WDPA 2017). From 2004 to 2008, major research, planning, and negotiation contributed to establishing a better balance between the exploitation of resources and the conservation of the natural heritage of Wemindji Cree territory.

Setting up the reserve was a process of discovery for all parties. Although it may appear straightforward and legitimate for a community to propose that a protected area be established on its traditional territory, the experience proved complex since Wemindji had to manage differing, and often unexpected, interests triggered by the initiative. The community was largely uninformed regarding the mining rights and interests in the territory, so it did not fully anticipate the associated legal and political problems that could arise. For their part, mining interests were not fully aware of the community's potential influence in encouraging or discouraging projects on its traditional territory.

Government procedures did not follow a pre-established path. The official presentation of the proposal was challenged, and some outcomes required intra-governmental and inter-governmental negotiations with organizations that represented other interests on the territory. In general, the community and the university partners who supported the proposal were well aware of mining exploration and sport hunting and fishing activities on the territory, in addition to the ongoing transformations produced by hydro-electric development. Proponents of the project believed that competing interests would not greatly oppose it. However, when they submitted the proposal for a biodiversity reserve, their information regarding mineral prospecting claims was a few months old. It indicated that few claims existed in the watersheds proposed for protection.

The steps involved in obtaining a mining claim are easy, rapid, and highly sensitive to recent mining success, as demonstrated by the example of the major gold deposit discovered at Éléonore in 2004, about forty kilometres east of Old Factory Lake (Chapter 3, this volume). Unbeknownst to Wemindji, in 2006, as it prepared to present its reserve proposal to MDDEP, a large number of claims were filed in crucial watersheds, encouraged by the success at Éléonore. They surrounded a large part of Old Factory Lake, which lay at the heart of the Wemindji proposal.

Mining proponents did not fully appreciate the legal and political problems that could arise if they encouraged mining activities and granted claims in a territory of high cultural, historical, and ecological importance, such as the Paakumshumwaau-Maatuskaau region. These watersheds are sites of key cultural activities, with traditional practices conducted on a regular basis; resource management remains anchored in the Cree system of family hunting territories and territory leaders. Clearly, the proponents of the biodiversity reserve and the mining interests were poorly informed regarding each other's goals and realities. Once each party perceived that the objectives of the other could jeopardize its own rights and interests, a dialogue rapidly ensued.

The JBNQA of 1975 provides for broad political and administrative autonomy for James Bay Cree communities and grants them exclusive hunting, fishing, and trapping rights on territories (Category I and II lands) covering approximately 150,000 square kilometres. It also provides for financial compensation over the short and medium terms when development occurs on those lands.[6] In return, the Quebec government obtained the right to develop the hydraulic, mineral, and forestry resources of northern Quebec insofar as the rights of the communities are guaranteed in agreement with the environmental and social assessment process (provincial and federal). The Paix des braves (Agreement Concerning a New Relationship between the Government of Québec and the Crees of Québec) (Quebec 2002) consolidated relations between the Cree Nation and the Quebec government.

The La Grande hydro-electric project that occurred under the JBNQA affects significant portions of Wemindji territory. The annual electricity production of its eight generating stations represented approximately 40 percent of the electricity consumed in Quebec in 2003. Wemindji has benefitted from certain economic and political spin-offs related to this industrial-scale exploitation. Nevertheless, the large portion of territory that was flooded in the 1970s and 1980s to create the reservoirs necessary for hydro-electric production left a legacy of cultural trauma that is still palpable today – more than forty years later.

DIALOGUE

Clearly, the scheme for a biodiversity reserve was potentially in direct competition with the interests of the mining companies. However, the fact that the reserve was proposed by a James Bay Cree community and would be on lands subject to the JBNQA demanded multilateral consultation. In addition, a major portion of the reserve was on Category II land, where any development project would come under particular scrutiny through consultations with Cree authorities. This is because Category II lands that are alienated for development must be replaced with lands of comparable ecological value from Category III, to compensate for traditionally harvested resources lost to the project.

A multi-party dialogue was thus established, involving the Grand Council of the Crees (which had provided resolutions of support for the reserve proposal), MDDEP and MRNF, and Wemindji – with its university partners – to discuss the future of this portion of the territory. Afterward, the dialogue was continued at Wemindji, with tallymen and the affected families, under the aegis of Rodney Mark, who was then chief of Wemindji. Representatives of the Cree Mineral Exploration Board joined the discussions,[7] as did university researchers from the Wemindji Protected Areas Project, who provided scientific support to the community (Chapter 3, this volume).

MRNF sketched the potential economic importance of mining spin-offs for Wemindji and estimated that the exploration activities on the proposed protected area would be of relatively low density. It warned against the hasty creation of a biodiversity reserve, noting that the knowledge of the area's mineral deposits was incomplete. To remedy this, MRNF suggested carrying out preliminary exploration work. For its part, MDDEP presented the recognized ecological and cultural value of the area for Wemindji, explained the protected area creation process, and examined the possible options for breaking the deadlock between exploitation and protection. The Cree Mineral Exploration Board outlined its mandate, its activities on the territory, and the importance of mining for the Crees. At the same time, it pointed out that it was not opposed to Wemindji's proposal for a biodiversity reserve. Finally, university researchers associated with the Wemindji Protected Areas Project presented the cultural, historical, and archaeological importance of the project (as reflected in contributions to this volume). These presentations were interspersed with numerous and supportive interventions by tallymen and members of Wemindji Cree families.

Chief Rodney Mark explained that the community used the two watersheds in question to ensure the continuity of traditional activities. The frequent use of the rivers and lakes, especially Old Factory Lake, helped to maintain links with the territory and land-based social institutions and cultural knowledge (see Chapter 10, this volume). No industrial project or financial compensation could ever replace this. That is why, after all the discussions, Mark reiterated his wish to create a biodiversity reserve comprising the two watersheds. His statement was supported by a consensus of community voices.

A few weeks later, MRNF imposed a moratorium on the acquisition of new mining claims on almost 80 percent of the proposed reserve. However, this did not apply to areas covered by claims that had already been granted.

The enterprises that held the claims were able to continue their mineral exploration. This situation alarmed the Wemindji community, particularly because some of the claims nearly encircled Old Factory Lake, whose ecological, cultural, and social features it greatly valued.

Wemindji continued to express its concerns to MRNF, to MDDEP, and to the companies holding the claims. It resolved to fight any mining development in the protected watersheds. Its ability to offer legal resistance was credible because, due to the scale of hydro-electric impacts elsewhere in Wemindji territory, it would have been impossible to replace Old Factory Lake – Category II lands – with an area of comparable ecological value, as required by the JBNQA. Finally, in the fall of 2009, the mining exploration company decided not to renew its claims. Old Factory Lake is now subject to a moratorium on the acquisition of new claims and has been included in the projected biodiversity reserve.

CONCLUSION

Over the past twenty years, the Quebec government has committed itself to building a network of protected areas, the result of which is that approximately 10 percent of the province's land area is currently under protection. Quebec has also adopted a strategy to ensure the future of the mining sector. This strategy banks on increased mining activities and the growth of spin-offs for Quebec and its regions. Further, Quebec adopted an energy strategy, based on a relaunching and acceleration of the development of hydro-electricity. At the heart of these strategies lies community participation in the orchestration of environmental use and protection. These development objectives are integral to Plan Nord, which explicitly commits to striking a balance between industrial and non-industrial activities on the northern territory.

The creation of the Paakumshumwaau-Maatuskaau Biodiversity Reserve illustrates the essential role that a community can play in the protection of natural and cultural heritage, to the benefit of society at large. Political and scientific structures and resources are required for striking such a balance between environmental, economic, and social needs. Without the participation of the Wemindji community, without its proactive political and research partnerships, without the guarantees and consultation processes required by the JBNQA and by Supreme Court rulings, such as those in *Haida Nation* and *Taku River Tlingit*, it is unlikely that the reserve would have been created. This highlights the importance for the Quebec government and for Indigenous communities to work on improving relations and on an enhanced understanding of their respective and mutual needs and realities.

NOTES

1 This chapter was drafted when Julie Hébert and François Brassard were working with the Ministère du Développement durable, de l'Environnement et des Parcs and addresses how the biodiversity reserve was developed in the years leading up to its formal acceptance in 2008.

2 MDDEP, with whom Wemindji originally conferred, is now divided between the Ministère de l'Énergie et des Ressources naturelles and the Ministère des Forêts, de la Faune et des Parcs.

3 An eleventh Eeyou Istchee Cree village is situated near the mouth of the Moose River in northeastern Ontario.

4 In 2012, the *Agreement on Governance in the Eeyou Istchee James Bay Territory between the Crees of Eeyou Istchee and the Gouvernement du Québec* (Quebec 2012) replaced the Municipalité de la Baie-James with new regional governance arrangements that assure Crees a greater voice.

5 Non-Crees, principally francophones who live in the James Bay municipality, refer to themselves as Jamesians.

6 It should be noted that the Crees also enjoy exclusive rights to harvest several wildlife species on Category III lands, which extend over the larger part of the territory not included in Category I and II lands.

7 The board is a joint Cree-Quebec committee whose general mandate consists of promoting and fostering the knowledge of the Crees and their participation in mining.

WORKS CITED

Berkes, Fikret, Johan Colding, and Carl Folke. 2000. "Rediscovery of Traditional Ecological Knowledge as Adaptive Management." *Ecological Applications* 10 (5): 1251–62.

CQFB (Cree-Québec Forestry Board). 2017. "2015–2020 Sustainable Development Action Plan." http://www.ccqf-cqfb.ca/wp-content/uploads/2017/03/2017-03-14-CQFB-2015-2020-SDAP.pdf.

Cree Nation Government. 2015. *Cree Regional Conservation Strategy.* https://www.eeyouconservation.com/wp-content/uploads/2018/02/cree-regional-conservation-strategy-e.pdf.

Grand Council of the Crees. 2011. "Cree Vision of Plan Nord." Nemaska, Grand Council of the Crees (Eeyou Istchee). https://www.worldcat.org/title/cree-vision-of-plan-nord/oclc/760992657.

Groupe interministériel de soutien sur la consultation des Autochtones. 2008. Guide intérimaire en matière de consultation des communautés autochtones. Gouvernement du Québec. http://numerique.banq.qc.ca/patrimoine/details/52327/ 1905549.

MDDELCC (Ministère du Développement durable, de l'Environnement et de la Lutte contre les changements climatique). 2017. "Protected Areas in Québec." http:// www.mddelcc.gouv.qc.ca/biodiversite/aires_protegees/aires_quebec-en.htm.

Quebec. 2012. *Agreement on Governance in the Eeyou Istchee James Bay Territory between the Crees of Eeyou Istchee and the Gouvernement du Québec.* Grand Council of the Crees/Gouvernement du Québec. https://www.cngov.ca/governance -structure/legislation/agreements/.

WDPA (World Database on Protected Areas). 2017. "Réserve de biodiversité projetée Paakumshumwaau-Maatuskaau in Canada." https://protectedplanet.net/reserve -de-biodiversite-projetee-paakumshumwaau-maatuskaau-proposed-biodiversity -reserve.

Wellen, Christopher, and Colin Scott. 2006. Proposed Paakumshumwaau-Maatuskaau Protected Area [map]. McGill University School of Environment.

Whiteman, Gail, and William H. Cooper. 2000. "Ecological Embeddedness." *Academy of Management Journal* 43 (6): 1265–82.

12

A Responsibility to Protect and Restore
Advancing the Tawich (Marine) Conservation Area

MONICA E. MULRENNAN and COLIN H. SCOTT

This chapter documents the experience of Wemindji Eeyouch in proposing the creation of the Tawich (Marine) Conservation Area (hereafter Tawich) along the central east coast of James Bay.[1] Larger than twenty thousand square kilometres, Tawich includes the waters, islands, rocks, and shoals of Wemindji's traditional territory, as well as parts of the territories of the neighbouring Eeyouch communities of Chisasibi, Eastmain, and potentially Waskaganish. The idea for the creation of Tawich came originally from the campaign to launch the Paakumshumwaau-Maatuskaau Biodiversity Reserve, itself the product of consultations between Wemindji elders and other community members, in which the importance of balance for good living was identified.[2] This notion of balance – between a salaried job and life in the bush, between development and environmental protection – gave rise to the idea that certain parts of the territory should be completely protected, whereas others could remain open to development (see the Introduction, this volume). A consensus quickly emerged that the Paakumshumwaau (including both Old Factory River and Old Factory Lake) and Maatuskaau (Poplar River) watersheds in the southern part of the territory would be protected, equivalent to 20 percent of Wemindji's total land territory. Early discussions were premised on the assumption that the protected area regime would extend into the offshore. Thus, protecting the watersheds of these two rivers and their estuaries would automatically protect the contiguous marine area (Figure 12.1). However, this turned out not to be the case. Due to complex jurisdictional arrangements at federal and provincial levels, protected area designation for the marine component was to be pursued separately from the land component (see Chapter 11, this volume). As this approach was not in line with Eeyouch understandings of the interface between land and sea, Wemindji Eeyouch turned their attention to the potential of a marine conserved territory to fulfill their land-sea stewardship responsibilities in a comprehensive and integrated fashion. The initiative snowballed into an ambitious proposal that extended over, and beyond, Wemindji's entire marine territory to include other coastal Cree communities and potentially Inuit communities to the north (Figure 12.1).

This chapter traces efforts to advance the Tawich project. These began in earnest in 2007, once the Paakumshumwaau-Maatuskaau Biodiversity Reserve was secured. In January 2009, the creation of the Tawich (Marine) Conservation Area was formally proposed to Parks Canada (Mulrennan, Bussières, and Scott 2009). The Wemindji Protected Areas Project provided research and community consultations in support of the proposal. This was a partnership involving members of the Wemindji First Nation and a team of academic researchers, with additional partners from various organizations at local, regional, provincial, and national levels, including the Wemindji Community Council, the Cree Trappers Association, the Cree Nation Government, the Grand Council of the Crees (Eeyou Istchee), the Ministère du Développement durable, de l'Environnement et des Parcs du Québec (MDDEP), Parks Canada, and la Société pour la nature et les parcs du Canada (SNAP Québec; Mulrennan, Mark, and Scott 2012).[3]

We begin by clarifying the context in which the regime of enhanced protection was proposed and by identifying some key constraints that hamper efforts to establish a marine conserved territory in James Bay. After discussing the opportunities associated with the Tawich initiative, we outline the steps taken to advance it and conclude with some of the lessons, achievements, and disappointments to date. It is worth noting that though we focus on the role of the Wemindji First Nation and partners in developing the Tawich proposal, the views and interests of other coastal Cree Nations must be considered if and when Tawich moves beyond the proposal stage.

Our analysis is intended to be instructive by providing a behind-the-scenes account of the experience of advancing a marine conservation

FIGURE 12.1

Proposed Paakumshumwaau-Maatuskaau Biodiversity Reserve and Tawich (Marine) Conservation Area

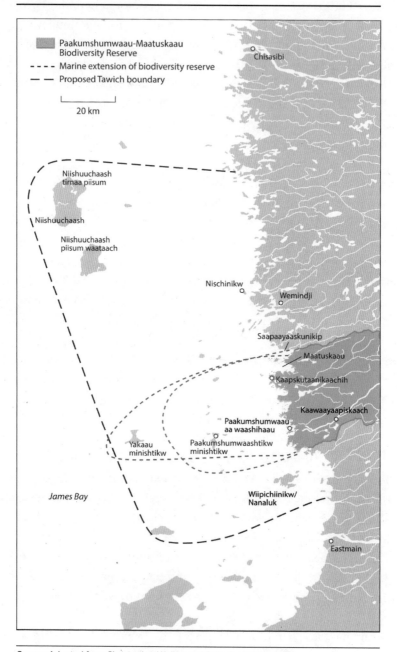

Source: Adapted from Christopher Wellen and Colin Scott (2007).

proposal. Of particular significance is that the vision, which is ambitious in size and innovative in intent, was initiated and has been driven by an Indigenous community, with support from neighbouring communities and the regional leadership. The case also speaks to the role of university and other partners in supporting a revised vision for marine protection and the possibilities that can emerge from the integration of varying and shared visions of environmental protection. The experience with Tawich is also a sober reminder of the extent of the challenge that lies before Indigenous communities in Canada if, even in contexts where opportunities and support clearly outweigh constraints and concerns, the creation of a marine conserved territory can take so long to achieve.

CONTEXT OF MARINE PROTECTED AREA CREATION IN JAMES BAY

The initial vision for the Wemindji Protected Areas Project, as articulated by *uchimaauch* (family hunting territory stewards, also known as tally-men), was for a marine conserved area centred on the Paakumshumwaau and Maatuskaau watersheds that would extend seamlessly beyond tide lines to incorporate the adjacent offshore. This integrated vision is consistent with Eeyouch ontologies that view land and sea as a continuum, as reflected in the customary tenure regime of Eeyou Istchee. The latter comprises multi-family hunting territories overseen by uchimaauch. Wemindji has twenty such territories, seven of which include coastal/marine components, extending inland as far as two hundred kilometres and seaward to encompass the inner islands (Mulrennan and Scott 2000).[4] This tenure arrangement supports the seasonal rotation of harvesting strategies across the land-sea interface – notably the hunting of geese and other wildfowl in the spring and fall (see Chapter 9, this volume), coastal fishing in the summer, berry picking in late summer and early fall (see Chapter 10, this volume), and inland hunting of fur bearing animals through the winter. It has contributed to the resilience of this social-ecological system over generations.

This integration of coastal land- and seascapes is affirmed in place names,[5] phenological indicators, and stories. For example, in the well-known story of Wolverine and Giant Skunk, the half-blinded Wolverine slowly makes his way through the shifting forest ecology to the shores of James Bay, where upon washing himself clean he releases some of his special power (and essence of Giant Skunk) into the sea, causing the waters of the bay to become salty (see also Chapters 6 and 10, this volume). The story connects land and sea as an ecological and cultural continuum, an understanding that underpins Eeyouch use, knowledge, and stewardship of the land-sea interface. Wemindji's decision to pursue a complementary rather than an integrated strategy for land-sea protection was taken reluctantly for political expedience but also because there were no other options. However, Eeyouch remain committed to maintaining the land-sea continuum as a defining feature of their customary tenure and land management practices.

Eeyouch have occupied and used this area for thousands of years (see Chapter 5, this volume). Their survival has depended on their ability to adapt, not only to the harsh winters, low biological productivity, and unpredictability of the area (see Chapter 7, this volume), but also to the reality of living on a highly dynamic shoreline associated with rapid coastal uplift (see Chapter 9, this volume). Crees have also had to contend with successive intrusions upon the broader region, from the early days of Hudson's Bay Company interest in furs during the 1630s to the massive James Bay hydro-electric project of the 1970s and 1980s (see Chapter 4, this volume). Crees have reproduced their institutions of knowledge and authority in territorial stewardship across this history, through a politics of resistance and negotiated accommodation that has induced Euro-Canadian companies and governments, often reluctantly, to engage them on their own terms (see Chapter 2, this volume).

The signing of the James Bay and Northern Québec Agreement (JBNQA) in 1975 established a regime of three categories that apply to lands above the high-water mark, along with a commitment from the federal government to resolve Cree rights and interests in areas below that mark in due course (Buchanan 1974). This meant that the terms and provisions of the JBNQA are limited to terrestrial zones.[6] Meanwhile, the offshore, until the ratification of an offshore agreement (Quebec 2010), was an area where Eeyouch held unextinguished Aboriginal title, although the hunting, fishing, and trapping regime established under the JBNQA applied to migratory birds and marine mammals (section 24.14), and the Cree Hunters and Trappers Income Security Program (section 30) included coastal and marine resource harvesting (Quebec 1975).

Ironically, Hudson's Bay Company contact and fur trade colonization were sustained by sea for nearly three centuries, so that major travel routes and trading centres of the James Bay region were focused on the coast. Yet, with the advent of railways, roads, and hydro-electric grids from the south, the coast became hinterland and backwater compared to southern inland areas that by the 1960s were already subject to various demands by Euro-Canadians for lands and resources. The relative inaccessibility of the James Bay coast, until recently at least, contributed to a sense of the offshore as beyond the reach of outsiders. Indeed, the James Bay coastal corridor now represents the only contiguous stretch of uninvaded Cree homeland and sea, extending nearly the length of the eastern Cree territory. Realizing that this situation may not last has been an impetus for some Crees to contemplate protected area status and to negotiate settlement of their rights and interests in the offshore.

At the same time, it has become increasingly clear to Wemindji Eeyouch and neighbouring coastal Cree communities that the ecological integrity of the James Bay coast and marine environment has been compromised in recent decades. The decline of eelgrass, possibly a result of the James Bay hydro-electric project, has serious implications for fish habitat quality and the availability of Canada geese along the coast (Scott and Mark forthcoming). Uchimaauch who have responsibilities for coastal areas see the creation of Tawich as an opportunity to protect but also as a chance to research and restore areas that are degraded or compromised, such as that adjacent to the La Grande estuary.

CONSTRAINTS ON MARINE PROTECTED AREA CREATION IN JAMES BAY

Preliminary conversations initiated in 2007 between members of the Wemindji Protected Areas Project, particularly with senior officials at Parks Canada, the Department of Fisheries and Oceans, and SNAP,[7] identified several constraints and challenges to the recognition of a marine conserved territory in James Bay. For example, the jurisdictional arrangements in the region were complex, the federal government had not prioritized James Bay for marine protection, many Eeyouch felt ambivalence and suspicion about parks and protected areas, and the public was unaware of the ecological and/or cultural significance of the region.

Complex Jurisdictional Arrangements

The extent and complexity of legal and jurisdictional interests in James Bay – unparalleled elsewhere in Canada – were not fully known to many Wemindji Eeyouch.[8] At the local level, there was general awareness that Cree claims to unextinguished Aboriginal title in the offshore conflicted with federal government jurisdictional assertions over the same area. A few people also knew of unfulfilled commitments made by the federal government during the negotiation of the JBNQA to resolve this impasse. Less widely known was that the Province of Quebec also asserted an interest in the James Bay intertidal zone, defining the shoreline in accordance with a "headland to headland" (or jaws of the land) rule, which Ottawa contested but was reluctant to confront. Furthermore, because James Bay had been designated a Special Management Area under the Nunavut Land Claim Agreement, Nunavut had administrative authority over offshore islands, including responsibility for wildlife regulations related to the Twin Islands Wildlife Sanctuary.

These divisions of externally imposed jurisdictional authority in the offshore did not connect with the realities of Eeyouch customary use and did not acknowledge the primacy of their rights as original occupants. Wemindji Eeyouch support for the the creation of a marine conserved territory therefore sprang from their commitment to retaining its economic and cultural value, jurisdictional challenges to de facto Cree control of the area, and a growing awareness of the potential for intrusive industrial development. A proposal to establish a protected area, whether in the context of unextinguished Aboriginal rights or in the settlement of those claims, raised a number of questions from the outset: What would a marine conserved territory founded upon customary management institutions look like? How could Cree hunting, fishing, and trapping be fully accommodated in the conserved territory? How might a marine conserved territory contribute to resolving current jurisdictional anomalies in the offshore? And what precedents were available to support the establishment of a marine conserved territory on Cree terms and with what level of legal protection? It was widely anticipated, at the regional Cree leadership level, that answers to these questions would be resolved with the settlement of the Eeyou Marine Region Land Claims Agreement (EMRLCA),[9] the negotiation of which was already well under way in 2007. In the meantime, consultations and supporting research continued, resulting in a formal proposal to Parks Canada in January 2009 (Mulrennan, Bussières, and Scott 2009). However, negotiation of Tawich was then suspended until an offshore agreement with settled provisions for marine protected area creation was in place.

Exclusion from National Planning Priorities

Early discussions with federal scientists and public servants involved in marine protected area creation mentioned the exclusion of the Hudson Bay and James Bay bioregion from the Large Ocean Management Areas (LOMAs) planning process.[10] This federal-level process identified areas that should receive special protection. Although this exclusion barely registered at the local level, federal partner organizations saw it as a significant constraint on the likelihood of federal co-operation. Ottawa's poor record in marine protected area development – less than 1 percent of Canada's marine territory was then protected, compared with 9 percent of its terrestrial area – was also identified as a constraint (Jessen, Morgan, and Bezaury-Creel 2016). This lack of political commitment translated into limited financial investment and limited expertise or experience in federal government agencies of partnering with Indigenous groups.

Legacy of Parks and Protected Areas

Despite their openness to employing a marine protected area as a potential strategy in enhancing the protection of James Bay, representatives from Wemindji's coastal hunting territories also expressed ambivalence. People told stories of the negative experiences of other Indigenous groups – that they were removed from or restricted in their use of areas once these were designated parks or wildlife areas (Tawich Community Workshop 2007). Others spoke of their own experience of co-management boards established under the JBNQA, pointing out that so-called power-sharing arrangements with external government agencies rarely supported Cree management perspectives (Tawich Community Workshop 2007). There were counter-narratives too and a strong awareness that parks and protected area agencies had been admonished for their treatment of Indigenous peoples, particularly by the influential International Union for Conservation of Nature (IUCN). On the strength of Aboriginal title, several arctic and subarctic Indigenous groups had employed their land claim agreements, some dating from the 1970s, to negotiate a series of large protected areas to safeguard their subsistence interests and stewardship roles. Nonetheless, there were few positive precedents in regard to marine protected areas, and Cree Eeyouch frequently voiced their concern about the potential for a protected areas regime to prevent or limit their ability to hunt, fish, and trap (see Chapter 1, this volume).

Public Perception of the Region

It became apparent that the exclusion of James Bay from the LOMA process was symptomatic of a larger problem of public perceptions. Much like subarctic environments in general, the James Bay region had not attracted the same level of environmental interest or concern from the public, media, or scientific community as more northern arctic latitudes (see Chapter 4, this volume). Indeed, for the Quebec and wider Canadian public, James Bay tended to evoke more associations with the massive hydro-electric project of the 1970s and the more recent Plan Nord (Quebec 2011, 2015) than with its rich, distinctive, and diverse ecological and cultural character. The significance of this was most clearly acknowledged by partners from SNAP,[11] who recognized the critical influence of public support for mobilizing political action on marine protected area creation.

OPPORTUNITIES FOR MARINE PROTECTED AREA CREATION IN JAMES BAY

Despite concerns and constraints, the notion of a marine conserved territory defined by and for Cree interests quickly gained traction among Wemindji Eeyouch. Several reasons, or opportunities, accounted for this. Eeyouch and their project partners agreed that the area had significant intrinsic natural values that warranted enhanced protection. The ecological decline in the coastal and marine environment spurred local leadership to protect the Tawich area. As a site of long-term Indigenous use and occupancy, it aligned Eeyouch aspirations for ecological and cultural heritage protection with emerging international discourses of biocultural diversity protection. Recent legal and political developments regarding the settlement of the EMRLCA could be harnessed to redefine jurisdictional arrangements in James Bay and generate favourable conditions for marine conserved territory creation. And finally, Wemindji Eeyouch had the social and political capacity, as evidenced by their success in establishing the Paakumshumwaau-Maatuskaau Biodiversity Reserve, to advance an ambitious agenda for protection of their marine territory. This capacity was reinforced by co-operation and policy alignment at regional, national, and international levels, and by the support of bridging organizations, especially those linked to the Wemindji Protected Areas Project.

Intrinsic Natural Values

The offshore islands and waters of Tawich support various species that are of particular interest from a conservation point of view. This diversity and distinctiveness can be explained in terms of the ecological edges that overlap in the area (Chapter 7, this volume). For example, terrestrial species typical of the tundra (such as polar bears) come into contact with boreal forest species (such as black bears), and the features of the more northerly and exposed islands are typical of arctic ecosystems, whereas more southerly and inshore locations are defined by subarctic conditions (Chapter 6, this volume). Similarly, several arctic species of birds co-occur with bird species that are at the northern limit of their range. The region also hosts both marine and freshwater fish, including anadromous species of whitefish, trout, and cisco (Mulrennan, Bussières, and Scott 2009). Marine and vegetative surveys conducted by members of the Wemindji Protected Areas Project in support of the Tawich proposal provide further evidence of the ecological significance of the area (Bussières et al. 2008; Scott et al. 2009).

The ecological and biological significance of Tawich is reflected in numerous official designations. For example, Twin Islands were declared a wildlife sanctuary in 1939 in recognition of their significance for polar bear breeding and denning. The contribution of James Bay as "a refuge or a reservoir" for eastern Hudson Bay beluga whales is also acknowledged (Gosselin et al. 2002, 11). The importance of numerous offshore islands for waterfowl is reflected in various designations, including an International Biological Program site, several Important Bird Area sites, and a key migratory terrestrial bird habitat. None of these designations bestow any form of legal protection, but they affirm the intrinsic natural values of the area.

James Bay is also one of nine marine regions in arctic Canada that Parks Canada acknowledges as meriting identification as a Natural Area of Canadian Significance (NAC) under the National Marine Parks Policy (Stewart and Lockhart 2005). However, since no single site of medium size was found to adequately represent the distinctive regional marine features of the area, four separate NACs were recommended for consideration, including one at Rivière du Castor estuary in eastern James Bay (Stewart, Dunbar, and Bernier 1993).[12] Tawich is not representative of all marine features in the region, but eastern James Bay offers a particularly rich diversity of coastal and marine habitats. Furthermore, the potential contribution of Tawich – as a single large area versus smaller, non-contiguous ones – is greater in terms of enhanced regional ecological connectivity, most notably for polar bears moving between Tawich, Polar Bear Provincial Park on the northwest shore of James Bay, and Wapusk National Park on the western shore of Hudson Bay.

Indigenous Use and Occupancy

Eeyouch have used and occupied the coastal and offshore portions of their traditional territory for thousands of years. Archaeological and palynological evidence (see Chapter 5, this volume) supports the long-term settlement history of this highly dynamic coast and the central importance of the James Bay offshore to Eeyouch survival and culture. Several dimensions of their past and ongoing connections – all primarily associated with resource harvesting and social interaction – inform and are integral to Eeyouch motivation and commitment to create Tawich.[13]

The creation of Tawich will protect and promote coastal archaeological sites and stories and will contribute to a more place-based understanding of the long-term human history of the area (see Chapter 5, this volume). It will also bring recognition to the historical significance of this coast during the fur trade era (late 1600s to the 1960s), particularly the hub of activities centred on coastal trading posts, as a record of Eeyouch experience of, and interactions during, that period. Indeed, a collective memory has prevailed among many Eeyouch, particularly older ones, of the coast as a focal point for seasonal social interactions. In the past, inland and coastal families gathered at coastal trading posts during the summer, following extended periods inland as isolated family groups (see Chapter 10, this volume). Today, this link persists in an annual gathering at Akwaanaasuukimikw, adjacent to the Old Factory post. Connections to the coast were further enriched by interactions with an Inuit community on Cape Hope Islands at the southern end of Tawich (Abbott 2013), some of whom still reside in Wemindji and Eastmain. Tawich is seen locally as a vehicle for protecting these connections but also as a way of celebrating and sustaining them.

The significance of the coast and offshore islands for hunting Canada geese and other waterfowl was also identified during our consultations with community members. The spring and fall goose hunt is the most important subsistence harvesting activity of coastal Eeyouch hunters, so its maintenance is vitally important for cultural and economic reasons. Concerns about its decline in recent years add further support for Tawich. Community members expressed similar anxieties about the availability and quality of fish in the bay, particularly given a formerly unified sense of the coast as a healthy and healing place, including a preference for fish caught on the coast. In this respect, Tawich is seen as an opportunity to protect valued resources but also to help restore them.

The coast, and its many associated activities, knowledge, and memories, contributes to the collective identity of coastal Eeyouch communities. Many Eeyouch believe that the coast and offshore present an opportunity to sustain a level of autonomy in their own affairs (including resource harvesting, stewardship responsibilities, and low-impact economic development) that is no longer possible on inland portions of the territory. This is due to various extractive resource developments and associated infrastructures, the increasing presence of non-Cree hunters on the land, and the influence of non-Cree interests in governance arrangements.

These considerations enrich and extend the purpose of marine protected areas from a focus on biodiversity conservation (which is often their sole or primary justification) to open up possibilities for cultural heritage protection and sustainable livelihood options. In doing so, Tawich demonstrates the importance and value of an integrated biocultural seascape approach.

Recent Political and Policy Developments

Interest in creating Tawich came at a time when Indigenous community-driven protected areas were uncommon in Canada. There were, however, indications that a change was coming (see Chapter 1, this volume). The 2003 IUCN endorsement of Indigenous governance in protected areas, along with formal recognition of Indigenous and Community Conserved Areas (ICCAs), opened up a discussion of designation options for Tawich. Some initial consideration was given to a uniquely Cree designation, which would build upon traditional and customary institutions of tenure, knowledge, and practice. However, the limited ability of such options to prohibit externally imposed development reduced their appeal.[14] At the same time, a willingness to explore state-sanctioned designations was buoyed by the recent success of Wemindji Eeyouch in creating the Paakumshumwaau-Maatuskaau Biodiversity Reserve. This was reinforced by the positive experience of Mistissini Eeyouch, with their newly created Albanel-Témiscamie-Otish Park.[15]

In early discussions, public servants at Parks Canada were enthusiastic about the potential fit of Tawich with the then seldom used National Marine Conservation Area (NMCA) program. Introduced in the mid-1990s, NMCAs are marine areas that are managed for sustainable use, with smaller areas dedicated to higher levels of protection. Large-scale resource extraction is prohibited, but hunting, fishing, and trapping by Indigenous people are permitted, as are low-impact forms of development. By 2008, only two NMCAs had been formally approved.[16] However, the Gwaii Haanas National Marine Conservation Area Reserve was well advanced and well received by the Haida.[17] There was also a sense, communicated by Parks Canada representatives, that Tawich furnished an opportunity to define the terms of NMCAs, particularly as they might apply in an Indigenous context.

A meeting in November 2008, organized by SNAP in collaboration with members of the Wemindji Protected Areas Project, brought together high-level officials from the Grand Council of the Crees, Parks Canada, Fisheries and Oceans, Environment Canada, and MDDEP to discuss Tawich. All parties expressed an interest in the project, along with a willingness to collaborate and move quickly. The meeting agreed that a Bilateral Working Group on MPAs was the best available mechanism to support Quebec and federal government co-operation. Creative solutions to jurisdictional complexities were also put forward, including the suggestion that a provincial aquatic reserve be established in the intertidal zone, thus bypassing the contentious matter of how to define Quebec's territorial limits.[18] Another suggestion, glossed as "a Swiss Cheese approach" to NMCA creation, was proposed in response to Eeyouch concerns that the NMCA would undermine soon-to-be affirmed Cree Aboriginal title over the islands. Instead, federal officials suggested that the Crees could retain title to the islands, whereas the NMCA would extend to adjacent waters and the seabed. In response, the Cree leadership signalled its openness to the possibility of a memorandum of understanding that would ensure consistency and co-ordination of management across the islands and waters.

The negotiation of federal-Cree relations in the offshore provided an important context for these early discussions. Negotiations of the EMRLCA (Quebec 2010) were launched at about the time that Tawich was first proposed but advanced quickly, benefitting from the experience and templates of two recently signed offshore agreements and a context unhindered by any large-scale development interests.[19] The EMRLCA was ratified in 2010, a year after Tawich was formally proposed to Parks Canada. It encompassed a marine region of approximately 61,270 square kilometres and a land area of 1,650 square kilometres, most of which would be islands under Cree title. The EMRLCA covered harvesting and associated rights, resource revenue sharing arrangements, and the creation of institutions of public government to co-manage wildlife, land-use planning, and development impact issues. A separate chapter of the agreement addressing provisions for marine protected area establishment prompted the regional Cree leadership to delay further discussions on Tawich until the EMRLCA, as the supporting framework for creating such an area, was formally approved. It was ratified in April 2010.[20] Legislation to implement the EMRLCA was signed into federal law on November 29, 2011, and came into force on February 15, 2012, with the expectation that the new regulatory regime would soon follow. Unfortunately, several years elapsed before the EMRLCA Commission and Boards were established and began their work. In the interim, support for a more extensive marine conserved territory increased, with other coastal Cree Nations confirming their interest in participating.

Social and Political Capacity

When asked to identify a particular challenge or frustration he had experienced with the Wemindji Protected Areas Project, Rodney Mark, then joint director (with Colin Scott) of the community-university research partnership, cited his difficulty in convincing Crees and non-Crees alike that the biodiversity reserve and Tawich were truly initiated and driven by Wemindji. Ironically, it is this community-driven dimension of the project that probably accounts, more than any other, for its success. This includes the strength and commitment of the local leadership, provided by the chief, the coastal uchimaauch, the community council, and the Cree Trappers Association, including their ability to present a coherent vision and strategy for balancing development and environmental protection. Community support for that vision was also essential and was reflected in the enthusiastic participation of residents in meetings, research activities, and various events associated with the project. The institutional support, and the quality of the physical infrastructure, at the community level was also significant, including contributions of time, energy, and ideas from the Cultural Department and the availability of the community centre and its resources for team activities. The latter also indicated the capacity and readiness of Wemindji to provide opportunities tied to cultural heritage protection and visitor experiences that might arise from Tawich.

The interest and support from the leaders of neighbouring communities have also been remarkable, including their positive response to invitations to meetings and conferences outside Eeyou Istchee. Similarly, the backing and strength of the regional Cree leadership has been essential, particularly for negotiations with Parks Canada on Tawich. In this respect, the development of the Cree Regional Conservation Strategy in 2015, which envisions various types of conservation areas and mechanisms to support Cree cultural heritage and way of life as well as to sustain biodiversity, reinforces the EMRLCA as an instrument for pursuing Tawich (Cree Nation Government 2015).

The Eeyouch leadership has also engaged in a larger national and international alliance of Indigenous peoples and has drawn on international discourses of human rights and environmental protection. In doing so, it has established relationships that have proved useful in the development of strategies, particularly those of resistance, against intrusions on their rights and interests. These include the Cree Nation Government's recent membership in the IUCN (Grand Council of the Crees 2014), as well as the Grand Council of the Crees' partnership with the Centre for Indigenous Conservation and Development Alternatives, a research team (led by Colin

Scott) that has strong continuity with the Wemindji Protected Areas Project.

DEVELOPMENT OF THE TAWICH (MARINE) CONSERVATION AREA

A commitment to protect at least some of Wemindji's marine territory was articulated from the earliest days of the Wemindji Protected Areas Project. However, consultations and supporting research for Tawich did not get under way until early 2007, when the establishment of the biodiversity reserve was confirmed. The latter followed a model of community-based participatory research, which was underpinned by a commitment to a community-defined research agenda, a collaborative research process, and meaningful research outcomes (see Mulrennan, Mark, and Scott 2012). The research involved extensive community consultations, a synthesis of previous studies on the James Bay marine environment, evaluation of various options for protected area designation, and a program of field research to address gaps between existing science/knowledge and the requirements for an NMCA proposal. Three summer field seasons were dedicated to survey work (2008–10), as well as ongoing research on the cultural and ecological significance of the area.

As mentioned earlier, jurisdictional divisions between Quebec and Ottawa required that marine protection be pursued in parallel with, rather than as part of, the protection of the Paakumshumwaau and Maatuskaau watersheds. Ironically, this constraint enabled a more ambitious proposal to emerge that would extend marine protection to an area of more than twenty thousand square kilometres. Consultations involved one-on-one conversations with tallymen, hunters, and elders, as well as formal and informal meetings with groups of hunters and residents, including a community workshop in Wemindji in July 2007. Informal consensus, later confirmed by formal resolutions at both the community and regional levels, was reached for a core protected zone. This included the estuaries between the La Grande and Eastmain Rivers; North and South Twin Islands, where Wemindji's interests overlapped with those of Chisasibi, its northern neighbour; the Cape Hope Islands, where Wemindji's interests overlapped with those of Eastmain to the south; and numerous other offshore islands that were important for migratory and nesting birds, polar bears, and a diversity of flora, including several threatened species. The active traditions of knowledge, use, and management that Cree coastal communities have sustained were also integral to the protection of these areas. Support for this extended area required collaboration between Wemindji, Eastmain, and Chisasibi because all three share interests in it. The possibility of enlarging the area has been discussed, particularly to the south to incorporate Charlton Island, with the support of the community of Waskaganish. Consensus was also achieved on the overarching management objectives of Tawich, which should

- Be framed by Cree knowledge, values, and practices for resource management and environmental protection, in dialogue and collaboration with relevant federal and Nunavut mandates and responsibilities
- Contribute to the enhanced protection and improved understanding of the natural habitats and associated resources in the area
- Support continued and enhanced local management embedded in customary institutions, particularly the hunting territory/tallyman system
- Provide Crees with a leadership role in the management, monitoring, surveillance, enforcement, and research activities associated with Tawich, with material and informational support from governments and university researcher partnerships
- Support the continuance of sustainable subsistence resource harvesting and assist in the creation of opportunities for locally managed low-impact economic development, such as ecotourism and cultural tourism
- Prohibit high-impact development activities that would damage ecosystem health and productivity, customary resource harvesting, and sites of historical importance
- Be flexible and adaptive, to facilitate an appropriate response to any technological, environmental, and social changes that may occur in the area
- Inform and foster public awareness about the significance of this area, both within and beyond James Bay

A formal fifty-page proposal document was submitted to Parks Canada in January 2009 (Mulrennan, Bussières, and Scott 2009), following a November 2008 meeting of high-level officials from the Grand Council of the Crees, Parks Canada, Fisheries and Oceans, and MDDEP. The proposal outlined the vision for protection and reviewed the environmental and ecological features of Tawich, as well as its cultural, social, and economic characteristics. The document drew on previous research, as well as research conducted by the Wemindji Protected Areas Project. It prompted written communications between Minister of Environment Jim Prentice and the grand chief of the Crees, who was initially Matthew Mukash (2005–09), followed by Matthew Coon Come (2009–17). Both parties consistently expressed a desire to advance the project. However, a series of delays ensued, occasioned by the 2009 change in grand chief, postponements in the ratification of the EMRLCA in 2009–10, Prentice's unexpected resignation in 2010, and the Harper government's lack of interest in both environmental protection and Indigenous-state relations.[21] Thus, the Tawich proposal made little progress for several years. Local commitment to the project did not wane, however, and research and knowledge building continued. In August 2016, following a clear mandate from the newly installed Trudeau government to set objectives and timelines for the creation of protected areas in northern seas, Parks Canada announced its interest in moving Tawich to the feasibility stage.

CONCLUSION

The proposed Tawich (Marine) Conservation Area has major cultural, economic, and historical significance to the Crees and is also internationally recognized for its high intrinsic ecological value. Local and regional Cree support for protection of Tawich is grounded in a growing awareness of the long-term and cumulative damage caused by industrial development, stemming from many parts of the James Bay/Hudson Bay watershed. It also builds upon earlier advances by the Cree Nation in establishing a larger network of terrestrial-based protected areas centred on Cree institutions of tenure and knowledge. Since the experience of Tawich may be instructive for other Indigenous groups, we close this chapter with some observations about Tawich.

First, Tawich is an example of an Indigenous-led marine conserved territory that employs a state-mandated form of protection rather than a uniquely Cree-informed ICCA (see Chapter 1, this volume). This reflects the limited uptake of ICCAs in Canada, as well as Cree preference, informed by decades of negotiations with the state, for a more robust legal framework – such as the EMRLCA provides – to define the terms of the partnership with the state. The certainty and confidence provided by an Aboriginal title agreement seemed more likely to support a level of innovation and "thinking outside the box" than a less secure negotiating context would provide.

A further explanation seems necessary for why Crees might want or trust a government-mandated protected area. In discussing Cree relations with outsiders, Harvey Feit (2004, 98, 107) refers to "aboriginal life-projects" and explains that due to their "tie to the land and all that has occurred on the land," Crees desire to engage with outsiders, including government and developers, and express "a willingness to respect the needs of others and an expectation that this will be reciprocated." For Cree people, this approach began with the earliest days of the fur trade. Their confidence in it was sorely tested during the early 1970s, in the initial stages of the massive James Bay hydro-electric project, which commenced without any consultation with the Cree people. The signing of the JBNQA in 1975 established the basis for a more constructive relationship, albeit one that did not come to fruition without successive demonstrations of Cree political acumen and determination (Scott 2008). The terms of a more consensual relationship were set forth in the Agreement concerning a New Relationship between Government of Quebec and the Crees of Quebec (Paix des braves) (Quebec 2002) and in other agreements with provincial and federal governments since.

In the context of protected areas, the intersection of Cree goals and agendas with those of provincial and federal governments presents a further opportunity for Crees to reinforce their role in territorial stewardship. Their willingness to engage with the state depends on their ability to significantly shape the terms of that engagement. Respectful relations are possible but must extend to a reciprocal respect for local customary systems, knowledge, and practices. Protected areas, which raise reasonable concerns as possible instruments of state intrusion, can also bring benefits on the right negotiated terms. The Tawich project, in tandem with adjacent terrestrial protected areas, presents the challenge and the opportunity of resolving complex trans-jurisdictional protection across the land-sea interface. It will fulfill Eeyouch conceptions of this environment as a continuous ecological and cultural region integral to Cree national territory.

Tawich challenges us to rethink the purpose of marine protected areas, a designation that is so often reserved for intact "pristine" locations. For southern Canadians, James Bay and Hudson Bay conform well to preconceptions of remote and pristine places, complete with charismatic fauna and exotic cultural histories, but the Cree people who live there know that care and attention are needed to remedy significant cumulative damage to the coastal and marine ecosystem. The Cree ethic of responsibility to protect all territory extends to places and processes that are deteriorating or degraded – indeed, as one uchimaau pointed out, responsibility to care for damaged areas is greater than for those that are thriving. If the Tawich (Marine) Conservation Area is implemented, it will be precious not as a "trophy" of unproblematic ecological integrity, but as a project of care sacred to the Cree Nation and – though many Canadians are unaware of the fact – of huge significance to us all. Tawich is the downriver recipient of potential influences from myriad development projects in the vast James Bay/Hudson Bay watershed, spanning portions of five provinces and two territories. It stands at the heart of a unique ecosystem, a vulnerable driver of bio-productivity and biodiversity for the continental and circumpolar North. And for all the ambivalent sentiment and relationship that this entails, it symbolizes and embodies a deep history of colonial relationships at the origin of Canada, going back well over three centuries.

NOTES

1 The Cree word *tawich* translates as marine, deep water, or, more specifically, the bay.
2 These consultations were led by then chief Rodney Mark, who spoke in 2000–01 with 128 members of the Wemindji community on a range of issues tied to Eeyouch values.
3 SNAP is the Quebec chapter of the Canadian Parks and Wilderness Society.
4 The outer islands, which are less accessible, are communally held.
5 For example, the place name Chiimaan aapimipiyit aashiipwaayaat translates as "where ships used to go through in old days," a reference to the effect of the land rising due to isostatic uplift.
6 There is no recognition in the JBNQA, or the more recent Eeyou Marine Region Land Claims Agreement (EMRLCA), that the Cree system of hunting territories extends from the land to the offshore.
7 Nelson Boisvert and Francine Mercier (Parks Canada Agency), Elaine Albert (Fisheries and Oceans), and Sylvain Archambault and Patrick Nadeau (SNAP) were particularly supportive during the early years (2007–10) of proposal development.

8   These arrangements were explained to and discussed with community members at various meetings and events (2007–10) organized by the Wemindji Protected Areas Project.

9   The EMRLCA is an agreement between the Crees of Eeyou Istchee and the Government of Canada concerning an offshore area of approximately 61,270 square kilometres in eastern James Bay and southern Hudson Bay (Government of Canada 2011). Under the EMRLCA, Canada owns the waters, tide lines, seabed, and some islands (200 square kilometres), whereas Crees retain exclusive ownership of the majority of islands (1,050 square kilometres) and joint ownership with Nunavik Inuit of some islands (400 square kilometres) in the overlap area in southern Hudson Bay. The EMR Wildlife Board (EMRWB) manages and regulates wildlife and harvesting. Crees have exclusive rights to harvest certain species and the right to harvest any wildlife species (subject only to conservation restrictions) to fulfill their economic, social, and cultural needs.

10 LOMAs were initiated by Fisheries and Oceans in 2005 as a collaborative ocean and coastal planning process.

11 As the Quebec chapter of the Canadian Parks and Wilderness Society (CPAWS), SNAP is an environmental non-governmental organization with a reputation for scientific expertise, particularly in marine conservation. SNAP/CPAWS brought expertise in communication and outreach to the Wemindji Protected Areas Project, as well as enhanced access to excellent networks in research and policy realms.

12 The other sites are Chickney Point (Ontario) in western James Bay, Richmond Gulf in the Hudson Bay Arc, and the Belcher Islands in southern Hudson Bay.

13 During a Marine Workshop held in Wemindji on July 28, 2007, one-on-one interviews with local Eeyouch identified some of the reasons for establishing Tawich.

14 A parallel situation had arisen with the decision to accept a biodiversity reserve rather than a Cree designation or ICCA in the southern part of Wemindji territory. The former prohibited large-scale resource extraction, whereas the latter did not have the legal teeth to do so.

15 The endorsement by Kathleen Wootton, deputy chief of Mistissini, of the potential of protected areas was an important development. Wootton was consulted because of her direct experience with the Albanel-Témiscamie-Otish Park project, one of several initiatives by the Crees and the Quebec government to support the development of parks, using a combination of social and ecological objectives. The latter included the maintenance and protection of the Cree system of hunting territories and tallymen.

16 Fathom Five Marine Park in Georgian Bay, Ontario, and Saguenay–St. Lawrence Marine Park in Quebec.

17 The Gwaii Haanas National Marine Conservation Area Reserve and Haida Heritage Site (3,500 square kilometres) in British Columbia was established in June 2010 under Canada's National Marine Conservation Areas Act. SNAP, as a partner of the Wemindji Protected Areas Project, connected Wemindji Eeyouch and the regional Cree leadership to the Haida, with whom CPAWS had worked closely in the establishment of Gwaii Haanas reserve.

18 Quebec has produced "administrative maps," which use headland-to-headland rules for defining limits. This is a principle of international law, but Canada contends that it does not apply to inland seas (Alan Penn, Memo to Colin Scott and Peter Brown concerning Old Factory Bay and the protected areas legislation in Quebec, April 14, 2003).

19 A preliminary mineral and energy resources assessment confirmed the limited potential for large-scale resource extraction in the offshore (McCourt, Larbi, and Stewart 2009).

20 The ratification process had to be delayed for several months due to community health concerns linked to the H1N1 flu virus.

21 Following the ratification of the EMRLCA, negotiations were initiated by SNAP with Makivik concerning possible collaboration between Crees and Inuit in creating a marine protected area where the territorial interests of the two groups overlapped. Makivik ultimately decided to pursue a marine protected area designation with Fisheries and Oceans, rather than an NMCA. These negotiations resulted in further delays.

Prentice's resignation on November 4, 2010, coincided with the Dare to Be Deep Canada Oceans tour of leaders and ceremonial dancers from the Haida Nation, organized by CPAWS. The tour included a scheduled stop in Montreal, where CPAWS, in collaboration with the Wemindji Protected Areas Project, arranged for the leadership of the Crees of Eeyou Istchee and the Haida Nation to formally meet and affirm their joint commitment to marine conservation. It was widely anticipated that Prentice would publicly announce the commitment of Parks Canada to Tawich at that meeting. Unfortunately, he announced his resignation at 4:00 p.m. and didn't make it to the event.

WORKS CITED

Abbott, Louise. 2013. *Nunaaluk: A Forgotten Story*. Documentary Video, 29 min, VOSTA.

Buchanan, Judd. 1974. Ministerial letter to Mr. Charlie Watt, President, Northern Québec Inuit Association, November 15.

Bussières, Véronique, Katherine Scott, Jessica Dolan, Henry Stewart, Monica Mulrennan, and Colin Scott. 2008. "Wemindji Marine and Island Surveys, 2008." Montreal, Wemindji Protected Area Project. http://wemindjiprotectedareapartner ship.weebly.com/uploads/1/3/9/6/13960624/bussieres_et_al_2008_marine__ island_survey.pdf.

Cree Nation Government. 2015. *Cree Regional Conservation Strategy*. https://www. eeyouconservation.com/wp-content/uploads/2018/02/cree-regional-conservation -strategy-e.pdf.

Feit, Harvey A. 2004. "James Bay Crees' Life Projects and Politics: Histories of Place, Animal Partners and Enduring Relationships." In *In the Way of Development: Indigenous Peoples, Life Projects and Globalization*, ed. Mario Blaser, Harvey A. Feit, and Glenn McRae, 92–110. London: Zed Books.

Gosselin, Jean-François, Véronique Lesage, Mike O. Hammill, and Hugo Bourdages. 2002. *Abundance Indices of Belugas in James Bay and Eastern Hudson Bay in Summer*

2001. Canadian Science Advisory Secretariat Research Document 042. Ottawa: Department of Fisheries and Oceans.

Government of Canada. 2011. "Backgrounder: The Eeyou Marine Region Land Claims Agreement." https://www.canada.ca/en/news/archive/2011/12/backgrounder -eeyou-marine-region-land-claims-agreement.html.

Grand Council of the Crees. 2014. "The International Union for the Conservation of Nature (IUCN) Welcomes the Cree Nation Government as a New Member of the Union." September 3. https://www.eeyouconservation.com/the-international -union-for-the-conservation-of-nature-iucn-welcomes-the-cree-nation -government-as-a-new-member-of-the-union/.

Jessen, Sabine, Lance Morgan, and Juan Bezaury-Creel. 2016. "Dare to Be Deep: SeaStates Report on North America's Marine Protected Areas." Ottawa, Canadian Parks and Wilderness Society and Marine Conservation Institute. http://cpaws. org/uploads/CPAWS-Oceans-Report-2016.pdf.

McCourt, George, Youcef Larbi, and Henry Stewart. 2009. "A Geologic and Geo-morphologic Survey of Coastal Islands of the Tawich National (Marine) Conservation Area, Eastern James Bay." Report prepared for Parks Canada, December.

Mulrennan, Monica E., Véronique Bussières, and Colin H. Scott. 2009. "Tawich (Marine) Conservation Area, Eastern James Bay." Proposal to the National Marine Conservation Program, Parks Canada, January.

Mulrennan, Monica E., Rodney Mark, and Colin H. Scott. 2012. "Revamping Community-Based Conservation through Participatory Research." *Canadian Geographer* 56 (2): 243–59.

Mulrennan, Monica E., and Colin H. Scott. 2000. "Mare Nullius: Indigenous Rights in Saltwater Environments." *Development and Change* 31 (3): 681–708.

Quebec. 1975. *The James Bay and Northern Quebec Agreement*. Quebec City: Editeur officiel du Québec.

–. 2002. *Agreement concerning a New Relationship between Le Gouvernement du Québec and the Crees of Québec – Paix des braves*. Quebec City: Editeur officiel du Québec.

–. 2010. *Agreement between the Crees of Eeyou Istchee and Her Majesty the Queen in Right of Canada concerning the Eeyou Marine Region*. Quebec City: Editeur officiel du Québec.

–. 2011. *Plan Nord: Faire le Nord ensemble: le chantier d'une generation/Plan Nord: Building Northern Québec Together; The Project of a Generation*. Quebec City: Gouvernement du Québec. http://numerique.banq.qc.ca/patrimoine/details/ 52327/2420757.

–. 2015. "Le Plan Nord, à l'horizon 2035: Plan d'action 2015–2020"/"The Plan Nord, toward 2035: 2015–2020 Action Plan." Quebec City, Secrétariat au Plan Nord. http://plannord.gouv.qc.ca/wp-content/uploads/2015/04/Synthese_PN_FR.pdf.

Scott, Colin H. 2008. "James Bay Cree." In *Handbook of North American Indians*. Vol. 2, *Indians in Contemporary Society*, ed. Garrick Bailey, 252–60. Washington, DC: Smithsonian Institution.

Scott, Colin H., and Rodney Mark. Forthcoming. "Responding to Environmental Decline in Eastern James Bay: Collaborative Approaches in Indigenous Knowledge and Trans-Disciplinary Research." In *Indigenous Stewardship of Environment and Alternative Development,* ed. C. Scott, E. Silva-Rivera, and K. Sinclair. Toronto: University of Toronto Press.

Scott, Katherine, Véronique Bussières, Sylvain Archambault, Wren Nasr, Jim Fyles, Karen Whitbeck, Henry Stewart, and Colin Scott. 2009. "Augmenting Information for a Proposed Tawich National Marine Conservation Area Feasibility Assessment, James Bay Marine Region: Cultural and Bio-Ecological Aspects." Ottawa, Parks Canada. http://wemindjiprotectedareapartnership.weebly.com/uploads/1/3/9/6/13960624/scott_et_al_2009_report_for_parks_canada.pdf.

Stewart, D.B., M.J. Dunbar, and L.M.J. Bernier. 1993. "Marine Natural Areas of Canadian Significance in the James Bay Marine Region." Report by Arctic Biological Consultants, Winnipeg, for Canadian Parks Service, Ottawa.

Stewart, D.B., and W.L. Lockhart. 2005. *An Overview of the Hudson Bay Marine Ecosystem.* Canadian Technical Report of Fisheries and Aquatic Sciences 2586. Winnipeg: Fisheries and Oceans Canada. https://www.researchgate.net/profile/D_Bruce_Stewart/publication/281536631_An_Overview_of_the_Hudson_Bay_Marine_Ecosystem/links/55ed0d3908aeb6516268cf4c.pdf.

Tawich Community Workshop. 2007. Workshop with community members on Tawich organized by Monica Mulrennan and Véronique Bussières, Wemindji Protected Areas Project and Concordia University, Wemindji, 28 July.

Wellen, Christopher and Colin Scott. 2007. Proposed Paakumshumwaau-Maatuskaau Protected Area [map]. McGill University School of Environment.

Conclusion

MONICA E. MULRENNAN, KATHERINE SCOTT,
and COLIN H. SCOTT

This concluding chapter is not the end of the story; it's early days yet for the Paakumshumwaau-Maatuskaau Biodiversity Reserve and the Tawich (Marine) Conservation Area proposal, although we started some years ago. But we take stock here and look to the future by reconsidering the partnerships, politics, and perspectives that shaped the Wemindji Protected Areas Project and inform the title of this volume. One benefit of time elapsed is the vantage point it can offer. Eighteen years have passed since the idea of this partnership came into being, and eight years since the main academic projects wrapped up, although related research continues to emerge and evolve, as the political process for completing the protected area initiatives continues. This closing chapter is a place to review the shifting mindscape and relationships that this book has brought into focus.

PARTNERSHIP

As we walked, talked, and paddled together in the early stages of this project, we began to consider the partnership we were building. Dorothy Stewart suggested we think about it in relation to a Cree word, *aa wiichaautuihkw*, which can be translated as "coming together to walk together." The word evokes the priority given to the relational aspects of our partnership. As we look back over our many years of working together, we recognize the significance of our partnership at several levels. The partnership was anchored by strong local leadership and a coherent vision for the future that had wide community support. It arose from the trust and respect that Rodney Mark and Colin Scott had already established for one another and from their agreement to work together to advance a collaborative research program in support of community-led protected areas creation. The research team that subsequently came together was composed of natural and social scientists who were committed to working with the community on terms that went beyond the standard requirements of most academic research projects (Mulrennan, Mark, and Scott 2012).

The partnership was also unprecedented in terms of its membership and almost certainly because it was initiated and led by an Indigenous community. Partners included representatives from Cree regional government bodies, federal and provincial government agencies responsible for protected areas, and an environmental non-governmental organization. The openness and willingness of government partners to acknowledge past failings, hear alternative views, and look for innovative ways forward were no less notable, especially given the constraints that affected them, including travel restrictions and budgetary limits imposed on Parks Canada staff during the nine years when the Harper Conservatives were in office. A turnover of government partner representatives was inevitable in that time, as indeed were changes in Cree leadership at both the community and regional levels. Graduate students too came and went, although a number of them sustained their involvement through subsequent doctoral studies. The addition of new people brought fresh ideas and energy, whereas the ongoing engagement of others ensured continuity. Along the way, we were welcomed, hosted, guided, and instructed by Wemindji elders and community members. We learned, for example, that though formal presentations seemed an efficient way to provide feedback to community partners, more time devoted to informal discussions, working in small groups, and outings on the land in the company of knowledge holders yielded the best results. Opportunities to reach the wider community included feasting, karaoke nights (thanks to Sam's Video and Coffee Shop), Summer Science Camp, and Traditional Skills Workshops.

Finally, the partnership flourished through its focus on protected area creation. The goal of protecting certain portions of Wemindji's land and sea territory provided a tangible shared project. With the value of hindsight, we now see these efforts as part of a movement that includes numerous other initiatives (occurring at various scales) to address how protected areas might be reimagined and remade to serve the rights, interests, and responsibilities of Indigenous peoples by enhancing their stewardship of their lands and waters. The report and recommendations of the Indigenous Circle of Experts (ICE 2018, 3) underscore how far this rethinking has come; they present Indigenous protected and conserved areas as "a more hopeful vision of the future – a future where Indigenous Peoples decide what conservation and protection means to them and to the lands and water and are given space to lead its implementation in their territories." This view of Indigenous protected and conserved areas as sites of reconciliation supported by partnerships in conservation is partially inspired by the vision and innovation of communities such as Wemindji. The building of trust and mutually beneficial relationships through partnerships such as the Wemindji Protected Areas Project affirm the possibilities for reconciliation in the context of conservation and environmental protection.

POLITICS

Our journey has required us to navigate the complex politics of protected area development through the turbulence of contested knowledges, Indigenous rights, and extractive resource development. Protected areas are sites of entanglement in a dark history of dispossession and marginalization of Indigenous peoples (Chapter 1, this volume). Despite shifting paradigms that are expanding the role of Indigenous (and local) communities in conservation and environmental protection, protected areas remain heavily implicated in the extension of state-authorized neo-liberal development agendas (Chapter 2, this volume).

One catalyst for the Wemindji Protected Areas Project was the threat of mining on two watersheds of particular importance to the community that had not been disrupted and degraded by the James Bay hydro-development project (Chapter 3, this volume). The creation of the Paakumshumwaau-Maatuskaau Biodiversity Reserve addressed this threat by providing a moratorium on large-scale resource extraction. There are other examples too, both within Eeyou Istchee (such as the Broadback Watershed Protected Area) and beyond, where protected areas are curtailing unwanted development and large-scale resource extraction. But they are not necessarily antithetical to development; indeed, they have become increasingly embedded in a "sustainable development" logic that seeks to legitimate resource extraction through the set-aside of protected areas.

In the case of Wemindji, the creation of protected areas was a core strategy in a two-pronged approach, centred on finding a balance between protection and development. The openness of the community to major development projects in certain parts of its territory while keeping other parts off-limits is heavily dependent on the ability of protected areas to restrict development. At the wider provincial level, protected area networks have become something of a handmaid to development. Under Plan Nord (Quebec 2011, 2015), a regional economic development strategy for northern Quebec, seemingly ambitious commitments to exclude large areas from industrial development were announced. The Cree Nations of Eeyou Istchee (2011) expressed fears, however, that protected areas could be relegated to the spaces leftover from large-scale resource extraction, an empty gesture aimed at increasing the acceptability of Plan Nord. In significant measure, the "Cree Vision of Plan Nord" (ibid.) declares an inverse prioritization, insisting that environmental protection must take precedence over development, especially in view of the cumulative broad-scale industrial impacts already affecting Eeyou Istchee. The Cree vision is nevertheless intimately entangled in neo-liberal development politics. For its part, the federal government seems intent on meeting its international commitments to biodiversity conservation, including Aichi Target 11, which requires that at least 17 percent of terrestrial areas and 10 percent of coastal and marine areas be set aside as protected areas (and other effective area-based means) by 2020. Although Quebec's engagement of the federal pathway to achieve this target is limited, its commitment to establish protected areas in 20 percent of the province's northern portion that is not targeted for development will probably secure its position as a leader in protected area creation. Fortunately, the recommendations of the Indigenous Circle of Experts point to a better vision – beyond the arbitrary targets and compromised politics of mainstream approaches to protected area development.

Consistent with the approach outlined by the Indigenous Circle of Experts, our Cree partners articulate an alternative vision of environmental protection, which centres on enhancing their authority and responsibilities for environmental stewardship. This is informed by what Rodney Mark describes as a Cree "concept of the world as a network of relations between all living things. Protecting the environment is about protecting the animals, a way of life, and all that nature gives" (page 6). Our challenge then was to develop a protected area agenda that fulfilled this vision. This required an approach that was grounded in Cree family hunting territories and the knowledge and authority of territory leaders *(nituuhuu uchimaauch)* as core institutions for decision making within and beyond Eeyou Istchee. This approach had the further value of leveraging the recognition and authority of these institutions under the James Bay and Northern Québec Agreement (JBNQA) and subsequent agreements, and reinforcing them in emerging discourses of protected area development.

Wemindji's openness to collaboration and partnership was in part a recognition that, despite its strengths, local stewardship alone would be ill-equipped to deal with external threats to protected areas. Cree partners acknowledged that a solely Cree-declared protected area was unlikely to find acceptance, due to the constraints of governance arrangements. But the willingness of Cree leaders to explore alternative options may also have a historical dimension (Alan Penn, pers. comm., October 25, 2017). The idea of a protected land base to safeguard the Cree hunting economy was of interest back in the 1970s, when the Cree leadership entered into talks with Quebec and Hydro-Québec. Those talks led eventually to the establishment of Category II lands under the JBNQA, which included provisions for Cree participation in management, exclusive rights to harvest certain species, and replacement of any lands used for development by lands of equivalent quality. Fortunately, Cree interest in collaborative protected areas creation seems to be matched by the willingness of state agencies to pursue this approach, on terms that recognize Cree stewardship responsibilities (see Chapters 11 and 12, this volume). Nevertheless, an ongoing challenge will be to ensure that these protected areas are not subsumed by the priorities of the state.

We are also alert to the "pitfalls and possibilities in the dance of hegemonic and counter-hegemonic discourses that make up the politics of TEK" (page 76). Again, our insistence on the engagement and social reproduction of Cree knowledge in the institutional and practical contexts of the hunting territories has helped to resist "appropriation of TEK by external authority and to negotiate more equitable conditions for knowledge exchange in collaborative decision making" (page 77). In other words, by foregrounding hunters and their relationships to the land, together with their knowledge and practices, we have sought to convey a sense of the ontological significance, ethical orientation, and political vitality of their relationships in Cree projects of self-determined cultural continuity. At the same time, we came to appreciate how Cree responsibilities of environmental stewardship can inform, enrich, and expand conventional understandings of conservation and environmental protection.

PERSPECTIVES

More than a hundred years have passed since researchers first journeyed to the James Bay area to collect scientific knowledge for their governments and universities back home. They depended on their Cree guides for logistical support and, from what we can tell, often developed positive relationships with community members but rarely acknowledged their significant contributions in published works (Chapter 4, this volume). We can only imagine how those interactions changed the participants and shifted their views. Certainly, the journey we shared through this project has altered and enlarged our perspectives, particularly our ideas about the possibilities of collaborative research, the politics of knowledge and environmental conservation, and the potential of protected areas.

One of the richest gold mines in eastern Canada is located on Wemindji territory, with the consent of the Wemindji leadership. Support for that project was contingent on enhancing measures to ensure that lands and waters would be available to sustain Cree traditional ways of life. Protected area creation offered one potential strategy to achieve this balance by enhancing the ability of Wemindji people to protect certain parts of their territory while leaving others open to development. Our partnership was born out of an invitation to explore the potential of this strategy, as reflected in the division of this volume into three parts – "context," "what to protect," and "how to protect."

Our research partnership has supported the intersection of Cree knowledge with multiple disciplinary perspectives at a range of temporal and spatial scales. This dialogue articulated a broad understanding of "what to protect," which brought natural science into conversation with local knowledge, in Chapters 5 to 7. Here, the assessments of local experts and biologists align in our exploration of priorities for protection, revealing that "a real and pragmatic intersection of conservation interests emanates from the distinct ontological and institutional standpoints of hunters and scientists, which we have sought to advance in negotiations with responsible state authorities" (page 75). The importance of social, cultural, and ecological connections and adaptations, and the agency of Cree hunters expressed in that conversation, is considered in Chapters 8 to 10, to give a fuller appreciation of the complex interconnections that shape this social-ecological system.

The final section on "how to protect" brings attention to the challenges and possibilities that protected areas offer as strategies for environmental protection (see Chapters 11 and 12). Here, the separation of terrestrial from marine environments stands in stark contrast to the integrated approach that Crees take in caring for territory. Cree responsibility, we have learned, extends to all parts of the territory, including those beyond the biodiversity reserve (and outside Category I and II lands) and into places that have been compromised by development. Indeed, local interest in the capacity of protected areas to assist environmental restoration, though at odds with Western conventions of privileging and preserving more pristine areas, is entirely consistent with Cree responsibilities of stewardship. Cree elders insist that their obligations to restore damaged parts of the territory, much like a parent's responsibility to a sick child, cannot be neglected. Our Cree partners have long known that a world in which all lives are interwoven is one where all life is respected. As Rodney Mark tells us, "Protecting the environment is about protecting the animals, a way of life, and all that nature gives." This centring of relationship and reciprocity with all beings has changed in fundamental ways our perspectives on the possibilities of protected areas.

THE RED CANOE

Wemindji elders have taught us that stories can provide instructive guidance. In keeping with this tradition, we end with a story about a red canoe that captures some of the excitement and anticipation around the creation of the Paakumshumwaau-Maatuskaau Biodiversity Reserve and proposed

Tawich (Marine) Conservation Area, as well as some of the anxieties and uncertainties.

Five years ago, a single red canoe with two paddlers was spotted on Old Factory Lake. People in the community wondered who might be in the canoe and whether the paddlers knew how to take care on the river. A day later, the red canoe was seen passing the place where the youth expedition group, on its annual trip, was camped. Everyone saw them and hailed them, waving their arms. Strangely, they did not respond. They kept on going, leaving the expedition leaders feeling puzzled and a bit affronted. Before long, radio calls reached the annual gathering, which was under way at the river's mouth on the coast, informing it about the strange red canoe and its unsociable occupants. Why didn't they stop? What were they hiding? Who doesn't stop at least to say hello and trade a story or two? It was entirely suspicious behaviour. And the question eventually arose: Was the protected area designation bringing in people who didn't know how to behave as guests?

When the red canoe reached the coast a day or two later, people gathered on the beach. The paddlers were tanned, outdoorsy white men, who looked very pleased with themselves. They came ashore, shook hands with everyone, and then, prompted by a greeter, sought out George Stewart as the person they should speak to. They shook his hand, said that they'd been planning a trip to James Bay for a couple of years, and explained why they had not stopped along the way; it seems they didn't have time and didn't want to intrude on a big group trip. They had come down the river in three days and wanted to get back home as soon as possible. When the youth expedition arrived a few days later, its members were full of questions about the red canoe.

Wemindji Crees – like a growing number of Indigenous groups across Canada and internationally – are using protected areas as an invitation to outsiders to enter into a new relationship with them and their lands and waters. These are not projects meant to establish anonymous "wilderness" spaces for curious and adventuresome individuals; rather, such projects welcome strangers into existing communities of life, on terms of respect and relationship. Protected areas have the potential to be places of reconciliation if newcomers understand their roles as guests, friends, and allies in networks of care for the living land.

WORKS CITED

Cree Nations of Eeyou Istchee. 2011. "Cree Vision of Plan Nord." Nemaska.

ICE (Indigenous Circle of Experts). 2018. "We Rise Together: Achieving Pathway to Canada Target 1 through the Creation of Indigenous Protected and Conserved Areas in the Spirit and Practice of Reconciliation." Indigenous Circle of Experts' Report and Recommendations, March.

Mulrennan, Monica E., Rodney Mark, and Colin H. Scott. 2012. "Revamping Community-Based Conservation through Participatory Research." *Canadian Geographer* 56 (2): 243–59.

Quebec. 2011. *Plan Nord: Faire le Nord ensemble: le chantier d'une generation/Plan Nord: Building Northern Québec Together; The Project of a Generation.* Quebec City: Gouvernement du Québec. http://numerique.banq.qc.ca/patrimoine/details/52327/2420757.

–. 2015. "Le Plan Nord, à l'horizon 2035: Plan d'action 2015–2020"/"The Plan Nord, toward 2035: 2015–2020 Action Plan." Quebec City, Secrétariat au Plan Nord. http://plannord.gouv.qc.ca/wp-content/uploads/2015/04/Synthese_PN_FR.pdf.

Contributors

Fikret Berkes is a distinguished professor emeritus at the Natural Resources Institute, University of Manitoba. His earliest work with the James Bay Cree goes back to the 1970s. He holds a two-term Tier 1 Canada Research Chair in Community-based Resource Management. His research bridges social sciences and natural sciences, and deals with commons theory, social-ecological resilience, and traditional ecological knowledge. He has numerous publications, including three major books (*Navigating Social-Ecological Systems,* 2003; *Linking Social and Ecological Systems,* 1998; and *Sacred Ecology,* 4th ed., 2018). His honours and awards include the International Union for Conservation of Nature CEESP Inaugural Award for Meritorious Research, the Elinor Ostrom Award for Senior Scholar, and the Ecological Society of America Sustainability Science Award for the book *Sacred Ecology.*

Jennifer Bracewell is an archaeologist and PhD candidate in the Department of Anthropology at McGill University. She has worked on community-oriented archaeology projects across Quebec, in Nunavik, James Bay, and in the Montreal area. Her PhD research focuses on post-sedentary hunter-gatherers in prehistoric northern Finland.

François Brassard is a forestry engineer and sustainable development adviser to the Ministère des Transports du Québec. He co-ordinated the development of multi-purpose protected areas in Quebec from 2011 to

2017 and contributed to the Quebec government's commitment to allocate 30 percent of the territory of Plan Nord to biodiversity conservation and adapted development. He holds a master's in forest science and geography of Southeast Asia from Laval University.

Véronique Bussières is currently a PhD candidate in the Department of Geography, Planning and Environment at Concordia University. She holds a bachelor's from McGill University and a master's in public policy and administration from Concordia University. She has worked in collaboration with the Cree Nation of Wemindji since 2003. Her research focuses on conservation, governance, and Indigenous stewardship, particularly at the land-sea interface. She has been actively involved in the development of the Tawich proposal.

Gail Chmura is an associate professor in the Department of Geography at McGill University. She uses paleo-ecological tools to examine dynamics and change of wetlands, particularly tidal salt marshes. Her research also documents the ecological services (such as carbon storage and global biodiversity) of salt marshes and how these can be regained through restoration of tidal flow to drained marshes.

Andre Costopoulos is an archaeologist who is interested in human adaptation to environmental change. Since July 2016, he has been vice-provost and dean of students at the University of Alberta, prior to which he served as dean of students and associate professor in the Department of Anthropology at McGill University. He conducts field research in Finland, in the James Bay area, and in southern Quebec.

James W. Fyles holds the Tomlinson Chair in Forest Ecology in the Department of Natural Resource Sciences, McGill University. His research revolves around understanding the interrelationships between vegetation and soils in the context of natural disturbance, human activity, and changing climate. Involvement in the Wemindji project fuelled his interest in how we see, understand, and explain the complex systems in which we are embedded.

Julie Simone Hébert is community liaison officer with the Quebec government's Société du Plan Nord (Territorial and Governmental Relations), based in Chibougamau. She previously served as biologist and wildlife habitat co-ordinator for the Ministère des Forêts, de la Faune et des Parcs, and before that she worked in protected area development with the Ministère du Développement Durable, de l'Environnement et des Parcs. She holds a master's in forest science from Laval University, as part of the Waswanipi Cree Model Forest initiative.

Eva Hulse is a geoarchaeologist with Archaeological Investigations Northwest, in Portland, Oregon. She has researched past human-environment interactions in James Bay, Finland, the Russian Far East, and the Pacific Northwest of the continental United States.

Murray M. Humphries is an associate professor of wildlife biology in the Department of Natural Resource Sciences at McGill University. He holds the McGill Chair in Northern Research and is the academic director of McGill's Centre for Indigenous Peoples' Nutrition and Environment. He leads the McGill–Alberta NSERC CREATE graduate training program, Environmental Innovation, focused on impact assessment, monitoring, and management in northern Canada. His research concentrates on wildlife and environmental contributions to the traditional food systems of Indigenous peoples. This has led him to studies of participatory approaches in natural sciences research, the nature of community-university research partnerships, and documentation of the food knowledge of northern Indigenous peoples.

Grant R. Ingram was a professor in the Department of Atmospheric and Oceanic Sciences at McGill University and subsequently the founding principal of St. John's College and the principal of the College for Interdisciplinary Studies, both at the University of British Columbia. He was an expert on the effects of physical processes on oceanic biological production. He contributed to the early stages of the Wemindji Protected Areas Project but passed away before seeing its completion.

Dustin Keeler is archaeological resources manager at Sapphos Environmental, in Pasadena, California. His research interests include maritime adapted hunter-gatherers, circumpolar archaeology, and GIS intrasite spatial analysis.

Ugo Lapointe works as the Canada program co-ordinator for Mining Watch Canada. He has a bachelor's in geological engineering from Queen's University, and at the time of writing his chapter, he was a master's candidate at the Institut des sciences de l'environnement, Université du Québec à Montréal. Ugo has over twenty years of diverse experiences in the Canadian mining sector. He has worked as a consultant on mining issues for non-governmental organizations, government agencies, research centres, and communities. He co-founded the Coalition Québec Meilleure Mine in 2008, a not-for-profit organization promoting improved social, environmental, and economic practices in the mining industry.

Rodney Mark is the director of the Social and Cultural Development Department of the Cree Nation Government. He was deputy grand chief of the Grand Council of the Crees (Eeyou Istchee) from 2013 to 2017, with responsibility for mining, justice, and protected area creation. Rodney is a proud member of the Wemindji Cree community, where he served as chief (2004–13), deputy chief (1999–2004), and youth chief (1993–96). He was co-director of the Wemindji Protected Areas Project and a key player in negotiations related to Goldcorp's Éléonore Mine. He is dedicated to the development of sustainable local economies, the protection of Cree lands and waters, and the promotion of Cree culture and language.

Gregory M. Mikkelson is an associate professor at the McGill School of Environment and Department of Philosophy. He studies the causes, and the intrinsic and instrumental value, of diversity and equality in nature and society.

Heather E. Milligan is a biologist with the Fish and Wildlife Branch, Department of Environment, Government of Yukon. She completed a master's in wildlife biology in the Department of Natural Resource Sciences at McGill University in 2008. Her current research focuses on invasive species in Yukon.

Monica E. Mulrennan is an associate professor in the Department of Geography, Planning and Environment, and associate vice-president of research (development and outreach) at Concordia University, Montreal. She has worked with the James Bay Cree since the mid-1990s and was a lead researcher on the Tawich (Marine) Conservation Area proposal. Her research addresses Indigenous-led strategies of conservation and environmental stewardship that draw on Indigenous institutions of knowledge and practice. She is particularly interested in local adaptations to environmental change, protected area creation, and local knowledge and use of coastal and marine territories. She also sustains a research partnership with Indigenous islanders in the Torres Strait, northern Australia, and more recently with seaweed collectors along the Atlantic coast of Ireland.

Wren Toombs Nasr is an organic farmer, research consultant, and PhD candidate at McGill University. His doctoral research with Cree hunters and administrators in Wemindji looks at how people are caring for, protecting, and sharing their traditional hunting territories through nested ideas and practices mediating the inclusion and exclusion of various land uses and actors. More broadly, his research interests, which are located at the interface of physical, cultural, and emotional landscapes, include the history and future of Indigenous-settler relationships, regenerative farming, and hunting/gathering in the Anthropocene.

Jari Okkonen is a university lecturer in the Department of Archaeology at the University of Oulu, Finland. He is interested in the archaeology of northern areas from various periods, as well as in issues related to the collection of archaeological information and the generation of data.

Claude Péloquin is a geographer who specializes in the politics of environmental management. He holds a bachelor's from McGill University, a master's in natural resource management from the University of Manitoba, and a PhD in geography from the University of Arizona. He has taught at the University of Arizona, Macalester College, and Concordia University, and has worked as a researcher and policy adviser for the James Bay Advisory Committee on the Environment, the Cree Nation Government, and the Government of Canada. His research, teaching, and policy work examine how institutions decide what to do about environmental things that evade certainty and predictability, and how these decisions relate to economic development strategies.

Florin Pendea is an associate professor in the Departments of Sustainability Sciences and Geography at Lakehead University, Thunder Bay. His research focuses on the human dimensions of environmental change and uses the North's high-resolution paleoenvironmental archives, such as peatlands and coastal sediments, to reconstruct the rate and magnitude of environmental change at centennial and millennial scales.

Jason Samson is a biologist at the Ministère des Forêts, de la Faune et des Parcs in Quebec. He holds a bachelor's in biology from Université Laval, a master's from the University of Guelph, and a PhD from McGill University. He was a post-doctoral researcher at the Université du Québec à Montréal. His current research involves endangered species and climate change.

Jesse S. Sayles is currently a postdoctoral researcher. He holds a bachelor's from McGill University, a master's in geography, planning, and environment from Concordia University, and a PhD in geography from Arizona State University. He conducts research on a broad range of environment and sustainability issues, including Indigenous resource management, landscape governance, social-ecological fit, and capacity building for activities such as environmental restoration in multi-level governance settings. He often uses network science concepts and tools, and is involved in their development to advance human-environment science and practice. His focus is on coastal and watershed systems.

Colin H. Scott is a McGill University anthropologist. Through an engaged anthropology, his research focuses on Indigenous ecological knowledge, relational ontologies, land and sea tenure, conservation governance, and the political and legal process of Indigenous rights among hunting and fishing peoples in northern Canada and northern Australia. He directs the Centre for Indigenous Conservation and Development Alternatives, with Indigenous and academic partners on five continents. He is also director of the centre's Indigenous Stewardship of Environment and Alternative Development research program, with Indigenous and academic partners throughout the Americas. Both involve university researchers, Indigenous groups, and non-governmental organizations in relations and processes of knowledge co-production aimed at improving conditions for the cultural and ecological diversity and integrity of Indigenous territories and associated land- and sea-based livelihoods.

Katherine Scott is a research consultant and a PhD candidate in anthropology at McGill University. From 2003 to 2012, she was the project administrator for the Wemindji Protected Areas Project. Her research focuses on traditional knowledge and language revitalization, Indigenous museums, and the planning and creation of an Eeyou Knowledge Center in Wemindji. She has been based in Wemindji since 2014, working as heritage research co-ordinator for the Cree Nation of Wemindji's Culture and Wellness Department.

Dorothy Stewart is a member of the Cree Nation of Wemindji. She holds a bachelor's in history and a bachelor's in education (primary/junior) from York University, Toronto. She served as community liaison to the Wemindji Protected Areas Project over several years, providing valuable guidance and advice to research team members. Fluent in Cree, English, and French, she is currently the regional Cree language co-ordinator for the Cree Nation Government in Eeyou Istchee.

Samuel Vaneeckhout is a professor in the Faculty of Humanities/ Archaeology at the University of Oulu, Finland. He holds a PhD in archaeology from the University of Oulu. His research interests include the impact of isostatic land uplift and shoreline displacement on mobility/ sedentism on the Finnish coast.

Kristen Whitbeck is an ecologist for the US Forest Service and an adjunct faculty member in the Department of Forest Ecosystems and Society at Oregon State University. She lived and worked with the James Bay Crees of Wemindji during her doctoral studies at McGill University. She continues to enjoy studying nature and mapping uncharted territory.

Colin D. Wren is an assistant professor in the Department of Anthropology at the University of Colorado, Colorado Springs. He participated in the archaeological investigations in the Wemindji region as an undergraduate and doctoral student, and continues to study the way in which people adapted to their natural environments in the past. His research focuses on mobility and human-environment interactions in case studies from several parts of the world, including northern Quebec, northern Finland, South Africa, and Western Europe.

Index

Note: Page numbers with (f) refer to figures; (t) refers to tables.

resiliency of, 214; usage of name, 190–91
Paakumshumwaau-Maatuskaau Biodiversity Reserve, 14; archaeological digs and, 333; and Azimut claim, 106; beaver in, 237; complexity of, 192; creation as new chapter, 317; creation of, 337–38; development "neutrality" of watersheds, 332–33; diversity of mammals, 240–41; establishment of, xvii; exclusion of land with mineral extraction potential, 105; exclusion of portion of lake, 318n6; family hunting territories and, 13f; fire and, 199, 232; future of Paakumshumwaau in, 214; government procedures and, 334; hunting territory boundaries shaping design of, 236; implications of natural history for, 213–14; as initiated/driven by Wemindji, 353–54; James Bay hydro-electric project, and creation of, 366; Jamesians' attitudes toward, 331; under JBNQA, 335; mammals and, 240–41; map, 324f, 342f; MDDEP and, 104, 105, 325, 333, 334, 335–37; mining and, 16–17, 85, 104, 334, 335, 336–37; Ministère des Ressources naturelles et de la Faune (MRNF) and, 105; moose in, 237; multi-party dialogue, 335–37; Municipalité de la Baie James and, 332; Old Factory Lake at heart of, 16; Old Factory River as motivation for establishment of, 298; partnership leading to creation of, 17; plotting future of, 192; process of discovery in setting up, 333–34; proposal for creation submitted to MDDEP, 103; and protection of marine territory, 348–49; Quebec government process in establishment, 19; red canoe story and, 370; relations with Municipalité de la Baie James and, 329; relationship between mammals and people in, 240–41; research, 355; size of, 333; spatial compartments/integrations protected by, 230; spatial compartments/interactions included in, 231f; spatial/temporal scale in, 241; stewardship and, 333; support of Indigenous way of life/resources linked to protection/sustainable use, 49; and Tawich (Marine) Conservation Area, 340, 341, 352, 355; temporal compartments/interactions included in, 234f; and threats to mammals, 221; university partners as intermediaries in community-government discussions, 328. *See also* Maatuskaau (Rivière du Peuplier/Poplar River)
Paaschikin minishtikw (Gun Island), 307
paaschimaau uchimaau (goose boss/tallyman). *See* tallymen
Paix des braves (Agreement Concerning a New Relationship between the Government of Québec and the Crees of Québec): about, 79n4, 86; and customary tenure in land use, 77; and hunting territories, 79; and hydro-electric development, 111n3; and mining, 101–2, 107; Moses and, 74; and relations between Cree Nation and Quebec government, 335, 358
paleogeography, 161–65
palynology/pollen, 17, 179–81, 182, 211, 350
parks: Aboriginal issues and, 31; activity of northern vs. southern Indigenous peoples in planning, 42; conservation and, 30; dispossession and, 16; ecological integrity as priority, 30, 31; federal government policy of joint management with Indigenous peoples, 30; Indigenous positive/negative experiences with, 347–48; Indigenous rights/title and, 30; land claims and, 32, 42; marginalization and, 16; and misrepresentations of Indigenous peoples, 30; partnerships with Indigenous peoples in development/maintenance, 31–32; privileging of science in management